Advances in Applied Mechanics
Volume 37

Editorial Board

Contributors to Volume 37

ADVANCES IN
APPLIED MECHANICS

Edited by

Erik van der Giessen

DELFT UNIVERSITY
OF TECHNOLOGY
DELFT, THE NETHERLANDS

Theodore Y. Wu

DIVISION OF ENGINEERING AND APPLIED SCIENCE
CALIFORNIA INSTITUTE OF TECHNOLOGY
PASADENA, CALIFORNIA

VOLUME 37

ACADEMIC PRESS
San Diego   London   Boston
New York   Sydney   Tokyo   Toronto

ACADEMIC PRESS
A Harcourt Science and Technology Company
525 B Street, Suite 1900, San Diego, California 92101-4495, USA
http://www.academicpress.com

ACADEMIC PRESS
Harcourt Place, 32 Jamestown Road, London, NW1 7BY, UK
http://www.academicpress.com

International Standard Serial Number: 0065-2165

International Standard Book Number: 0-12-002037-8

Printed in the United States of America

00 01 02 03 04 QW 9 8 7 6 5 4 3 2 1

Contents

A Unified Theory for Modeling Water Waves

Theodore Yaotsu Wu

Coastal Hydrodynamics of Ocean Waves on Beach

Jin E. Zhang, Theodore Y. Wu, and Thomas Y. Hou

Onset of Oscillatory Interfacial Instability and Wave Motions in Bénard Layers

Manuel G. Velarde, Alexander A. Nepomnyashchy, and Marcel Hennenberg

Role of Cryogenic Helium in Classical Fluid Dynamics: Basic Research and Model Testing

Katepalli R. Sreenivasan and Russell J. Donnelly

Recent Advances in Applications of Tensor Functions in Continuum Mechanics

Josef Betten

List of Contributors

Numbers in parentheses indicate the pages on which the authors' contributions begin.

JOSEF BETTEN (277), Department of Mathematical Models in Materials Science, Technical University of Aachen, Aachen, Germany D-52062

RUSSELL J. DONNELLY (239), Cryogenic Helium Turbulence Laboratory, Department of Physics, University of Oregon, Eugene, Oregon 97403

MARCEL HENNENBERG (167), Instituto Pluridisciplinar, Universidad Complutense, Madrid, Spain 28 040, and Microgravity Research Center, Faculte des Sciences Appliquées, Université Libre de Bruxelles, Brussels, Belgium 1 050

THOMAS Y. HOU (89), Applied Mathematics, California Institute of Technology, Pasadena, California 91125

ALEXANDER A. NEPOMNYASHCHY (167), Instituto Pluridisciplinar, Universidad Complutense, Madrid, Spain 28 040, and Department of Mathematics, Technion-Israel Institute of Technology, Haifa, Israel 32000

KATEPALLI R. SREENIVASAN (239), Mason Laboratory, Yale University, New Haven, Connecticut 06520

MANUEL G. VELARDE (167), Instituto Pluridisciplinar, Universidad Complutense, Madrid, Spain 28 040

THEODORE Y. WU (1, 89), Engineering Science, California Institute of Technology, Pasadena, California 91125

JIN E. ZHANG (89), Economics and Finance, City University of Hong Kong, Kowloon, Hong Kong, and Engineering Science, California Institute of Technology, Pasadena, California 91125

Preface

This volume of *Advances in Applied Mechanics* consists of five articles addressing a variety of topics of basic interest and current activity.

One of the most noticeable areas of recent activity is the application of mechanics to coastal phenomena involving evolution of ocean waves and current in coastal waters under various conditions approaching that in nature. Intended to facilitate scientific and engineering research on this general subject, a new theoretical model is introduced in the article by T. Y. Wu for modeling unsteady, fully nonlinear, fully dispersive, three-dimensional gravity-capillary waves on water of variable depth. The emphasis is on achieving both analytical resources and computational efficiency for solving problems. Reduction of the full theory to approximate versions is illustrated for a number of issues of classical and new interest.

The article by J. E. Zhang, T. Y. Wu, and T. Y. Hou addresses a systematic approach to developing a comprehensive theoretical model for ocean waves on coastal waters and beaches to embrace a scope of diversity and range of the physical factors involved. The authors investigate the three-dimensional effects, first for a simple representative case of topographical and wave features, together with the nonlinear and dispersive effects playing their exact roles in regions including the moving waterline on beaches. These important effects are examined in a series of comparative studies. For this task, a hybrid Lagrangian–Eulerian method is developed by incorporating the local Lagrangian determination of moving waterline with the Eulerian field computation of the interior flow domain, to constitute an efficient exact method. This preliminary study may serve to assist further research over a vast scope of practical interest.

The article by M. G. Velarde, A. A. Nepomnyashchy, and M. Hennenberg is devoted to a review and exposition of the literature about theory and experiments dealing with the excitation of oscillatory interfacial Bénard–Marangoni convection. Emphasis is paid to the role played by surface tension gradients (the Marangoni effect) in shallow liquid layers subjected to thermal gradients and having a deformable surface open to air or an interface to another liquid, when the outcome instability is in one form or another of wave motion. It is argued that the oscillatory surface-tension-gradient instability offers a field of research with a high potential for development in various disciplines, both theoretical and experimental.

The use of helium as a working fluid, known for its very low kinematic viscosity and its rather sensitive variations in physical properties near the critical point with changes in pressure, has attracted strong interest in generating flows at ultra-high Reynolds and Rayleigh numbers. Such flows can create new opportunities for conducting hydrodynamic experiments and for pursuing turbulence research over some extreme parametric regimes otherwise unattainable. On the other hand, there remains a host of uncertainties to be examined and overcome before helium can be used with familiarity and confidence. The article by K. R. Sreenivasan and R. J. Donnelly presents an expository review on some relevant considerations and an in-depth discussion of the opportunities and challenges ahead.

Among the advances in the mechanics of fluids and solids during the last decades is the continued development of constitutive models, both with enhanced predictive power and for a growing class of materials actually used in modern technology. While the formulation of constitutive equations is essentially a matter of physical insight and experimental determination, to have a theoretical continuum mechanics framework is nevertheless of great value. The article by J. Betten gives an exposé about how general mathematical results on tensor functions can be used in a constitutive framework to satisfy objectivity and material symmetry conditions. After a presentation of the general principles, the survey focuses on recent applications to describe anisotropic plasticity and creep, supplemented with experimental validation.

A Unified Theory for Modeling Water Waves

THEODORE YAOTSU WU

California Institute of Technology
Pasadena, California

ADVANCES IN APPLIED MECHANICS, VOL. 37
ISBN 0-12-002037-8
ISSN 0065-2165/01 $35.00

I. Introduction

The phenomena of water waves have fascinated observers of all ages and the subject's extent and diversity are enormous. The different types of waves in water, varying from ripples on a placid pond, breaking of shoaling waves on a beach, billows on a stormy sea and in the ocean interior, to geophysical waves and devastating tsunamis, are truly extensive. These wave phenomena attract interest and stimulate curiosity because the salient features of wave properties can often be perceived, qualitatively at least, with attentive naked eyes. Of the various wave phenomena found in nature, water waves have the distinction of exhibiting some basic properties with a range and strength rarely matched by other kinds of waves. First, water waves possess a wide range of variation in dispersive effects that make wave (phase) velocity vary with wavelength, which in turn may differ considerably from the (group) velocity at which wave energy propagates. In deep water, the group velocity varies from three-halves of phase velocity for ripples to one-half the phase velocity for gravity waves. For long waves, of length large relative to water depth, the dispersive effects, though slight in this case, can accumulate, in time, to a noticeable margin as shown by the length increment of tsunami waves propagating across the Pacific. In addition, water waves can give conspicuous displays of nonlinear effects (making linear superposition of solutions no longer a solution) as is evident in breaking roll waves on beaches. However, with dispersive (wave-separating) and nonlinear (wave-focusing) effects kept in appropriate balance, moderately steep waves are found to propagate on shallow water and to be permanent in shape. This was first discovered by John Scott Russell (1844) who in August 1843 observed a heap of water formed in front of the prow of a channel boat, which suddenly stopped from running under tow by a pair of horses and became a "grand wave of translation" progressing forward along the channel with uniform speed as a free wave. In the recent 35 years of the colorful modern history of pure and applied mathematics, physics, engineering science, biology, and other disciplines, the same type of weakly nonlinear and weakly dispersive wave phenomena, now known under the blanket term of *solitons*, seem to occur almost universally.

The different fundamental ideas that have been developed to interpret and predict the properties of water waves and the applications of different theories and methods to various problems can be delineated according to the specific regimes of the key physical parameters involved and on which regime a theory is based. Of the rich ensemble of physical parameters involved, two key parameters are essential, namely,

$$\alpha = a/h, \qquad \varepsilon = h^2/\lambda^2, \tag{1}$$

which represent, respectively, the nonlinear and dispersive effects for characterizing waves of amplitude a and typical length λ in water of rest depth h. Theoretical modeling of water waves becomes greatly more difficult when two additional geometric parameters are extended from two to three spatial dimensions and from uniform ($h = $ const.) to nonuniform medium (with h varying in space and time). In fact, the states of the art are decades apart between the simple and more general cases. Nevertheless, such generalizations are needed for various reasons. In this respect, the nonlinearity factor α, the dispersion factor ε, and the medium geometry (nondimensional h) can be termed the *three primary parameters* for modeling water waves. In addition, there are other parameters including the Fruode number $F = U/\sqrt{gh}$ (for measuring the gravity effect $\rho g h$ relative to the inertial effect ρU^2 for motions with typical velocity U of the fluid of density ρ under gravity acceleration g), the Weber number $U/\sqrt{\gamma/l}$ (for scaling the capillary effect $\rho \gamma/l$ for typical length l relative to inertia effects, with γ being the kinematic surface tension), the specific fluid density variation $\Delta\rho/\rho$ for density-stratified fluids, and also the frame rotation effect for rotating flows and so on. These parameters need be considered when their effects play a role in problems of concern. It is in this framework and scope that modeling of water waves is constituted, as has been much illuminated by C. C. Lin and A. Clark (1959), J. D. Cole (1968), G. Whitham (1974), and Sir James Lighthill (1978).

For infinitesimal gravity waves on deep water, with the two parameters α and ε merging (with h eliminated) into one regime (see Figure 1) characterized by $\tilde{\alpha} = \alpha\sqrt{\varepsilon} = a/\lambda \ll 1$ for an infinitesimal bound of wave slope, much useful knowledge can be attained on linear theory since the dispersive effects are fully represented in governing the evolution of wave patterns, making waves of different lengths spatially further separated and thus hindering

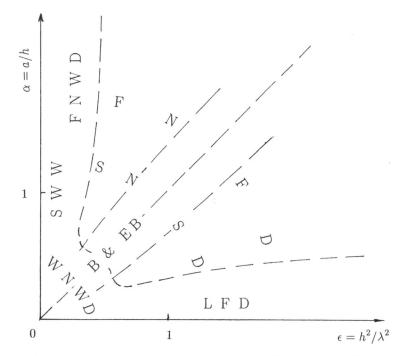

FIG. 1. The parametric domain spanned by the nonlinear effect $\alpha = a/h$ and dispersive effect $\varepsilon = h^2/\lambda^2$ for characterizing waves of amplitude a and typical length λ in water of typical depth h. The domain comprises various regimes for specialized modeling of waves ranging from linear through weakly, strongly, and fully nonlinear (L, WN, SN, FN), and from nondispersive through weakly, strongly, and fully dispersive (ND, WD, SD, FD). in grades. The regime about the $\alpha = \varepsilon$ line is the Boussinesq and extended Boussinesq family. Line $\varepsilon = 0$ locates the classical linear and nonlinear shallow-water wave (SWW) theories.

nonlinear growth accumulation. Furthermore, such linear theories provide analytical convenience for producing useful results and lend analytical resources to finding the basic mechanism underlying (linear) phenomena of interest. Tacitly, these approximate theories assume that there exists an exact solution that is being approximated. Nevertheless, for inviscid irrotational waves restricted to periodic motion in deep water, proof of existence has been given by Nekrasov (1921) assuming a series expansion of his nonlinear integral equation, and by Levi-Civita (1925) using a different series expansion, both indicating constructive algorithms for evaluating periodic waves up to a certain small amplitude. (For an expository survey, see Wehausen and Laitone, 1960.)

For irrotational waves of finite amplitude in water of uniform depth, the classical linear theory has been extended by the pioneering works of Boussinesq (1871, 1872), Rayleigh (1876), and Korteweg and de Vries (1895) who developed weakly nonlinear weakly dispersive (WNWD) wave models for interpreting the solitary wave discovered by Russell (1844). For the WNWD parametric regime (see Figure 1), Boussinesq (1872) found that the assumption of $\alpha = O(\varepsilon) \ll 1$ provides well-balanced roles between the nonlinear and dispersive effects for solitary waves to exist. For modeling long waves of finite amplitude in layered media, Green and Naghdi (1976) assumed $\alpha = O(1)$ and $\varepsilon \ll 1$, so that ε is the only small parameter adopted in deriving the Green–Naghdi model. This class of theory can be classified for the fully nonlinear weakly dispersive (FNWD) waves. In this direction, recent advances have been made by Ertekin *et al.* (1984, 1986), Choi (1995), Choi and Camassa (1996), and others for modeling nonlinear and dispersive wave motions in a single-layer or two-layer fluid, with intent to achieve higher accuracy than existing models. This demonstrates that various theories can be sought by making analysis based on different parametric regimes to obtain different approximations to serve as differing theoretical models, provided they are duly validated under the original premises.

For unidirectional waves of the Korteweg–de Vries (KdV) family in particular, the original KdV equation has been shown by Zabusky and Kruskal (1965) to possess remarkable properties, implying that the KdV equation admits not only one, but N solitary waves, for $N = 1, 2, \dots$, of arbitrary amplitudes, propagating and interacting on one real line and conserves all their entities. The term *soliton* is coined for solitary waves having such properties. This in fact implied that the KdV equation possesses an infinite number of conserved quantities, which was later proven to be the case. From a model making viewpoint, it is of great significance to reckon that model equations, usually derived by methods devised for reducing an original system of often large and intractable sets of partial differential equations, are supposedly in an appropriate simpler version, generally reduced in dimension by one or two, that can be handled more easily, both for analytical and computational purposes. This is the primary objective of modeling flow motion. While it may not be a rule, it nevertheless seems commonplace that the reduced system (e.g., the KdV, the modified KdV, the nonlinear Shrödinger (NLS) equation, etc.) has an infinite number of conserved integrals, whereas the underlying full system probably has only three, in mass, momentum, and energy. A philosophical query is therefore proper at this point: Why can reduction methods bring out extra symmetries that were not there, or at best hidden in the first place? We shall come back to this later.

Needs nevertheless do exist for modeling fully nonlinear fully dispersive water wave phenomena, as needed for further development of the subject and for practical applications, that would require the nonlinear and dispersive effects retained in the model to play their full exact physical roles. In a series of studies, we have developed an exact inviscid theory for evaluating fully nonlinear fully dispersive (FNFD) incompressible, irrotational water waves in water of *uniform* depth in a single layer (Wu, 1997, 1998a,b,c, 1999a) and in double layers (Wu, 1999b). In the present work, the theory is further extended in Section II for evaluating three-dimensional, time-evolving, FNFD gravity-capillary waves in water of *variable* depth. This exact theory is first built on two basic equations, one being the free-surface kinematic condition, and the other the horizontal momentum equation projected onto the free surface. These two partial differential equations are both exact but involve three unknowns, the horizontal velocity, $\hat{u}$, and the vertical velocity $\hat{w}$ at the free surface (in two horizontal dimensions) and the water surface elevation, ζ. Closure of the system is accomplished either in differential form by series expansion of the velocity potential or in integral form by adopting a boundary integral equation for the velocity field. The emphasis is focused on achieving both analytical facilities and computational efficiency that a theoretical model can provide for resolving nonlinear water wave problems. In versatility, this theory can be approximated to various degrees of validity to agree with existing theories as special cases for predicting nonlinear dispersive water wave phenomena pertaining to their specific parametric regimes, as shown in later sections. Reduction of the full theory to an approximate version is illustrated in Section V for Stokes waves in water of arbitrary depth and in Section VIII.B for the third-order solitary wave theory. In the latter task, the scaling perturbation method is presented structurally to elucidate the method known as the *asymptotic reductive perturbation method*, which is in general use for deriving weakly nonlinear evolution equations. Further applications of these model equations are briefly reviewed in later sections with exposition of the salient behaviors of bidirectional nonlinear waves in wave–wave interactions, generation of solitons in nonlinear dispersive systems forced at resonance, and waves evolving in a nonuniform medium such as in variable channels of arbitrary shape and in the related processes of transport of mass and energy.

This article is intended to present a systematic approach to modeling water waves over a scope as comprehensive as appropriate, although dissipative effects are beyond the present consideration. In view of the vast diversity of the subject, further reviews of the literature are left with other resources.

II. Fully Nonlinear Fully Dispersive (FNFD) Waves in Water of Variable Depth

We adopt Euler's equations to describe three-dimensional, incompressible, inviscid long waves on a layer of water of variable depth $h(\mathbf{r}, t)$ which may vary with the horizontal position vector $\mathbf{r} = (x, y, 0)$ and possibly also with the time t for representing submerged moving body, submarine landslides, drifting sandbars, or seismic activities along the seafloor. The fluid moves with velocity $(\mathbf{u}, w) = (u, v, w)$ in the flow field bounded below by the seabed at $z = -h(\mathbf{r}, t)$ and above by the free water surface at $z = \zeta(\mathbf{r}, t)$ measured from its rest position at $z = 0$. [The bathymetry $h(\mathbf{r}, t)$ is tacitly assumed to be gradual in variation, and will be further qualified later.] For evaluating flows assumed incompressible and inviscid, we have the Euler equations of continuity, horizontal, and vertical momentum as follows:

$$\nabla \cdot \mathbf{u} + w_z = 0, \tag{2}$$

$$\frac{d\mathbf{u}}{dt} = \mathbf{u}_t + \mathbf{u} \cdot \nabla \mathbf{u} + w\mathbf{u}_z = -\frac{1}{\rho}\nabla p, \tag{3}$$

$$\frac{dw}{dt} = w_t + \mathbf{u} \cdot \nabla w + ww_z = -\frac{1}{\rho}p_z - g, \tag{4}$$

where $\nabla = (\partial_x, \partial_y, 0)$, $(\partial_x = \partial/\partial x$, etc.) is the horizontal gradient operator, p is the pressure, ρ the constant fluid density, and g the gravitational acceleration. Here, the subscripts t and z denote partial differentiation. The boundary conditions are

$$\hat{w} = \hat{D}\zeta \quad (\hat{D} = \partial_t + \hat{\mathbf{u}} \cdot \nabla, \quad \text{on } z = \zeta(\mathbf{r}, t)) \tag{5}$$

$$p = p_a(\mathbf{r}, t) + \rho\gamma\nabla \cdot \mathbf{n} \quad (\text{on } z = \zeta(\mathbf{r}, t)), \tag{6}$$

$$\check{w} = -\check{D}h \quad (\check{D} = \partial_t + \check{\mathbf{u}} \cdot \nabla, \quad \text{on } z = -h(\mathbf{r}, t)), \tag{7}$$

where $p_a(\mathbf{r}, t)$ is an external pressure disturbance gauged over the ambient pressure (which is set to zero), $\rho\gamma$ is the uniform surface tension, $\mathbf{n}$ is the outward unit vector normal to the water surface, and $\hat{\mathbf{u}}$, $\check{\mathbf{u}}$, $\hat{w}$, $\check{w}$ use the definition that $\hat{f}(r, t)$ and $\check{f}(r, t)$ denote the value of an arbitrary flow variable $f(r, z, t)$ at the free surface and at the seabed, namely,

$$f(r, \zeta(r, t), t) = \hat{f}(r, t), \qquad f(r, -h(r, t), t) = \check{f}(r, t). \tag{8}$$

The momentum equations (3) and (4) can be projected under boundary conditions (5) and (6) onto the free surface to obtain an equation for $(\hat{u}, \zeta)$. First, using notation (8) for $\hat{f}(r, t)$ and $\check{f}(r, t)$, we have for their derivatives, by the chain rule, the relations

$$\partial_t \hat{f}(r, t) = \left(\partial_t f + \frac{\partial f}{\partial z} \partial_t \zeta\right)\Big|_{z=\zeta}, \qquad \nabla \hat{f}(r, t) = \left(\nabla f + \frac{\partial f}{\partial z} \nabla \zeta\right)\Big|_{z=\zeta}, \tag{9a}$$

$$\partial_t \check{f}(r, t) = \left(\partial_t f - \frac{\partial f}{\partial z} \partial_t h\right)\Big|_{z=-h}, \qquad \nabla \check{f}(r, t) = \left(\nabla f - \frac{\partial f}{\partial z} \nabla h\right)\Big|_{z=-h}. \tag{9b}$$

From this we deduce, under conditions (5) and (7), the following theorems:

$$\frac{df}{dt}\Big|_{z=\zeta} = \hat{D}\hat{f} \qquad (\hat{D} = \partial_t + \hat{u} \cdot \nabla); \tag{10a}$$

$$\frac{df}{dt}\Big|_{z=-h} = \check{D}\check{f} \qquad (\check{D} = \partial_t + \check{u} \cdot \nabla). \tag{10b}$$

They assert that the material derivative of $f(r, z, t)$ evaluated at a boundary surface and its projection onto that boundary surface are equal.

We can then project the horizontal momentum equation (3) onto the free surface so that

$$\hat{D}\hat{u} = -\frac{1}{\rho} \nabla p|_{z=\zeta} = -\frac{1}{\rho}\left(\nabla\hat{p} - \frac{\partial p}{\partial z}\Big|_{z=\zeta} \nabla\zeta\right), \tag{11}$$

from which we use the vertical momentum equation (4) and conditions (5) and (6) to yield

$$\hat{D}\hat{u} + [g + \hat{D}^2\zeta]\nabla\zeta = -(1/\rho)\nabla p_a - \gamma\nabla\nabla \cdot n. \tag{12}$$

Similarly, we project the momentum equation onto the seabed under condition (7) to give

$$\check{D}\check{u} - [g - \check{D}^2 h]\nabla h = -(1/\rho)\nabla\check{p}. \tag{13}$$

Here, eq. (12) provides an equation for $(\hat{u}, \zeta)$ at the water surface while (13) relates $\check{p}$ with $\check{u}$ at the seabed, $z = -h(r, t)$. Although (12) and (13) involve, superficially, only the variables pertaining to their own specific boundaries, they actually have incorporated the vertical momentum equation as well as the kinematic and dynamic conditions at the free surface or at the seabed to yield these equations of an overall equilibrium. Moreover, they are exact.

It is significant to point out that eqs. (12) and (13) hold regardless of whether the flow is irrotational or contains vorticity distributions. In the present work, however, we consider only irrotational water waves so that there exists a velocity potential, $\phi(\mathbf{r}, z, t)$, such that $\mathbf{u} = \nabla\phi$, $w = \partial\phi/\partial z$ and it satisfies the Laplace equation, $\nabla^2\phi + \phi_{zz} = 0$.

We have now two exact equations that may facilitate modeling FNFD unsteady water waves, one being the free surface kinematic condition (5) involving $(\hat{\mathbf{u}}, \hat{w}, \zeta)$, and the other the surface-projected momentum equation (12) for $(\hat{\mathbf{u}}, \zeta)$. Closure of the system can be accomplished by further finding for the velocity a third exact equation that relates the three unknowns, either in differential form by series expansion of the velocity potential or in integral form by adopting a boundary integral equation for the velocity field. The series solution in differential form has advantages because it can provide various approximate model equations, by series truncation, that can yield analytical approximate solutions for specific parametric regimes, and it can open the field for rich analytical resources to develop. In practice, models do exist that are capable of providing good predictions of target phenomena even with only the leading term of such reductions (for specific parametric regimes) and that are highly accurate with one or two higher-order terms amended. On the other hand, the velocity solution in surface integral form can be used directly for numerical computation of exact solutions.

To proceed, we note that since the two basic equations (5) and (12) are exact, we can ignore the nonlinearity parameter α by regarding it as arbitrary and consider first the special case of inviscid long waves in shallow water by tacitly assuming that only the dispersion parameter $\varepsilon = h^2/\lambda^2$ is small. (It turns out that this assumption can also be relaxed eventually; see the concluding passage of Section II.A.)

For simplicity, we adopt the following dimensionless variables:

$$r_* = r/\lambda, \quad z_* = z/h_0, \quad t_* = c_0 t/\lambda, \quad \zeta_* = \zeta/h_0, \quad h_*(x_*) = h/h_0, \quad \phi_* = \phi/c_0\lambda,$$

$$\varepsilon = h_0^2/\lambda^2, \quad u_* = u/c_0 = \nabla_*\phi_*, \quad w_* = w/c_0\sqrt{\varepsilon} = \varepsilon^{-1}\partial\phi_*/\partial z_*, \quad p_* = p/\rho g h_0,$$

$$\tag{14}$$

where h_0 is a typical or the mean water depth, λ a typical wavelength, $c_0 = \sqrt{gh_0}$ is the linear wave speed. With the $*$ omitted as understood, we note that eqs. (2), (5), and (7) remain intact in the dimensionless form, while

$u = \nabla\phi$, $w = \varepsilon^{-1}\partial\phi/\partial z$, and the dimensionless Laplace equation for the velocity potential ϕ involves the parameter $\varepsilon = h_0^2/\lambda^2$ as

$$\phi_{zz} + \varepsilon\nabla^2\phi = 0 \qquad (-h(r, t) \leqslant z \leqslant \zeta(r, t)). \qquad (15)$$

The solution ϕ of (15) may assume a series expansion of the form (with the * omitted)

$$\phi(r, z, t; \varepsilon) = \sum_{n=0}^{\infty} \varepsilon^n \Phi_n(r, z, t), \qquad (16a)$$

$$(\Phi_0)_{zz} = 0, \qquad (\Phi_n)_{zz} = -\nabla^2\Phi_{n-1} \qquad (n = 1, 2, 3, \ldots), \qquad (16b)$$

which follow from (15) and (16a). This set of recurrence equations has the integral:

$$\Phi_0 = \Phi_0(r, t),$$

$$\Phi_n = \Phi_{n0}(r, t) + (z + h)\Phi_{n1}(r, t) + I\Phi_{n-1} \qquad (n = 1, 2, 3 \ldots), \qquad (17)$$

$$I\Phi_{n-1} = -\int_{-h}^{z} dz_1 \int_{-h}^{z_1} dz' \nabla^2\Phi_{n-1}(r, z', t)$$

$$= -\int_{-h}^{z} (z - z')\nabla^2\Phi_{n-1}(r, z', t)\, dz'. \qquad (18)$$

We note that the zeroth order integral Φ_0 cannot admit a term linear in $(z + h)$ like that in Φ_n $(n \geqslant 1)$ due to the small ε assumption with $w = \varepsilon^{-1}\partial\phi/\partial z$. Substituting this set of integrals into (16a) yields, after some rearrangement, the series solution as

$$\phi(r, z, t; \varepsilon) = \sum_{n=0}^{\infty} \varepsilon^n I^n \phi_0(r, t; \varepsilon) + \sum_{n=0}^{\infty} \varepsilon^{n+1} I^n [(z + h)\phi_1(r, t; \varepsilon)], \qquad (19)$$

$$\phi_0(r, t; \varepsilon) = \Phi_0(r, t) + \sum_{n=1}^{\infty} \varepsilon^n \Phi_{n0}(r, t), \qquad \phi_1(r, t; \varepsilon) = \sum_{n=1}^{\infty} \varepsilon^{n-1} \Phi_{n1}(r, t),$$

where I^n is the n-times successive application of the integral operator I defined in (18). Here the dependence of ϕ_0 and ϕ_1 on ε arises from the above regrouping of the complementary solutions of Φ_n such that $\phi_0 \to \Phi_0(r, t)$ and $\phi_1 \to \Phi_{11}(r, t)$ as $\varepsilon \to 0$. This regrouping is admissible provided the medium is horizontally unbounded, in the absence of any other boundary effects of specific orders in magnitude.

The n-tuple integral operators I^n can be carried out in closed form. More specifially, we can show, by induction, that $I^0 = 1$, and for $n = 1, 2, 3, \ldots$,

$$I^n f(r, z) = I[I^{n-1} f(r, z)] = -\int_{-h}^{z} (z - z') \nabla^2 [I^{n-1} f(r, z')] \, dz'$$

$$= (-1)^n \nabla^{2(n-1)} \int_{-h}^{z} \frac{(z - z')^{2n-1}}{(2n - 1)!} \nabla^2 f(r, z') \, dz'. \tag{20a}$$

In fact, (20a) agrees with definition (18) for $n = 1$. Clearly, $If(r, z) = O(z + h)^2$ as $z \to -h$ provided that $\nabla^2 f$ is everywhere bounded, a condition which we assume to hold in general. For $n = 2$, we proceed as follows:

$$I^2 f(r, z) = I[If] = -\int_{-h}^{z} (z - z_1) \nabla^2 [If(r, z_1)] \, dz_1$$

$$= -\nabla \cdot \int_{-h}^{z} \nabla \{(z - z_1)[If(r, z_1)]\} \, dz_1$$

$$= \nabla^2 \int_{-h}^{z} (z - z_1) \, dz_1 \int_{-h}^{z_1} (z_1 - z') \nabla^2 f(r, z') \, dz',$$

which yields (20a), for $n = 2$, upon interchanging the order of integration. In the above steps, the operator ∇^2 can be moved freely across the integral sign [with zero contribution from the variable integration limit at $-h(r, t)$] because of $If(r, z) = O(z + h)^2$ as $z \to -h$. Now suppose (20a) holds for $n = m(\geqslant 2)$; then we can similarly show that (20a) holds for $n = m + 1$ by observing that we can again slide the operator ∇^2 across the integral sign because $I^n f(r, z) = O(z + h)^2$, $\forall n \geqslant 1$ as $z \to -h$. This proves (20a). From (20a) it follows at once that for arbitrary differentiable $f(r, z)$ we have the general relationship

$$\frac{\partial^2}{\partial z^2} I^n f(r, z) = -\nabla^2 [I^{n-1} f(r, z)] \qquad (n = 1, 2, 3, \ldots). \tag{20b}$$

Hence (19) is readily seen to satisfy (15).

As shown above, $I^n[(z + h)\phi_1] = O(z + h)^2$ as $z \to -h$, $\forall n \geqslant 1$ (except for $n = 0$) (with h, ϕ_0, and ϕ_1 understood as analytic in r), hence by condition (7) and solution (19), the only contribution to $\check{w}$ at seabed comes from the leading term ($n = 0$) of the ϕ_1 series, giving

$$\check{w}(r, t) = \frac{1}{\varepsilon} \frac{\partial \phi}{\partial z}\bigg|_{z=-h} = \phi_1(r, t) = -\check{D}h = -(\partial_t + \check{u} \cdot \nabla)h, \tag{21}$$

which determines $\phi_1(r, t)$. This leaves $\phi_0(r, t)$ as the only unknown in solution (19).

Applying (20)–(21) to the two terms in (19) yields the solution in integrated form as

$$\phi(r, z, t; \varepsilon) = \sum_{n=0}^{\infty} \varepsilon^n I^n \phi_0(r, t; \varepsilon) + \sum_{n=0}^{\infty} \varepsilon^{n+1} I^n [(z + h)\check{w}(r, t; \varepsilon)], \qquad (22a)$$

$$I^n \phi_0 = (-1)^n \nabla^{2(n-1)} \left[\frac{(z + h)^{2n}}{(2n)!} \nabla^2 \phi_0 \right] \qquad (n = 1, 2, 3, \ldots), \quad (22b)$$

$$I^n [(z + h)\check{w}] = (-1)^n \nabla^{2(n-1)} \left[\frac{(z + h)^{2n}}{(2n)!} H_1(r, t) + \frac{(z + h)^{2n+1}}{(2n + 1)!} H_2(r, t) \right],$$

$$(22c)$$

$$H_1(r, t) = (\check{w} \nabla^2 h + 2\nabla h \cdot \nabla \check{w}), \qquad H_2(r, t) = (\nabla^2 \check{w}). \qquad (22d)$$

Here, the differential operators ∇^m ($m = 1, 2, \ldots$) have been so selectively positioned in our analysis as to render the final exact solution, eqs. (22a–d), in perhaps the simplest form.

From solution (22a) for ϕ we deduce for u and w these results:

$$u = \nabla \phi = \sum_{n=0}^{\infty} (\varepsilon^n u_n + \varepsilon^{n+1} U_n), \qquad u_n = \nabla I^n \phi_0, \qquad U_n = \nabla I^n [(z + h)\check{w}],$$

$$(23)$$

$$w = \frac{1}{\varepsilon} \phi_z = \check{w} + \sum_{n=1}^{\infty} (\varepsilon^{n-1} w_n + \varepsilon^n W_n),$$

$$w_n = \partial_z I^n \phi_0, \qquad W_n = \partial_z I^n [(z + h)\check{w}], \qquad (24)$$

where $I^n \phi_0$ and $I^n [(z + h)\check{w}]$ are given by (22b–d). Hence, at the seabed, $z = -h(r, t)$,

$$\check{u} \equiv u(r, -h, t) = u_0(r, t) - \varepsilon(\check{D}h)\nabla h, \qquad u_0 \equiv \nabla \phi_0(r, t; \varepsilon). \qquad (25)$$

Equation (25) indicates how $\check{u} - u_0$ varies for a wave advancing into shallower or deeper water, by a term of $O(\varepsilon)$. For water of uniform depth, $h = $ const., we have $\check{u} = u_0(r, t)$.

At the free surface, $z = \zeta(r, t)$, $u \to \hat{u}$ and $w \to \hat{w}$ in the limit of (23) and (24) as $z \to \zeta(r, t)$, giving for $\hat{u} = u(r, \zeta, t)$ and $\hat{w} = w(r, \zeta, t)$ the result in operator form as follows:

$$\hat{\boldsymbol{u}} - \hat{\boldsymbol{U}} \equiv \Delta\hat{\boldsymbol{u}} = A[\boldsymbol{u}_0] = \sum_{n=0}^{\infty} \varepsilon^n \hat{\boldsymbol{u}}_n = \sum_{n=0}^{\infty} \varepsilon^n A_n \boldsymbol{u}_0, \qquad \hat{\boldsymbol{U}} = \sum_{n=0}^{\infty} \varepsilon^{n+1} \hat{\boldsymbol{U}}_n, \quad (26)$$

$$\hat{w} - \check{w} - \hat{W} = C[\boldsymbol{u}_0] = \sum_{n=1}^{\infty} \varepsilon^{n-1} C_n \boldsymbol{u}_0, \qquad \hat{W} = \sum_{n=1}^{\infty} \varepsilon^n \hat{W}_n, \qquad (27)$$

$$A_0 = 1, \qquad A_n = (-1)^n \nabla_w^{2n-1} \frac{\eta^{2n}}{(2n)!} \nabla\cdot, \qquad (\eta = h(\boldsymbol{r}, t) + \zeta(\boldsymbol{r}, t)),$$

$$C_n = (-1)^n \nabla_w^{2(n-1)} \frac{\eta^{2n-1}}{(2n-1)!} \nabla\cdot \qquad (n = 1, 2, \ldots), \qquad (28a)$$

$$\hat{\boldsymbol{U}}_0 = \nabla[(z + h)\check{w}]|_{z=\zeta} = \nabla_w(\eta\check{w}) = \eta\nabla\check{w} + \check{w}\nabla h, \qquad (28b)$$

$$\hat{\boldsymbol{U}}_n(\boldsymbol{r}, t) = (-1)^n \nabla_w^{2n-1} \left[\frac{\eta^{2n}}{(2n)!} H_1(\boldsymbol{r}, t) + \frac{\eta^{2n+1}}{(2n+1)!} H_2(\boldsymbol{r}, t) \right], \qquad (28c)$$

$$\hat{W}_n(\boldsymbol{r}, t) = (-1)^n \nabla_w^{2(n-1)} \left[\frac{\eta^{2n-1}}{(2n-1)!} H_1(\boldsymbol{r}, t) + \frac{\eta^{2n}}{(2n)!} H_2(\boldsymbol{r}, t) \right], \qquad (28d)$$

where the new operator ∇_w is defined by

$$\nabla_w \zeta \equiv 0 \qquad \text{and} \qquad \nabla_w f(\boldsymbol{r}, t) = \nabla f(\boldsymbol{r}, t) \quad \forall f(\boldsymbol{r}, t) \neq \zeta(\boldsymbol{r}, t). \qquad (29)$$

The operator ∇_w is introduced to make the limit operator $\lim(z \to \zeta)$ and the operator ∇, originally noncommuting, to be *conditionally commutative* if ∇ is replaced by ∇_w to invoke the condition that of all the functions $f(\boldsymbol{r}, t)$, *only* the wave elevation $\zeta(\boldsymbol{r}, t)$ is ignored (or waived) as a variable, and *only* with respect to the operator ∇_w. Thus, using ∇_w helps retain for $\hat{\boldsymbol{u}}$ and $\hat{w}$ the same simple expressions as those in (22)–(24) for $\boldsymbol{u}$ and w by avoiding the lengthy expansions of $\nabla^m(z + h)^n$ as a prerequisite for taking their limit as $z \to \zeta(\boldsymbol{r}, t)$. This is exemplified above for $\hat{\boldsymbol{U}}_0$, and further

$$\hat{\boldsymbol{u}}_1 = A_1 \boldsymbol{u}_0 = -\frac{1}{2} \nabla_w \eta^2 \nabla \cdot \boldsymbol{u}_0 = -\frac{1}{2} \eta^2 \nabla\nabla \cdot \boldsymbol{u}_0 - \eta(\nabla h)\nabla \cdot \boldsymbol{u}_0, \qquad (30a)$$

$$\hat{\boldsymbol{U}}_1 = -\nabla_w \left[\frac{1}{2} \eta^2 H_1 + \frac{1}{3!} \eta^3 H_2 \right]$$

$$= -\frac{1}{2} \eta^2 \nabla H_1 - \frac{1}{3!} \eta^3 \nabla H_2 - \left(\eta H_1 + \frac{1}{2} \eta^2 H_2 \right) \nabla h, \qquad (30b)$$

etc. Such expansions become increasingly more complicated with increasing n.

For this series solution we can show that if $\breve{u}$ and ζ are analytic everywhere in the flow domain, the original series in (22)–(24) are all convergent, absolutely and uniformly within their radius of convergence, which is infinite (the radius being the maximum of the bounds of $\breve{u}$ and ζ and their derivatives).

Summing up, we see that by (25) and (21), u_o and $\breve{w}$ are unique linear functions of $\breve{u}$, $h(r, t)$ being given. Because both $\hat{U}$ and $\hat{W}$ contain only $(\breve{u}, \zeta)$ as basic unknowns, by (26)–(28), $\hat{u}$ and $\hat{w}$ are unique functions of $(\breve{u}, \zeta)$. By substituting these relationships for $\hat{u}$ and $\hat{w}$ in (5) and (12), we therefore obtain a set of model equations for evaluating fully nonlinear, fully dispersive, unsteady gravity-capillary waves on water of variable depth in terms of $(\breve{u}, \zeta)$ as the basic variables to give the following system.

A. The $(\breve{u}, \zeta)$ System: The Bottom Velocity Base

The $(\breve{u}, \zeta)$ system for modeling FNFD water waves comprises, straightforwardly, the following set of basic equations:

$$\zeta_t + \hat{u} \cdot \nabla \zeta = \hat{w}, \tag{5n}$$

$$\hat{D}\hat{u} + [g + \varepsilon\hat{D}^2\zeta]\nabla\zeta = -\nabla p_a - \gamma\nabla\nabla \cdot n, \tag{12n}$$

$$\hat{u} = \hat{u}[\breve{u}, \zeta], \tag{26n}$$

$$\hat{w} = \hat{w}[\breve{u}, \zeta]. \tag{27n}$$

Here, (5n) and (12n) are (5) and (12) in dimensionless form of (14). The only term involving ε in (5n) and (12n) is the one of $O(\varepsilon)$ in (12n), which is due to the effect of vertical fluid acceleration. Equations (26n) and (27n) stand for (26) and (27) with u_0 converted into $\breve{u}$ by using (25). This system of model equations is exact.

It is important to note that due to the uniform and absolute convergence of the series involved in model equations (26n) and (27n), it is no longer necessary to require ε to be small as originally assumed, and we can indeed set $\varepsilon = 1$ by rescaling all the lengths, vertical as well as horizontal, by a typical depth $h_0(=1)$. Physically, this implies that the original assumption for long waves on shallow water can be relaxed, even totally relaxed provided the series involved remain intact, i.e., without severe truncations unless ε is actually small. In this context, the present theory is shown in Section V to provide the third-order theory of the Stokes wave in water of arbitrary depth (even infinite depth). The parameter ε, however, will be retained for general reference and use.

For water depth variation, we assume $|h_t| \leqslant O(\varepsilon^{1/2})$ and $|\nabla h| \leqslant O(\varepsilon^{1/2})$ so as to comply with (7) and (21) since $w = O(\varepsilon^{1/2})$ and $\check{u} = O(1)$. However, this restriction can also be considerably relaxed in conjunction with the relaxation of ε.

B. The $(\hat{u}, \zeta)$ System: The Free-Surface Velocity Base

We next consider eliminating $\check{u}$ from (26n) and (27n) to achieve a functional relation as $\hat{w} = \hat{w}[\hat{u}, \zeta]$. By direct series inversion, (26) yields

$$u_0(r, t) = J[\Delta\hat{u}(r, t)] = \sum_{n=0}^{\infty} \varepsilon^n J_n \Delta\hat{u}, \qquad (\Delta\hat{u} \equiv \hat{u} - \hat{U}), \qquad (31)$$

$$J_0 = 1, \qquad J_n = -\sum_{n=0}^{n-1} A_{n-m} J_m \qquad (n = 1, 2, \ldots),$$

which determines all the J_n's in terms of A_n, with the leading few given by

$$J_1 = -A_1, \qquad J_2 = A_1^2 - A_2, \qquad J_3 = A_1 A_2 + A_2 A_1 - A_3 - A_1^3, \qquad \text{etc.}$$

Here we note that the operators A_n and J_m are in general noncommutative, i.e., $A_n J_m \neq J_m A_n$ $(m + n > 2)$; it is nevertheless true that the inversion is identical on u_0 or on $\hat{u}$, i.e., $AJ = JA = 1$, as can be easily shown. Substituting (31) into (27) to eliminate u_0 yields

$$\hat{w} - \check{w} - \hat{W} = C[u_0] = CJ[\Delta\hat{u}] = Q[\Delta\hat{u}] = \sum_{n=1}^{\infty} \varepsilon^{n-1} Q_n(\hat{u} - \hat{U}), \quad (32)$$

$$Q_n = \sum_{m=0}^{n-1} C_{n-m} J_m \qquad (n = 1, 2, \ldots);$$

$$Q_1 = C_1, \qquad Q_2 = C_2 - C_1 A_1, \qquad Q_3 = C_3 - C_2 A_1 + C_1(A_1^2 - A_2), \qquad \text{etc.}$$

Therefore we have jointly from (5n), (12n), (25), (31), and (32) the basic equations:

$$\zeta_t + \hat{u} \cdot \nabla\zeta = \check{w} + \hat{W} + \sum_{n=1}^{\infty} \varepsilon^{n-1} Q_n(\hat{u} - \hat{U}), \qquad (5s)$$

$$\hat{D}\hat{u} + [g + \varepsilon\hat{D}^2\zeta]\nabla\zeta = -\nabla p_a - \gamma\nabla\nabla \cdot n, \qquad (12s)$$

$$\check{u} + \varepsilon(h_t + \check{u} \cdot \nabla h)\nabla h = u_0 = \sum_{n=0}^{\infty} \varepsilon^n J_n(\hat{u} - \hat{U}). \qquad (31s)$$

where $\hat{U}$ and $\hat{W}$ are given in (28b–d), and Q_n by (32). These equations, expressed in terms of $(\hat{u}, \zeta)$, are exact for modeling FNFD waves in water of variable depth. (Note that $\hat{U}$ and $\hat{W}$ involve $\check{w}$, hence $\check{u}$ — this can perhaps be conveniently dealt with by iteration.)

C. The $(\bar{u}, \zeta)$ System: The Depth-Mean Velocity Base

Another system of model equations uses the depth-mean velocity, denoted by $\bar{u}$, and the surface elevation ζ as the basic variables. This base, adopted by Boussinesq (1871) in introducing the Boussinesq model, has been in common use, e.g., by Wu (1979, 1981) for developing a generalized Boussinesq model for simulating three-dimensional waves in water of variable depth under external forcing at resonance, and by Wu (1998a, 1999a) for this series of studies, among many others.

Taking the average of (2) over the instantaneous water column $(-h(r, t) < z < \zeta(r, t))$ under conditions (5) and (7), we obtain an exact depth-mean continuity equation as follows:

$$\eta_t + \nabla \cdot (\eta \bar{u}) = 0 \qquad (\eta = h(r, t) + \zeta(r, t)), \tag{33}$$

where $\bar{u}$ is the depth mean of $u(r, z, t)$ defined [for arbitrary function $f(r, z, t)$] by

$$\bar{f}(r, t) = \frac{1}{\eta} \int_{-h}^{\zeta} f(r, z, t)\, dz \qquad (\eta = h(r, t) + \zeta(r, t)). \tag{34}$$

Applying the averaging operation (34) to (23) and (22a–d), we obtain (in operator form)

$$\bar{u}(r, t) - \bar{U} = B[u_0] = \sum_{n=0}^{\infty} \varepsilon^n \bar{u}_n = \sum_{n=0}^{\infty} \varepsilon^n B_n u_0, \qquad \bar{U} = \sum_{n=0}^{\infty} \varepsilon^{n+1} \bar{U}_n, \tag{35a}$$

$$\bar{u}_n = (-1)^n \frac{1}{\eta} \int_{-h}^{\zeta} \nabla^{2n-1} \left[\frac{(z+h)^{2n}}{(2n)!} \nabla \cdot u_0 \right] dz = B_n u_0,$$

$$B_0 = 1, \qquad B_n = (-1)^n \frac{1}{\eta} \nabla_w^{2n-1} \frac{\eta^{2n+1}}{(2n+1)!} \nabla \cdot, \qquad (n = 1, 2, \ldots) \tag{35b}$$

$$\bar{U}_0(r, t) = \frac{1}{2} \eta \nabla \check{w} + \check{w} \nabla h, \tag{35c}$$

$$\bar{U}_n = \frac{(-1)^n}{\eta} \int_{-h}^{\zeta} \nabla^{2n-1} \left[\frac{(z+h)^{2n}}{(2n)!} H_1(r,t) + \frac{(z+h)^{2n+1}}{(2n+1)!} H_2(r,t) \right] dz$$

$$= \frac{(-1)^n}{\eta} \nabla_w^{2n-1} \left[\frac{\eta^{2n+1}}{(2n+1)!} H_1(r,t) + \frac{\eta^{2n+2}}{(2n+2)!} H_2(r,t) \right]. \qquad (35d)$$

Here, using the operator ∇_w, defined by (29), makes the integral operator and the operator ∇^{2n-1} conditionally commutative on the condition that ζ is ignored as a variable with respect to ∇_w. With ∇ replaced by ∇_w, the integral operator can clearly be moved across the operator ∇_w^{2n-1} since the upper integration limit ζ is ignored as a variable and the lower limit at $z = -h$ has no contribution due to the quantity in $[\cdot] = O(z+h)^{2n}$ as $z \to -h$. For $n = 1$, e.g.,

$$\bar{U}_1 = -\frac{1}{\eta} \nabla_w \left[\frac{\eta^3}{3!} H_1 + \frac{\eta^3}{4!} H_2 \right] = -\frac{\eta^2}{3!} \nabla H_1 - \frac{\eta^3}{4!} \nabla H_2 - \left(\frac{\eta}{2!} H_1 + \frac{\eta^2}{3!} H_2 \right) \nabla h.$$
$$(35e)$$

From the above expansions, we find the series inversion of (35d) as

$$u_0(r,t) = K[\Delta \bar{u}(r,t)] = \sum_{n=0}^{\infty} \varepsilon^n K_n \Delta \bar{u}, \qquad (\Delta \bar{u} = \bar{u} - \bar{U}), \qquad (36)$$

$$K_0 = 1, \qquad K_n = -\sum_{n=0}^{n-1} B_{n-m} K_m \quad (n = 1, 2, \ldots);$$

$$K_1 = -B_1, \quad K_2 = B_1^2 - B_2, \quad K_3 = B_1 B_2 + B_2 B_1 - B_3 - B_1^3, \quad \text{ctc.}$$

Substituting eq. (36) into (26) yields

$$\ddot{u} - \hat{U} = A\lfloor u_0(r,t) \rfloor = AK[\Delta u] = L[\Delta \bar{u}] = \sum_{n=0}^{\infty} \varepsilon^n L_n(\bar{u} - \bar{U}), \qquad (37)$$

$$L_0 = 1, \qquad L_n = \sum_{m=0}^{n} A_{n-m} K_m, \quad (n = 1, 2 \ldots);$$

$$L_1 = A_1 - B_1, \quad L_2 = A_2 - B_2 - L_1 B_1, \quad L_3 = A_3 - B_3 - L_1 B_2 - L_2 B_1, \quad \text{etc.}$$

The $(\bar{u}, \zeta)$ system therefore has the basic equations (33), (12), and the inverted series (37):

$$\eta_t + \nabla \cdot (\eta \bar{u}) = 0 \qquad (\eta = h(r,t) + \zeta(r,t)), \qquad (33m)$$

$$\hat{D}\hat{u} + [g(t) + \varepsilon \hat{D}^2 \zeta] \nabla \zeta = -\nabla p_a - \gamma \nabla \nabla \cdot n, \qquad (12m)$$

$$\hat{u}(r,t) = \bar{u} + (\hat{U} - \bar{U}) + \sum_{n=1}^{\infty} \varepsilon^n L_n(\bar{u} - \hat{U}), \qquad (37m)$$

where L_n is given by (37), together with (25) correlating $\hat{u}$ to $\check{u}$ involved in $\hat{U}$ and $\bar{U}$.

In addition, we cite the inverted series of (37), expressing $\bar{u}$ in terms of $\hat{u}$ as

$$\bar{u} - \bar{U} = B[u_0(r,t)] = BJ[\Delta\hat{u}] = M[\Delta\hat{u}] = \sum_{n=0}^{\infty} \varepsilon^n M_n(\hat{u} - \hat{U}), \quad (38)$$

$$M_0 = 1, \quad M_n = \sum_{m=0}^{n} B_{n-m} J_m, \quad (n = 1, 2, \ldots);$$

$$M_1 = B_1 - A_1, \quad M_2 = B_2 - A_2 - M_1 A_1,$$

$$M_3 = B_3 - A_3 - M_1 A_2 - M_2 A_1, \quad \text{etc.}$$

Substituting (38) in (33) gives a new continuity equation in terms of $(\hat{u}, \zeta)$ as

$$\eta_t + \nabla \cdot (\eta\hat{u}) = \nabla \cdot \eta \left[(\hat{U} - \bar{U}) - \sum_{n=1}^{\infty} \varepsilon^n M_n(\hat{u} - \hat{U}) \right], \quad (\eta = h + \zeta). \quad (39)$$

This equation can be used equivalently in place of (5s) for $(\hat{u}, \zeta)$ system analysis.

D. The (u_*, ζ) System: The Intermediate-Depth Base

Another approach of interest is to select an intermediate (constant) depth, at $z = -h_* = -\beta h_0 = -\beta$ $(0 \leqslant \beta < 1)$, at which depth $u = u_*$, say, for deriving a new system of model equations. Here the objective is to seek such an optimum β, including the mean free surface at $z = 0$ $(\beta = 0)$, as to provide the "best fit" with the linear wave dispersion relationship (over a specific range of ε; see, e.g., Section V.A) for a wave model represented by a finite series truncated of (23) and (24) to $O(\varepsilon^N)$ for some N, as pursued by Nwogu (1993), Madsen *et al.* (1991, 1992, 1996), Schäeffer and Madsen (1995), Wei *et al.* (1995), and Agnon *et al.* (1999), among others [e.g., in Madsen and Schäffer (1998), $N = 3$ with $\alpha = O(\varepsilon)$, in the present notation, for a range up to $\varepsilon = h_0^2/\lambda^2 = O(1)$]. We pursue this case based on the general solution (23)–(25) as follows.

For an intermediate depth β $(0 \leqslant \beta < 1)$, (23) gives the local velocity u_* as

$$u_* - U_* \equiv \Delta u_* = A^*[u_0] = \sum_{n=0}^{\infty} \varepsilon^n A_n^* u_0, \quad U_* = \sum_{n=0}^{\infty} \varepsilon^{n+1} U_n^*, \quad (40a)$$

$$A_0^* = 1, \qquad A_n^* = \frac{(-1)^n}{(2n)!} \nabla^{2n-1} \eta_*^{2n} \nabla \cdot, \qquad \eta_* \equiv h(r, t) - \beta, \qquad (40b)$$

$$U_n^* = (-1)^n \nabla^{2n-1} \left[\frac{\eta_*^{2n}}{(2n)!} H_1(r, t) + \frac{\eta_*^{2n+1}}{(2n+1)!} H_2(r, t) \right]. \qquad (40c)$$

The inversion of (40a) is given by

$$u_0 = J_* [\Delta u_*(r, t)] = \sum_{n=0}^{\infty} \varepsilon^n J_n^* (u_* - U_*), \qquad (41)$$

$$J_0^* = 1, \qquad J_n^* = - \sum_{n=0}^{n-1} A_{n-m}^* J_m^* \qquad (n = 1, 2, \ldots).$$

Substituting (41) into (26) and (27) yields

$$\hat{u} = \hat{U} + \sum_{m=0}^{\infty} \varepsilon^n P_n^* (u_* - U_*), \qquad \hat{w} = \check{w} + \hat{W} + \sum_{m=1}^{\infty} \varepsilon^{n-1} Q_n^* (u_* - U_*), \qquad (42)$$

$$P_0^* = 1, \qquad P_n^* = \sum_{m=0}^{n} A_{n-m} J_m^*, \qquad Q_n^* = \sum_{m=0}^{n-1} C_{n-m} J_m^* \qquad (n = 1, 2, \ldots).$$

Thus, (5s), (12s), and (42) constitute the set of model equations for the $\{u_*, \zeta\}$ system.

Summarizing, we have obtained four sets of theoretical models for describing FNFD unsteady gravity-capillary waves on water of variable depth in terms of four sets of basic variables. We have noted that if u_0 and ζ are analytic everywhere in the flow domain, all the original series in (23) and (24) converge uniformly and absolutely with an infinite radius of convergence. The inverted series (31), (32), and (35)–(38) then define the inverse functions of the original functions with the property that they are all analytic within the flow domain since these inverted series are noted to possess a finite radius of convergence. The same argument prevails as before on relaxing the assumption that ε be small and the waves long.

In principle, these four models, being all exact, are therefore entirely equivalent in modeling the class of FNFD water waves without any limitation to the order of magnitude of nonlinearity and dispersion. Possible differences may arise, however, when they are reduced to approximate sets (e.g., by consistent series truncation to equal order), as will be illustrated in Section VI.A for waves of Boussinesq's class. At low orders, it has been demonstrated (Mei, 1983; Madsen *et al.*, 1991) that the accuracy of the linear dispersion relationship for large wavenumbers [with $\varepsilon = O(1)$] is sensitive to the choice of velocity variables used in the basic equations.

For long waves in shallow water, variations in water depth give rise, at $O(\varepsilon^n)$, to various derivatives of $h(r, t)$ through the terms

$$A_n \boldsymbol{u}_0 = (-1)^n/(2n)! \nabla_w^{2n-1} \eta^{2n} \nabla \cdot \boldsymbol{u}_0 \quad \text{and} \quad \hat{U}_n(\boldsymbol{r}, t),$$

in which there are terms with $\nabla^m h$, $(m = 0, 1, \ldots, 2n - 1)$. All of these terms vanish at $\eta = h + \zeta = 0$, i.e., at a physical (moving) waterline.

We further note that the $(\breve{\boldsymbol{u}}, \zeta)$ system seems to be the most straightforward and efficient for computational applications, especially for finite and large water depths [$\varepsilon \geqslant O(1)$], since the series substitutions involved are all direct, as will be shown, for example, in Section V.A for application to the Stokes waves in deep water.

Finally, we realize that for determining solutions to these model equations for initial-boundary value problems with external forcing by surface pressure $p_a(\boldsymbol{r}, t)$ and seabed motion $h(\boldsymbol{r}, t)$, effective numerical schemes are useful. When approximate solutions are sought, with the series solution truncated to a certain low order, analytical solutions may exist, possibly in closed form. These possibilities will be illustrated later. After a solution is thus attained for a problem, the general solution is then known for the entire flow field by the original series (23)–(25) for $\boldsymbol{u}$ and w, and the pressure by the Bernoulli equation.

III. Boundary Integral Closure

Instead of using the series solution given above for closure, we may also adopt the solution in boundary integral form to close the basic equations (5n) and (12n), as will be derived below separately for the two- and three-dimensional waves on water of variable depth. These closure equations are useful for direct applications to numerical computations of FNFD waves.

A. Two-Dimensional Waves: Contour Boundary Integral Closure

For two-dimensional irrotational water waves in particular, we can derive an alternative closure relation by applying Cauchy's contour integral formula. Here we consider the class of two-dimensional irrotational water waves in the vertical xy plane (with a slight change in convention of coordinates), with the y axis pointing vertically upward from $y = 0$ at the rest water level, and with the x-, y-velocity components denoted by u and v. The flow is assumed to be incompressible, inviscid, and irrotational as before and the wave field is either periodic with period $2A$, in $-A < x < A$, or localized, with the fluid at infinity at rest. The complex velocity $w(z, t) = u(x, y, t) - iv(x, y, t)$ is an analytic function of the complex coordinate variable $z = x + iy$ at any time instant t, and by Cauchy's integral formula,

$$i\Omega(z)w(z, t) = \oint_{\partial \mathscr{D}} \frac{w(z_1, t)}{z_1 - z} \, dz_1, \tag{43}$$

where the contour is counterclockwise along the boundary $\partial \mathscr{D}$ of the flow domain $\mathscr{D}$, $\Omega = 2\pi$ for $z \in \mathscr{D}$, and for $z \in \partial \mathscr{D}$, $\Omega(z)$ is the two-dimensional interior angle subtended by $\partial \mathscr{D}$ at z so that if $\partial \mathscr{D}$ is a Jordon (smooth) contour, then $\Omega = \pi$ for $z \subset \partial \mathscr{D}$, and $\Omega = 0$ for $z \notin \mathscr{D} + \partial \mathscr{D}$. Further, for $z \in \partial \mathscr{D}$, the integral assumes its Cauchy principal value. The domain $\mathscr{D}$ will be taken as either a single cell in case of periodic flow, or otherwise the entire flow field, so that the boundary $\partial \mathscr{D}$ consists of four parts: $C_- + B_- + C_+ + B_+$, where $C_\pm (x = \pm A, -h < y < \zeta)$ are two vertical paths connecting $B_- (-A < x < A, y = -h(x, t))$ and $B_+ = (-A < x < A, y = \zeta(x, t))$ for periodic flow with period $2A$, or with $A = \infty$ for localized waves. The contributions from C_+ and C_- cancel by virtue of the flow being periodic or being at rest at infinity.

Along the boundary $\partial \mathscr{D}$, we denote $\hat{z} = x + i\zeta$, $\check{z} = x - ih(x)$, $\hat{z}_1 = x_1 + i\zeta(x_1)$, $\check{z}_1 = x_1 - ih(x_1)$, $\hat{w} = w(\hat{z}, t) = \hat{u} - i\hat{v}$, $\check{w} = w(\check{z}, t) = \check{u} - i\check{v}$, $\hat{w}_1 = w(\hat{z}_1, t)$, and $\check{w}_1 = w(\check{z}_1, t)$. Applying Cauchy's formula (43) to an arbitrary point $\hat{z}$ on the upper free surface and to another point $\check{z}$ on the seafloor and taking the real and imaginary parts, respectively, of the resulting equations, we obtain

$$\pi \hat{v} = \text{Re} \oint_{\partial \mathscr{D}} \frac{u_1 - iv_1}{z_1 - \hat{z}} \, dz_1, \tag{44}$$

$$\pi \check{u} = \text{Im} \oint_{\partial \mathscr{D}} \frac{u_1 - iv_1}{z_1 - \check{z}} \, dz_1, \tag{45}$$

where Re and Im denote the real part and imaginary part, respectively. These two equations join (5) and (12) to provide four relations between four boundary variables, $\hat{u}$, $\hat{v}$, $\check{u}$, and ζ [noting that by (7), $\check{v} = -\check{D}h$ involves only

$\breve{u}$ as an unknown], thereby closing the system to form a new set of exact model equations as follows:

$$\hat{v} = \hat{D}\zeta, \qquad (\hat{D} = \partial_t + \hat{u}\partial_x), \tag{5c}$$

$$\hat{u}_t + \hat{u}\hat{u}_x + [g + \hat{D}^2\zeta]\zeta_x = -\partial_x p_a - \gamma\partial_x\nabla\cdot\mathbf{n}, \tag{12c}$$

$$\pi\hat{v} = \text{Re}\left[\int_{-A}^{A} \frac{\breve{u}_1 + i\breve{D}_1 h}{\breve{z}_1 - \hat{z}} d\breve{z}_1 - \int_{-A}^{A} \frac{\hat{u}_1 - i\hat{v}_1}{\hat{z}_1 - \hat{z}} d\hat{z}_1\right], \tag{44c}$$

$$\pi\breve{u} = \text{Im}\left[\int_{-A}^{A} \frac{\breve{u}_1 + i\breve{D}_1 h}{\breve{z}_1 - \breve{z}} d\breve{z}_1 - \int_{-A}^{A} \frac{\hat{u}_1 - i\hat{v}_1}{\hat{z}_1 - \breve{z}} d\hat{z}_1\right], \tag{45c}$$

where $\breve{D}_1 = (\partial_t + \breve{u}_1\partial_x)$, (5c) and (12c) are the two-dimensional versions of (5n) and (12n), and (44c) and (45c) expand the integral equations (44) and (45) in detail, with condition (7) applied. They are Fredholm's integral equations of the second kind, well known for their convergence and efficiency in numerical computation for solution.

B. THREE-DIMENSIONAL WAVES: BOUNDARY INTEGRAL EQUATION

For three-dimensional irrotational water waves, several standard methods are available to express the solution to eq. (15) for the velocity potential in terms of a boundary integral. One of them is by applying Green's second identity to $\phi(\mathbf{x}, t)$ and the free-space Green's function $G(\mathbf{x} - \mathbf{x}_1)$ [in this section, the three-dimensional position vector $\mathbf{x} = (\mathbf{r}, z) = (x, y, z)$ is adopted for simplicity] to give, for $\mathbf{x} \in \partial\mathcal{D}$ (which is the boundary of the three-dimensional flow domain $\mathcal{D}$), the equation

$$\Omega(\mathbf{x})\phi(\mathbf{x}, t) = \int_{\partial\mathcal{D}} [\phi_n(\mathbf{x}_1, t)G(\mathbf{x} - \mathbf{x}_1) - \phi(\mathbf{x}_1, t)G_n(\mathbf{x} - \mathbf{x}_1)] dS_1, \tag{46a}$$

$$G(\mathbf{x} - \mathbf{x}_1) = -|\mathbf{x} - \mathbf{x}_1|^{-1} \tag{46b}$$

where the subscript n denotes the outward normal derivative to $\partial\mathcal{D}$ at $\mathbf{x}_1$, $\Omega(\mathbf{x})$ is the interior solid angle subtended by $\partial\mathcal{D}$ with respect to $\mathbf{x} \in \partial\mathcal{D}$, so that $\Omega = 2\pi$ if $\partial\mathcal{D}$ is a smooth boundary surface, and the Cauchy principal value of the singular integral is understood. With the free-space Green's function, this expression for ϕ represents a distribution of sources and dipoles on $\partial\mathcal{D}$. For $\mathbf{x} \in \partial\mathcal{D}$, it is straightforward to derive from (46a,b) an integral equation (of Fredholm's second kind) relating $\hat{w}$ with $\hat{\mathbf{u}}$, $\breve{u}$, and ζ so that the resulting integral equation can join (5) and (12) to form a closed set of basic equations for modeling three-dimensional FNFD water waves. Integration of (46a) can be conveniently executed, like in the two-dimensional case discussed above, over a single cell ($-A < x < A$, $-B < y < B$) for periodic flow, or with $A = \infty$, $B = \infty$ for localized waves.

Alternatively, we can represent ϕ in terms of a source, or dipole distribution over the boundary surface $\partial\mathscr{D}$, namely,

$$\phi(x, t) = \int_{\partial\mathscr{D}} \sigma(x_1, t)G(x - x_1)\, dS_1, \qquad (47)$$

$$\phi(x, t) = -\int_{\partial\mathscr{D}} \mu(x_1, t)\frac{\partial}{\partial n} G(x - x_1)\, dS_1, \qquad (48)$$

with source strength $\sigma(x, t)$, or dipole strength $\mu(x, t)$ per unit area of $\partial\mathscr{D}$. The analysis to obtain an integral equation with these representations, or with a vortex-sheet representation, is classical and the reader is referred to the literature (e.g., Tsai and Yue, 1996; Beale *et al.*, 1996; Hou *et al.*, 1996). Nevertheless, it is worth noting that the boundary integral equation (BIE) methods, actively developed and applied since the 1970s [see, e.g., the review by Yeung, 1982, with the mixed-Euler–Lagrange (MEL) approach], have been applied, if only for strongly nonlinear water wave problems, to almost all two-dimensional cases such as Longuet-Higgens and Cokelet (1976) for steep overturning waves, and Faltinsen (1977) for flows with floating bodies. The primary difficulties in extending them to three dimensions are thought to be largely computational in nature. The very large scale of computation in three-dimensional cases would normally be too large without having innovative algorithms with parallel computer facilities.

IV. Modeling FNFD Waves in Water of Uniform Depth

In the special case of a uniform medium, i.e., with $h = $ const., the general results obtained above are simplified considerably. In fact, with $h = $ const. $= 1$, all the terms U_n and W_n representing the topographic effects vanish. In addition, all the factors $(h + z)^m = (1 + z)^m$, now being independent of r, can be moved freely outside the operator ∇ in (22a–d) and all its related formulas, and so $(1 + \zeta)^m$ can be moved across ∇_w. Thus, the general results of (26)–(28), (31), and (35a–e) reduce to that previously obtained by

Wu (1998a, 1999a), namely,

$$\hat{u} = u(r, \zeta, t) = A[u_0] = \sum_{n=0}^{\infty} \varepsilon^n A_n u_0, \tag{49}$$

$$\bar{u} = B[u_0] = \sum_{n=0}^{\infty} \varepsilon^n B_n u_0, \tag{50}$$

$$\hat{w} = w(r, \zeta, t) = C[u_0] = \sum_{n=1}^{\infty} \varepsilon^{n-1} C_n u_0, \tag{51}$$

$$A_n = \frac{(-1)^n}{(2n)!} \eta^{2n} \nabla^{2n}, \quad B_n = \frac{1}{2n+1} A_n, \quad C_n = \frac{(-1)^n}{(2n-1)!} \eta^{2n-1} \nabla^{2(n-1)} \nabla \cdot, \tag{52}$$

in which $\eta(r, t) = h + \zeta(r, t) = 1 + \zeta(r, t)$. We note that now $u_0(r, t) = \check{u}(r, t)$, by (25).

With the above reduction, the present theory for the general case of variable $h(r, t)$ yields, for the special case of $h = 1$, the same sets of model equations as that of Wu's studies (1998a, 1999a), which we cite for later reference as follows:

A. The (u_0, ζ) system: the basic equations comprise (5n), (12n), (49), and (51) for $(\hat{u}, \hat{w}, \zeta, u_0)$ in explicit form.

B. The $(\hat{u}, \zeta)$ system: the basic equations are (5s), (12s) with $\hat{U} = \hat{W} = 0$, Q_n given by (32), and A_n and C_n given by (52).

C. The $(\bar{u}, \zeta)$ system: the basic equations are (33m), (12m), and (37m), with $\hat{U} = \bar{U} = 0$, L_n given by (37), and A_n and B_n by (52).

D. The (u_*, ζ) system: the basic equations are (5n), (12n), (42), with $\hat{U} = \hat{W} = U_* = 0$, and

$$J_n^* = E_n/(2n)! \, (1 - \beta)^{2n} \nabla^{2n}, \tag{53}$$

where the E_n's are Euler's numbers.

In this special case, the two-dimensional BI equation for closure can also be simplified due to the symmetry that the vertical velocity component vanishes at the horizontal bottom $y = -1$. The flow domain can therefore be <u>continued</u> analytically by Schwarz's principle of symmetry, $w(z+i) = \overline{w(z+i)}$, i.e., $u(x, y+1) = u(x, -y-1)$, $v(x, y+1) = -v(x, -y-1)$. Hence we can revise the original flow boundary to that of the extended flow domain $(-2 - \zeta < y < \zeta, -\infty < x < \infty)$ and thus curtail the need to calculate the velocity $u_0 = u(x, -1)$ on the horizontal seafloor. In fact, this extended contour integral is immediately reduced, by the symmetry, to only one line integral along the free surface as

$$\hat{v} = \frac{1}{\pi} \operatorname{Re} \int_{-A}^{A} \left[\frac{1}{\hat{z}_1 - x + i(2 + \zeta)} - \frac{1}{\hat{z}_1 - \hat{z}} \right] (\hat{u}_1 - i\hat{v}_1)\, d\hat{z}_1. \quad (54)$$

This equation, relating $\hat{v}$ to $\hat{u}$ and ζ, can therefore join (5c) and (12c) to comprise the basic model equations for the case of $h = 1$.

For three-dimensional flows over a horizontal plane, similar simplifications can be attained by invoking the analytic continuation with

$$\mathbf{u}(\mathbf{r}, z + 1) = \mathbf{u}(\mathbf{r}, -(z + 1)) \quad \text{and} \quad w(\mathbf{r}, z + 1) = -w(\mathbf{r}, -(z + 1)), \quad (55)$$

and by extending the flow domain to include its mirror image through reflection into $z = -1$, , and then by combining the contributions from the original free surface and its image.

A. Preliminary Numerical Results

A computational task has been pursued by Qu et al. (2000) to obtain numerical results for two-dimensional, time-evolving nonlinear waves in water of uniform depth based on the present FNFD wave model. Two separate courses are conducted to proceed in parallel but with interactive connections for fine course tuning to facilitate results efficiently. Along one course, the basic equations of the FNFD model, namely, (5), (12), and (54), are adopted to compute the evolution of a specified initial wave, through discrete steps of time marching, into its long-time asymptotic motion of a terminal wave of permanent shape. Along the other course, the corresponding permanent wave is calculated as a solution to the corresponding stationary wave problem formulated in the wave frame with applications of conformal mapping methods and analytic series expansion of the steady wave solution. The latter course, employing numerical methods with an algorithm specially devised, has succeeded in producing numerical results for solitary waves of all finite amplitudes, including the highest one with a cornered peak of interior angle of 120°, achieving the accuracy marked by a local maximum relative error of less than 10^{-5}. These solitary wave solutions are then furnished, invariably with data somewhat distorted to probe the robustness of the numerical code for the unsteady wave computation. It has been found that the convergence of the iterative procedures involved in the numerical program of our FNFD code depends moderately sensitively on the initial data entered; the closer the initial wave is to the permanent profile and the smaller the wave amplitude, the faster the convergence.

Figure 2a shows a time-sequence plot of the asymptotic solution obtained by Qu *et al.* (2000) for a solitary wave of amplitude $\alpha = 0.75815$ after it has achieved the final state within the local maximum error limit of 10^{-4}, with the corresponding wave velocity determined as $c = 1.291$. Figure 2b presents a comparison between the final asymptotic solution and the corresponding solution for a stationary wave of equal amplitude of $\alpha = 0.75815$, while the two results for wave velocity agree within the assigned error limit. The error of the numerical results for stationary wave solutions is less than 10^{-5}.

V. Weakly Nonlinear Fully Dispersive (WNFD) Water Wave Theory

The present theory was initially intended for evaluating long waves in shallow water based on small $\varepsilon = h^2/\lambda^2$. However, this assumption is noted to be not strictly binding (on observing that all the infinite series in ε converge absolutely and uniformly in any closed flow domain) and can therefore be considerably — even totally — relaxed provided that the series are not severely truncated unless ε is actually small. With ε regarded as arbitrary, this theory can also be used for modeling the category of weakly nonlinear fully dispersive (WNFD) waves in water of variable depth by expansion of the basic equations for small wave steepness in the parametric regime shown in Figure 1. For this category, we shall pursue in the next section the expansion methodology based on the $(\hat{u}, \zeta)$ system — in the spirit of Sir G. G. Stokes (1847), who founded the nonlinear theory of dispersive water waves.

A. The Stokes Gravity Waves in Water of Arbitrary Depth

To illustrate the full ε range spanned by the present theory, we consider the infinitesimal one-dimensional gravity waves progressing in water of arbitrary, but uniform (even infinite), depth characterized by

$$0 < \varepsilon = (h/\lambda)^2 < \infty \quad \text{and} \quad \tilde{\alpha} = a/\lambda = (a/h)(h/\lambda) = \alpha\sqrt{\varepsilon} \ll 1, \quad (56)$$

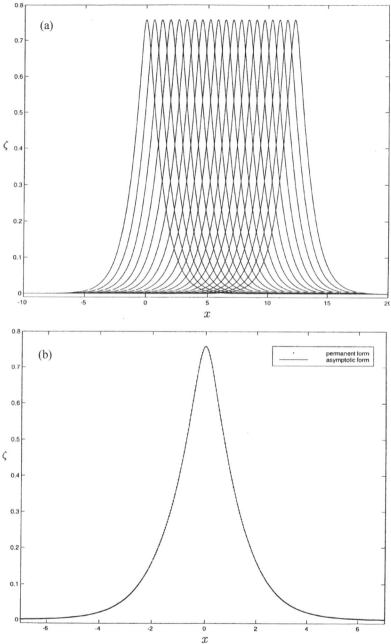

FIG. 2. (a) A time-sequence plot of progression of a large time asymptotic solitary wave profile computed using the present FNFD theory, ——; (b) comparison of the FNFD theory with the solitary wave profile based on permanent wave theory, ····; both for wave amplitude $\alpha = 0.75815$ and phase velocity $c = 1.291$, with maximum local error $<10^{-4}$ and $<10^{-5}$, respectively.

where λ is the wavelength and the water depth is eliminated in $\tilde{\alpha}$, so that the maximum wave slope ($\simeq a/\lambda$) is assumed to be small and h is free to vary from very shallow to very deep extremes. On this premise, we proceed to expand all the flow variables, $f(x, t)$, in power series in $\tilde{\alpha}$ and substitute them into the basic equations to obtain linearized equations to various orders. For brevity, we use

$$f(x, t) = \tilde{\alpha}f_1(x, t) + \tilde{\alpha}^2 f_2(x, t) + \tilde{\alpha}^3 f_3(x, t) + \cdots, \tag{57}$$

for $f = \zeta, u_0, \hat{u}, \hat{w}$. Here, we are interested solely in periodic right-going waves so that

$$f(x, t) = f(\theta), \qquad \theta = kx - \omega t + \delta, \tag{58a}$$

$$\text{or} \qquad \theta = \tilde{k}x_* - \tilde{\omega}t_* + \delta, \qquad \tilde{k} = k\lambda, \qquad \tilde{\omega} = \omega\lambda/c_0, \tag{58b}$$

where θ is the phase function, k is the wavenumber, ω is the circular frequency of the fundamental mode $\exp(\pm i\theta)$, δ being a phase constant, and (x_*, t_*) are given in (14). We omit the $*$ as before but retain $\tilde{k}$ and $\tilde{\omega}$ to distinguish them from the dimensional k and ω. Further, following Stokes, we also expand $\tilde{\omega}$ in a power series as

$$\tilde{\omega} = \tilde{\omega}_0(\tilde{k}) + \tilde{\alpha}\tilde{\omega}_1(\tilde{k}) + \tilde{\alpha}^2\tilde{\omega}_2(\tilde{k}) + \cdots. \tag{59}$$

Because trouble arises starting at the third order, we may put $\tilde{\omega}_1 = 0$ in advance.

Substituting (57)–(59) into (5c), (12c), (49), and (51) yields (with surface tension $\gamma = 0$) a hierarchy as follows:

$O(\tilde{\alpha})$ terms:

$$\tilde{\omega}_0\zeta_1' + \hat{w}_1 = 0,$$

$$\tilde{k}\zeta_1' - \tilde{\omega}_0\hat{u}_1' = 0,$$

$$\hat{u}_1 = \sum_{n=0}^{\infty} \varepsilon^n \frac{(-1)^n}{(2n)!} d^{2n}u_{01}(\theta) = \cos(\kappa d)u_{01}(\theta), \qquad (\kappa = kh, \, d = ' = d/d\theta)$$

$$\sqrt{\varepsilon}\,\hat{w}_1 = \sum_{n=1}^{\infty} \varepsilon^{n-1/2} \frac{(-1)^n}{(2n-1)!} d^{2n-1}u_{01} = -\sin(\kappa d)u_{01};$$

$O(\tilde{\alpha}^2)$ terms:

$$\tilde{\omega}_0\zeta_2' + \hat{w}_2 = \tilde{k}\hat{u}_1\zeta_1',$$

$$\tilde{k}\zeta_2' - \tilde{\omega}_0\hat{u}_2' = -\tilde{k}\hat{u}_1\hat{u}_1' - \varepsilon\tilde{k}\tilde{\omega}_0^2\zeta_1'\zeta_1'',$$

$$\hat{u}_2 = \cos(\kappa d)u_{02} - \zeta_1\kappa d \sin(\kappa d)u_{01},$$

$$\sqrt{\varepsilon}\,\hat{w}_2 = -\sin(\kappa d)u_{02} - \zeta_1\kappa d \cos(\kappa d)u_{01};$$

$O(\tilde{\alpha}^3)$ terms:

$$\tilde{\omega}_0\zeta_3' + \hat{w}_3 = \tilde{k}\hat{u}_1\zeta_2' + (\tilde{k}\hat{u}_2 - \tilde{\omega}_2)\zeta_1',$$

$$\tilde{k}\zeta_3' - \tilde{\omega}_0\hat{u}_3 = -\tilde{k}(\hat{u}_1\hat{u}_2)' - \varepsilon\tilde{k}\tilde{\omega}_0^2(\zeta_1'\zeta_2')' + \varepsilon\tilde{k}^2\omega_0(\hat{u}_1\zeta_1'\zeta_1')' + \tilde{\omega}_2\hat{u}_1',$$

$$\hat{u}_3 = \cos(\kappa d)u_{03} - \zeta_1\kappa d\,\sin(\kappa d)u_{02}$$
$$- [\zeta_2\kappa d\,\sin(\kappa d) + (\zeta_1^2/2)(\kappa d)^2\,\cos(\kappa d)]u_{01},$$

$$\sqrt{\varepsilon}\,\hat{w}_3 = -\sin(\kappa d)u_{03} - \zeta_1\kappa d\,\cos(\kappa d)u_{02}$$
$$- [\zeta_2\kappa d\,\cos(\kappa d) - (\zeta_1^2/2)(\kappa d)^2\,\sin(\kappa d)]u_{01},$$

where the prime denotes $f'(\theta) = df/d\theta \equiv df$, and the operator d in $\sin(\kappa d)$ and $\cos(\kappa d)$ stands for the same repeated $d/d\theta$ as indicated in their series expansion, and $\kappa = kh$.

From the first-order equations, we readily find

$$\zeta_1 = \cos\theta, \quad u_{01} = \beta_{01}\zeta_1, \quad \hat{u}_1 = u_{01}\cosh\kappa,$$

$$\hat{w}_1 = \tilde{\omega}_0\sin\theta, \quad \beta_{01} = (2\kappa/\sinh 2\kappa)^{1/2}, \tag{60}$$

$$\tilde{\omega}_0^2 = (\kappa/\varepsilon)\tanh\kappa, \quad \text{or} \quad \omega_0^2 = gk\tanh\kappa \quad (\kappa = kh). \tag{61}$$

Here, ζ_1 assumes the reference amplitude of unity to let $\tilde{\alpha}$ in (57)–(59) resume the original small amplitude a of ζ_1. The above relation for $\omega_0(k; h)$, which is the *frequency dispersion relation* on linear theory, with h as a parameter, is actually the solvability condition that makes the two differential equations of $O(\alpha)$ linearly dependent to yield the first-order solution. The second-order equations, with $\tilde{\omega}_1 = 0$, are regular and have the solution

$$\zeta_2 = \beta_2\cos 2\theta, \quad u_{02} = \beta_{02}\cos 2\theta, \quad \hat{u}_2 = u_{02}\cosh 2\kappa + \varepsilon\tilde{k}\tilde{\omega}_0\cos^2\theta, \tag{62}$$

$$\beta_2 = \frac{1}{2}\kappa\coth\kappa\left(1 + \frac{3}{2\sinh^2\kappa}\right), \quad \beta_{02} = \frac{3}{4}\frac{\varepsilon\tilde{k}\tilde{\omega}_0}{\sinh^4\kappa} \tag{63}$$

In the third-order equations for (ζ_3, u_{03}), the inhomogeneous terms on the right-hand side involving the lower-order solutions can be expressed as a linear combination of $\sin\theta$ and $\sin 3\theta$. The terms with $\sin 3\theta$ can be resolved by ζ_3 and u_{03} both taking the third harmonic term $\cos 3\theta$. However, the resonance (when the left-hand sides vanish) due to the secular terms in $\sin\theta$ can only be eliminated by imposing the solvability condition which has the third-order frequency term $\tilde{\omega}_2$ determined. When all this is carried through, with somewhat lengthy algebra, the final result is

$$\zeta = a \cos \theta + \beta_2 a^2 \cos 2\theta + \cdots, \tag{64}$$

$$u_0 = \beta_{01} a \cos \theta + \beta_{02} a^2 \cos 2\theta + \cdots,$$

$$\hat{u} = \beta_{01} a \cosh \kappa \cos \theta + \beta_{02} a^2 \cosh 2\kappa \cos 2\theta + (\kappa^3 \tanh \kappa)^{1/2} a^2 \cos^2 \theta + \cdots,$$

$$\hat{w} = \beta_{01} a \sinh \kappa \sin \theta + \beta_{02} a^2 \sinh 2\kappa \sin 2\theta$$

$$+ (1/2)(\kappa^3 \tanh \kappa)^{1/2} a^2 \sin 2\theta + \cdots,$$

$$\frac{\omega}{\omega_0} = 1 + \frac{1}{16}(9 \operatorname{csch}^4 \kappa + 8 \coth^2 \kappa) k^2 a^2 + \cdots. \tag{65}$$

These results agree with the literature (e.g., Wehausen and Laitone, 1960; Whitham, 1974). For long waves, $\varepsilon \ll 1$, these formulas readily reduce to expansions automatically in terms of α/ε, or the Ursell number. With increasing water depth, the bottom effect falls off rapidly for $\varepsilon > 1$ as u_0 decreases exponentially relative to $\hat{u}$, yet u_0 always retains its role exactly, however small it may become itself. For waves in deep water, the dispersion relation reduces to the classical result, $\omega^2 = \omega_0^2(1 + k^2 a^2)$.

The Stokes waves have been pursued in the literature to fifth order; however, it is important to note that these waves are unstable due to sideband disturbances, as shown by Benjamin and Feir (1967).

VI. Weakly Nonlinear Weakly Dispersive (WNWD) Water Wave Models

In this section, the FNFD wave models will be reduced to another approximate category for modeling weakly nonlinear and weakly dispersive (WNWD) waves characterized by the parametric regime with $\alpha \ll 1$ and $\varepsilon \ll 1$ (see Figure 1). By expansion of the FNFD model equations to a specific order in small α and ε, the resulting models with different base variables are found to exhibit slightly different properties of the predicted wave form and velocity, especially at very low orders, in spite of the contention that these models are supposed to be equivalent. Furthermore, these systems of model equations may even have different mathematical properties such as possessing different numbers of conservation laws and distinct Hamiltonian structures. It is therefore of great significance to find out the degree by which they differ and any qualitative distinctions between them. To investigate these features, our central interest is in a comparative study of the solitary-wave solutions in the (1 + 1) dimensions (one space and one time) given by these models of $O(\alpha) = O(\varepsilon) \ll 1$ (which we call the *Boussinesq family*) for the simple case with water depth $h = 1$. In other words, the particular study at hand is pursued at the lowest order at which both the nonlinear and dispersive effects first appear (see Wu and Zhang, 1996a).

A. Various Systems of Model Equations

In the current premise, we further elaborate the scaling of (14) with

$$\zeta = a\zeta' = \alpha\zeta', \quad u = \alpha u', \quad x = \lambda x' = \varepsilon^{-1/2}x', \quad t = \varepsilon^{-1/2}t', \quad \alpha = O(\varepsilon) \ll 1,$$
(66)

with $h = 1$ and with the prime (') omitted for brevity. Then the three sets of model equations reduce for waves on water of uniform depth, in the absence of external forcing and surface tension, to the following sets of equations (in which α and ε have been rescaled out):

(A_2)—the (u_0, ζ) model:

$$\zeta_t + [(1 + \zeta)u_0]_x = (1/6)u_{0xxx},$$
(67)

$$u_{0t} + u_0 u_{0x} + \zeta_x = (1/2)u_{0xxt}.$$
(68)

(B_2)—the $(\hat{u}, \zeta)$ model:

$$\zeta_t + [(1 + \zeta)\hat{u}]_x = -(1/3)\hat{u}_{xxx},$$
(69)

$$\hat{u}_t + \hat{u}\hat{u}_x + \zeta_x = 0.$$
(70)

(C_2)—the $(\bar{u}, \zeta)$ model:

$$\zeta_t + [(1 + \zeta)\bar{u}]_x = 0,$$
(71)

$$\bar{u}_t + \bar{u}\bar{u}_x + \zeta_x = (1/3)\bar{u}_{xxt}.$$
(72)

These sets of basic equations are all accurate up to $O(\alpha^2\varepsilon^{1/2}, \alpha\varepsilon^{3/2}, \varepsilon^{5/2})$, with an error of $O(\varepsilon^{7/2})$, provided that all the derivatives involved are smooth while observing the criterion $\alpha = O(\varepsilon) \ll 1$. Here, the $(\bar{u}, \zeta)$ model is actually the original *Boussinesq's two-equation model*. If these equations are regarded as absolute (i.e., ruling out further justifiable modifications of the highest order terms retained therein by using lower-order relations), and we seek only their exact solutions, then only model (B_2) is found to have a solitary wave solution in closed (but implicit) form. In contrast, models (A_2) and (C_2) are found also to possess solitary wave solutions, but only by numerical means.

In fact, for the solitary wave solution, $\zeta = \zeta(s)$ and $\hat{u} = \hat{u}(s)$, where $s = x - ct$ (c being the undetermined wave velocity), the (B_2) system has this first integral:

$$\zeta = c\hat{u} - \hat{u}^2/2 \equiv Z(\hat{u}; c), \tag{73}$$

$$\hat{u}_s^2 = (3/4)\hat{u}^2(2c + 2 - \hat{u})(2c - 2 - \hat{u}) \equiv G(\hat{u}; c). \tag{74}$$

Clearly, $G(\hat{u}; c)$ has a double zero as $\hat{u} \to 0$, so that $\hat{u}$ falls off exponentially at infinity, and a simple zero at $\hat{u} = \hat{u}_c$, say, at the wave crest where $\zeta = \alpha$. Hence wave speed $c = c(\alpha)$ is given jointly by

$$G(\hat{u}_c, c) = 0 \quad \text{and} \quad Z(\hat{u}_c, c) = \alpha, \tag{75}$$

which in this case yields

$$c = \pm(1 + \alpha/2) \tag{76}$$

with $+$ for the right-going and $-$ for left-going waves. And (74) has a solitary wave solution in implicit form (Wu and Zhang, 1996a) as

$$\left[\frac{1}{\hat{u}} - \frac{1}{2(c+1)}\right]^{1/2} + \left[\frac{1}{\hat{u}} - \frac{1}{2(c-1)}\right]^{1/2} = \frac{1}{\sqrt{(c^2-1)}} \exp\left[\sqrt{\frac{3}{4}(c^2-1)}\,|x-ct|\right]. \tag{77}$$

in which c is given by (76) and the wave profile $\zeta = \zeta(s)$ by (73).

With regard to the phase velocity c for a solitary wave of amplitude α, the $(\bar{u}, \zeta)$ model of (C_2) is found [*à la* (73) and (74)] by Teng and Wu (1992) to have the following explicit form:

$$c = \left[\frac{6(1 + \alpha)^2}{\alpha^2(3 + 2\alpha)}\{(1 + \alpha)\ln(1 + \alpha) - \alpha\}\right]^{1/2}, \tag{78}$$

while we have (76) for model (B_2), and as for model (A_2), we resort to numerical means. By comparison, the phase velocity $c = 1 + \alpha/2$ found for the $(\hat{u}, \zeta)$ model agrees with that found by Korteweg and de Vries (1895) for the KdV model and is numerically the fastest of the three models in question, whereas the phase velocity determined numerically for the (u_0, ζ) model is numerically the slowest. The phase velocity (78) found for the $(\bar{u}, \zeta)$ model is numerically intermediate of the three and is found to be in excellent agreement with the experimental results of Daily and Stephan (1952) and Weidman and Maxworthy (1978), up to amplitudes as high as $\alpha = 0.6$, as shown in Figure 3.

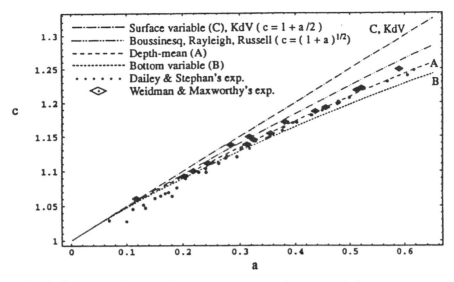

FIG. 3. Comparison between theory and experiment for wave velocity c versus wave amplitude a of solitary waves predicted by different models (A_2, B_2, C_2) (with subscripts omitted).

Regarding the solitary wave profile predicted by these models, Figure 4 (Wu and Zhang, 1996a) shows a comparison between the three models versus Daily and Stephan's experiment (1952) for $\alpha = 0.350$, 0.493, 0.593. The overall agreement is found to be quite good, with the $(\hat{u}, \zeta)$ profile slightly fatter, the (u_0, ζ) profile very slightly leaner than the experiment, and the $(\bar{u}, \zeta)$ profile fit being the closest.

In summary, we first note that the specific depth selected for adopting its local velocity as a basic variable dictates a definite distribution of the dispersive effects between the continuity equation (with a term au_{xxx}, say) and the momentum equation (with bu_{xxt}), to which the different results of A_2, B_2, and C_2 can be attributed. (Interestingly, the difference between these two terms seems to be invariably all giving $b - a = 1/3$.) From these results, we can conclude that the higher the level at which the flow velocity is adopted as a basic variable (in conjunction with ζ), the greater the predicted wave velocity [determined as a part of the solution; see (75)], and the fatter the predicted wave profile, both versus amplitude α as the reference parameter. Of the three models, Boussinesq's two-equation model excels in predicting both wave profile and wave velocity as compared with Daily–Stephan's experiments (1952). Further improvement can be achieved

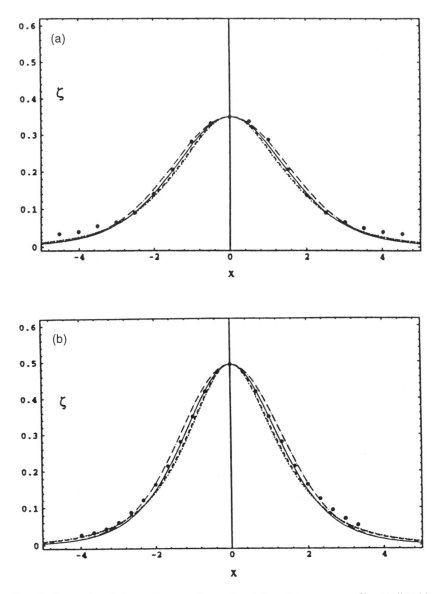

F IG. 4. Comparison between theory and experiment for solitary-wave profiles predicted by different models: (a) for $a = 0.350$; (b) for $a = 0.493$; (c) for $a = 0.593$ (after correcting the misprint $a = 0.61$ in Figure 9 of Daily and Stephan, 1952).

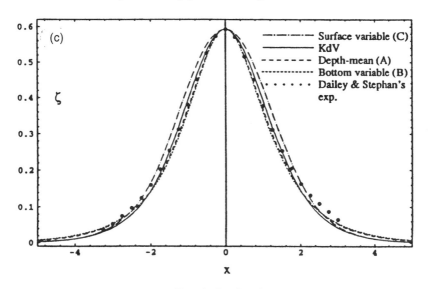

FIG. 4. Continued.

by including higher-order corrections, as will be shown later in Section VIII.B.

B. HAMILTONIAN STRUCTURES

In mathematical studies of nonlinear evolution equations, determination of their Hamiltonian structures is of fundamental importance, because these structures are intimately related to the question, and answer, concerning the integrability of the system.

For the present comparative study (here pursued at the lowest order of nonlinearity and dispersion), it is of great interest to note that of the three model systems at hand, only the $(\hat{u}, \zeta)$ model of (B_2) is known to have a Hamiltonian structure. According to Kaup (1975) and Kupershmidt (1985), the system of equations (69) and (70) has not only a Hamiltonian structure but a tri-Hamiltonian one. In other words, this system has the remarkable property that it can be written in Hamiltonian form in not just one, but three distinct ways.

Indeed, by a suitable rescaling, the coefficient $(1/3)$ of the term u_{xxx} in (69) can be converted to $(1/4)$ without altering (70). Then by the transformation

$$u = -v, \qquad 1 + \zeta = w - v_x/2,$$

(69) and (70) are converted to

$$v_t = \partial(v^2 + 2w - v_x)/2,$$

$$w_t = \partial(vw + w_x/2),$$

where $\partial = \partial/\partial x$. This system of equations has the following tri-Hamiltonian form:

$$V_t = B_1\delta H_3 = B_2\delta H_2 = B_3\delta H_1,$$

$$V = \begin{bmatrix} v \\ w \end{bmatrix}, \qquad B_1 = \begin{bmatrix} 0 & \partial \\ \partial & 0 \end{bmatrix}, \qquad B_2 = \begin{bmatrix} 2\partial & \partial v - \partial^2 \\ v\partial + \partial^2 & w\partial + \partial w \end{bmatrix},$$

$$B_3 = \begin{bmatrix} 2v\partial + 2\partial v & 2w\partial + 2\partial w + \partial(v - \partial)^2 \\ 2w\partial + 2\partial w + (v + \partial)^2\partial & (v + \partial)(w\partial + \partial w) + (w\partial + \partial w)(v - \partial) \end{bmatrix},$$

$$H_1 = \int \frac{1}{2} w \, dx, \quad H_2 = \int \frac{1}{2} wv \, dx, \quad H_3 = \int \left(\frac{1}{2} wv^2 + \frac{1}{2} w^2 + \frac{1}{2} vw_x \right) dx,$$

where

$$\delta H = \begin{bmatrix} \delta H/\delta v \\ \delta H/\delta w \end{bmatrix}$$

denotes the vector of variational differentials.

With this remarkable result, it then follows from the general theorem (e.g., Zakharov, 1968; Olver, 1986) that the system is completely integrable, possesses an infinite number of conservation laws, and integrals can be obtained by application of inverse scattering techniques (IST). Further, the system can support solitary waves, and interactions between solitary waves will always be elastic and remain clean.

In contrast to this conclusive result for system (B_2), no such findings are known for systems (A_2) and (C_2). Extensions of this first-order study to higher-order theories seem to have fundamental interest, basic importance, and broad ramifications in scope.

VII. Fully Nonlinear Weakly Dispersive (FNWD) Water Wave Models

The subject of fully nonlinear and weakly dispersive (FNWD) water waves has been of considerable interest in various aspects. In reality, there exists a parametric regime ($\alpha = O(1)$, $\varepsilon \ll 1$; see Figure 1) in which such waves are found to display salient features as noted in tidal bores, hydraulic jumps, tsunamis, roll waves, etc. Besides, the analogy between supercritical flow of shallow water and supersonic gas flow has an active history, as first noted by Prandtl (1931) and discussed by von Kármán (1938). This hydraulic analogy has been actively applied since the early 1950s in developing water-table experimental methods for investigating two-dimensional high-speed gas flow problems and shock wave dynamics. In addition, this category of wave models has been employed in the literature for investigating strongly nonlinear water wave phenomena, with results assessed in some publications as somewhat better than that of the Boussinesq family models discussed in the preceding section.

For modeling FNWD three-dimensional gravity waves on shallow water of variable depth, theoretical models can be derived directly from the present FNFD wave models by reduction for the pertinent parametric regime with $\alpha = O(1)$, $\varepsilon \ll 1$ (see Figure 1).

To the zeroth order in particular, i.e., with $\varepsilon = 0$, (25)–(28a–d) and (35a–e) reduce to

$$\hat{U} = \hat{W} = 0, \quad \hat{u} = u_0 = \bar{u} = \breve{u} \equiv u, \quad \text{say,}$$

while the basic equations (5) and (12) simplify to become

$$\zeta_t + \nabla \cdot [(h + \zeta)u] = -h_t, \tag{79}$$

$$u_t + u \cdot \nabla u + g\nabla\zeta = -\nabla p_a. \tag{80}$$

This is the set of classical (nonlinear) shallow-water wave equations for (u, ζ), also known as *Airy's model*, here extended to include forcing functions h_t and p_a. This long-wave model may have a certain range of validity in subcritical and supercritical regimes, but is well known for its failure to support solitary waves of permanent form in the transcritical regime, in sharp contrast with the experimental findings pioneered by John Scott Russell. This shortcoming is due to its mathematical characteristics dictating that the higher the wave elevation becomes, the faster its local velocity, hence the steeper the wave on the frontal side than on its lee side, much the same in analogy with the genesis of shock waves in gas dynamics. This particular case of mismatch may indeed cast light on how severe drawbacks can arise from making mismatched accounts of the nonlinear and dispersive effects for modeling water waves without scrutiny.

For modeling this category of FNWD waves to higher orders in ε, the $(\hat{u}, \zeta)$ system is perhaps the simplest of the four systems in providing a set of model equations:

$$\zeta_t + \nabla \cdot (\eta \hat{u}) = -h_t - \nabla \cdot \left\{ \eta \sum_{n=1}^{N} \varepsilon^n \left[M_n \hat{u} + \bar{U}_{n-1} - \hat{U}_{n-1} - \varepsilon \sum_{m=0}^{n-1} M_{n-m} \hat{U}_m \right] \right\},$$

(81)

$$\hat{D}\hat{u} + [g + \varepsilon \hat{D}^2 \zeta]\nabla \zeta = -\nabla p_a - \gamma \nabla \nabla \cdot n.$$
(82)

Here, (81) simply results from truncating the series in (39) to $N(\geqslant 1)$ terms, bearing an error of $O(\varepsilon^{N+1})$, while (82) recites (12s). We remark that the bottom velocity $\check{u}$ involved in $\check{w}$, $\hat{U}_n$, $\hat{W}_n$ can be handled by using (25) iteratively, namely,

$$\check{u} = u_0 - \varepsilon(\check{D}h)\nabla h = u_0 - \varepsilon(\partial_t h + u_0 \cdot \nabla h)\nabla h + O(\varepsilon^2), \quad \text{etc.,}$$
(83)

though a direct conversion is possible for two-dimensional flows, for by (25), we have

$$\check{u} = (u_0 - \varepsilon h_t h_x)/(1 + \varepsilon h_x^2).$$
(84)

More specifically, with $N = 2$ in (81), using the terms involved as defined gives

$$\zeta_t + \nabla \cdot (\eta \hat{u}) = -h_t - \varepsilon \nabla \cdot [\tfrac{1}{3}\eta^3 \nabla \nabla \cdot \hat{u} + \tfrac{1}{2}\eta^2 R_1] + \varepsilon^2 \nabla \cdot (\eta R_2) + O(\varepsilon^3),$$ (85)

$$R_1 = (\nabla h)\nabla \cdot \hat{u} - \nabla \hat{w} = \nabla[h_t + \nabla \cdot (h\hat{u})] - h\nabla \nabla \cdot \hat{u},$$

$$R_2 = \hat{U}_1 - \bar{U}_1 + M_1 \hat{U}_0 - M_2 \hat{u} + O(\varepsilon^3),$$

in which $\hat{U}_0$, $\hat{U}_1$, and $\bar{U}_1$ are given by eqs. (28a–d)–(30) and (35a–e), and $\check{u}$ in $\check{w}$ has been changed to $\hat{u}$ within the error bound.

Alternatively, the continuity equation (5s) can be used to replace (81); that is,

$$\zeta_t + \hat{u} \cdot \nabla \zeta = Q_1 \hat{u} + \check{w} + \sum_{n=1}^{N} \varepsilon^n \left[Q_{n+1} \hat{u} + \hat{W}_n - \sum_{m=0}^{n-1} Q_{n-m} \hat{U}_m \right],$$
(86)

for they are equivalent. However, it is of interest to note that the right-hand side of (81) contains the forcing function h_t [which is implicit in (86)] followed by a term of $O(\varepsilon)$ whereas the right-hand side of (86) is of $O(\varepsilon^0)$, a fact that may relate to computational efficiency, other things being equal. As regards N, it seems (Wu, 1998a) that $N = 2$ is consistent with the $\alpha = O(\varepsilon)$ balance that characterizes the Boussinesq family (see below), leaving the error term in (81) and (85) at $O(\varepsilon^3, \alpha\varepsilon^2)$.

Historically, the $\{\bar{u}, \zeta\}$ system was first used by Green and Naghdi (1976), and by Ertekin *et al.* (1984, 1986), Choi (1995), and Choi and Camassa (1996) among others in developing model equations under the same premise as assumed here. We now develop the theory in $(\bar{u}, \zeta)$ to one order higher than Green–Naghdi's. To proceed, we first take (37) truncated to three terms,

$$\hat{u} = \bar{u} + \varepsilon V_1 + \varepsilon^2 V_2 + O(\varepsilon^3), \tag{87}$$

$$V_1 = L_1\bar{u} + \hat{U}_0 - \bar{U}_0 = -[\tfrac{1}{3}\eta^2\nabla + \tfrac{1}{2}\eta(\nabla h)]\nabla \cdot \bar{u} + \tfrac{1}{2}\eta\nabla\check{w},$$

$$V_2 = L_2\bar{u} + \hat{U}_1 - \bar{U}_1 - L_1\bar{U}_0$$

For the basic equations, (8n) remains intact while the conversion of (12n) involves

$$\hat{D}\hat{u} - \bar{D}\bar{u} = \bar{D}(\hat{u}-\bar{u}) + (\hat{D}-\bar{D})\hat{u} = -\varepsilon(\tfrac{1}{3}\eta^2 G_{11} + \tfrac{1}{2}\eta G_{12}) + \varepsilon^2 G_2 + O(\varepsilon^3), \tag{88}$$

$$G_{11} = \bar{D}\nabla\nabla \cdot \bar{u} - \nabla(\nabla \cdot \bar{u})^2 + (\nabla\nabla \cdot \bar{u}) \cdot \nabla\bar{u},$$

$$G_{12} = [(\nabla h) \cdot \nabla\bar{u} - (\nabla h)\nabla \cdot \bar{u}]\nabla \cdot \bar{u} - [(\nabla\check{w}) \cdot \nabla\bar{u} - (\nabla\check{w})\nabla \cdot \bar{u}] + \bar{D}R_1,$$

$$G_2 = \bar{D}V_2 + V_2 \cdot \nabla\bar{u} + V_1 \cdot \nabla V_1,$$

where use has been made of (8n) and (32) (or, equivalently, $\bar{D}\eta = -\eta\nabla \cdot \bar{u}$ to convert $\bar{D}\eta$) for simplification, and R_1 is given in (85). In addition, we have

$$\varepsilon\hat{D}^2\zeta = \varepsilon\hat{D}[\bar{D}\zeta + (\hat{D} - \bar{D})\zeta] = -\varepsilon(\eta F_1 + \bar{D}^2 h) - \varepsilon^2 F_2 + O(\varepsilon^3), \tag{89}$$

$$F_1 = \bar{D}(\nabla \cdot \bar{u}) - (\nabla \cdot \bar{u})^2,$$

$$F_2 = V_1 \cdot \nabla(\eta\nabla \cdot \bar{u} + \bar{D}h) - \bar{D}(V_1 \cdot \nabla\zeta).$$

Accordingly, substituting (87)–(88) in (12n), we obtain the basic equations in $(\bar{u}, \zeta)$:

$$\zeta_t + \nabla \cdot (\eta\bar{u}) = -h_t \qquad (\eta = h + \zeta), \tag{90}$$

$$\bar{u}_t + \bar{u} \cdot \nabla\bar{u} + g\nabla\zeta = -\nabla p_a - \gamma\nabla K + \varepsilon[\eta F_1\nabla\zeta + \tfrac{1}{3}\eta^2 G_{11} + \tfrac{1}{2}\eta G_{12} + (\bar{D}^2 h)\nabla\zeta]$$
$$+ \varepsilon^2(F_2\nabla\zeta - G_2) + O(\varepsilon^3), \qquad (\kappa = \nabla \cdot n) \tag{91}$$

for modeling FNWD waves in shallow water of variable depth, with (90) being exact and (91) bearing an error of $O(\varepsilon^3)$. In the $O(\varepsilon)$ terms of (91), the last two are due to topographic variation; they vanish if $h = $ constant. The first two terms of $O(\varepsilon)$ can also be written as

$$\eta F_1 \nabla \zeta + \frac{1}{3} \eta^2 G_{11} = \frac{1}{3\eta} \nabla(\eta^3 F_1) + \frac{1}{3} \eta^2 \bar{\omega}_3 \times (\nabla \nabla \cdot \bar{u}) - \eta F_1 \nabla h,$$

upon observing the relation

$$G_{11} = \nabla F_1 + \bar{\omega}_3 \times (\nabla \nabla \cdot \bar{u}), \qquad \bar{\omega}_3 = (0, 0, \bar{v}_x - \bar{u}_y), \tag{92}$$

where $\bar{\omega}_3$ may be called the *virtual vorticity* or *apparent vorticity* vector that arises from the depth averaging under topographic variation effects and has only a vertical component, which is not necessarily zero unless $\nabla h \equiv 0$ (for a proof, see Wu, 1981), and $\times$ denotes the vector cross product. (Note that $\nabla \times \bar{u} = 0$ and $\nabla \times u = 0$ are entirely different in nature.)

This set of basic equations remains valid up to one order higher than Green–Naghdi (1976) and also conforms, to $O(\varepsilon)$, with the more general case for two-layer stratified flows studied by Choi and Camassa (1996). (Note that to extend Green–Naghdi to higher orders will require additional assumptions, if pursued in the original approach.) Finally, it can be argued (Wu, 1998a) that if weakly nonlinear waves are to be included as well, the $[\alpha = O(\varepsilon)]$ balance of Boussinesq's would require that the $O(\varepsilon^2)$ terms be retained. This contention for improvement (over $N = 1$) has been numerically verified by Liu and Wu (1998).

VIII. Higher-Order WNWD Long-Wave Models

Following the leading-order theory of WNWD water waves discussed in Section VI, we now develop a general theory to higher orders for water of variable depth. We deduce the result directly from the FNWD model systems attained above, since with the expansion in small dispersion parameter ε already accomplished, we merely need to expand the FNWD model equations for small α in addition, while observing $\alpha = O(\varepsilon) \ll 1$.

A. Higher-Order WNWD Long Waves in Water of Variable Depth

For water depth variation, we assume (in addition to $\alpha = O(\varepsilon) \ll 1$) that

$$|\nabla h| \leqslant O(\varepsilon^{1/2}), \quad |\nabla^2 h| \leqslant O(\varepsilon), \quad |h_t| \leqslant O(\alpha \varepsilon^{1/2}), \tag{93}$$

which is seen to be consistent with the scaling $u = O(\alpha)$, $w = O(\alpha \varepsilon^{1/2})$ and condition (7). For the $(\hat{u}, \zeta)$ base, applying scaling (93) to (82) and (85) for $\alpha = O(\varepsilon) \ll 1$ yields

$$\eta_t + \nabla \cdot [(1 + \alpha\zeta)\hat{u}] = -\varepsilon[\tfrac{1}{3}\eta^3\nabla^2\nabla \cdot \hat{u} + \tfrac{1}{2}\eta^2\nabla \cdot R_1 - (\eta^2\nabla\nabla \cdot \hat{u} + \eta R_1)\nabla h]$$

$$- \alpha\varepsilon[(\eta^2\nabla\nabla \cdot \hat{u} + \eta R_1)\nabla\zeta - \nabla \cdot (\eta R_2)] + O(\varepsilon^3), \quad (94)$$

$$\hat{u}_t + \alpha\hat{u} \cdot \nabla\hat{u} + \nabla\zeta = -\nabla p_a - \gamma\nabla\nabla \cdot n - \varepsilon\zeta_{tt}\nabla\zeta$$

$$- \alpha\varepsilon(2\hat{u} \cdot \nabla\zeta_t + \hat{u}_t \cdot \nabla\zeta)\nabla\zeta + O(\varepsilon^3), \quad (95)$$

where R_1 and R_2 are given by (85).

Similarly, for the $(\bar{u}, \zeta)$ base, we obtain from (33m), (12m), and (37m), with additional scaling in α, the resulting model equations as

$$\zeta_t + \nabla \cdot [(h + \alpha\zeta)\bar{u}] = -h_t \qquad (\eta = h + \zeta), \quad (96)$$

$$\bar{u}_t + \alpha\bar{u} \cdot \nabla\bar{u} + \nabla\zeta = -\nabla p_a - \gamma\nabla K + \varepsilon[\tfrac{1}{2}\eta\nabla[h_t + \nabla \cdot (\eta\bar{u})]_t - \tfrac{1}{6}\eta^2\nabla^2\bar{u}_t]$$

$$+ \alpha\varepsilon[\eta F_1\nabla\zeta + \tfrac{1}{3}\eta^2 G_{11}^{(2)} + \tfrac{1}{2}\eta G_{12}^{(2)} + (2\bar{u} \cdot \nabla\nabla\zeta + V_2)] + O(\varepsilon^3),$$

$$(97)$$

$$G_{11}^{(2)} = G_{11} - \tfrac{1}{3}\eta^2\nabla^2\bar{u}_t,$$

$$G_{12}^{(2)} = G_{12} - \tfrac{1}{2}\eta\nabla(h_t + \bar{u} \cdot \nabla h)_t.$$

The above higher-order WNWD model equations, with a relative error of $O(\varepsilon^3)$ (relative to the linear term), are one order higher than the existing models. The (96) and (97) system agrees with the gB equations of Wu (1979, 1981), but is higher by one order, and also higher in accuracy than the latter, especially near waterlines where $\eta = h + \zeta = 0$.

B. HIGHER-ORDER SOLITARY WAVES

The simplest case of higher-order WNWD wave theory is concerned with two-dimensional solitary (and cnoidal) waves progressing in a uniform medium. Since the pioneering works by Boussinesq (1872), Rayleigh (1876), and Korteweg-de Vries (1895) on solitary waves in the Euler context, the theory has been further developed to second order by Laitone (1960), and to third order by Chappelear (1962) and Grimshaw (1971). It is of interest to apply the present theory for a comparative study using a new base in $(\hat{u}, \zeta)$ to illustrate the solution construction method in uniformity.

For waves progressing in the $(1 + 1)$ dimension of (x, t) in a uniform medium, permanent in shape, the flow is characterized by variables of the form $f(x, t) = f(s)$, where $s = x - ct$, with c being the undetermined constant wave velocity. In terms of s, the basic equations (5s) and (12s) reduce (in the absence of external forcing and surface tension effect) to ordinary differential equations which can be integrated once to yield

$$-c\zeta + (1 + \zeta)u + \sum_{n=1}^{\infty} \varepsilon^n (1 + \zeta)M_n u = 0, \tag{98}$$

$$\zeta - cu + \tfrac{1}{2}u^2 + \tfrac{1}{2}\varepsilon(c - u)^2\zeta_s^2 = 0, \tag{99}$$

where $\zeta_s = d\zeta/ds$, and $\hat{u}$ is written as u for brevity. These equations are exact in representing the Euler equations for solitary waves in particular. Existence of a solution to this system would therefore imply that the Euler model supports solitary waves. While this contention remains unproven, it has nevertheless been shown by Friedrichs and Hyers (1954) that existence of an exact solution for solitary waves can be described with a specific expansion method in the Euler context, at least asymptotically. Here we attempt to illustrate such an asymptotic analytical construction of higher-order theory of solitary waves.

Assuming $\alpha = O(\varepsilon) \ll 1$, we introduce for u and ζ the expansion

$$u = \sum_{n=1}^{N} \alpha^n u_n, \qquad \zeta = \sum_{n=1}^{N} \alpha^n \zeta_n, \tag{100}$$

for some N $(< \infty)$. We further assume for the wave velocity c and the phase function $\theta = ks$ (k being the wavenumber) the expansion

$$c = c_0 + \alpha c_1 + \alpha^2 c_2 + \alpha^3 c_3 + \cdots, \tag{101}$$

$$\theta = ks = k_0 s(1 + \alpha k_1 + \alpha^2 k_2 + \alpha^3 k_3 + \cdots) \qquad (\theta_0 = k_0 s). \tag{102}$$

It is convenient to determine (u_n, ζ_n) as functions of θ, which we call the *phase function*. Substituting these expansions in (98) and (99) and the operators M_n [by (38) and (52)], we obtain the following hierarchy for various orders in α.

$O(\alpha)$ — for the leading order, we have

$$c_0\zeta_1 - u_1 = 0, \tag{103}$$

$$\zeta_1 - c_0 u_1 = 0. \tag{104}$$

For (u_1, ζ_1) to be solvable, the determinant of the coefficients must vanish; so

$$c_0^2 = 1 \quad \text{or} \quad c_0 = \pm 1; \tag{105}$$

$$u_1 = c_0 \zeta_1. \tag{106}$$

We shall work with $c_0 = 1$ for right-going waves (similarly with $c_0 = -1$ for left-going waves). The equations for solving (u_1, ζ_1) will come from the next step.

$O(\alpha^2)$ — for the second order, the perturbation equations are found to be

$$c_0 \zeta_2 - u_2 = -c_1 \zeta_1 + \zeta_1 u_1 + (k_0^2/3) u_1'', \tag{107}$$

$$\zeta_2 - c_0 u_2 = c_1 u_1 - u_1^2/2, \tag{108}$$

where the prime (') denotes $d/d\theta$. The solvability condition for (u_2, ζ_2) yields, with $c_0 = 1$ and $u_1 = \zeta_1$, the equation

$$(k_0^2/3)\zeta_1'' = 2c_1 \zeta_1 - (3/2)\zeta_1^2. \tag{109}$$

This equation can be integrated once, giving

$$(k_0^2/3)(\zeta_1')^2 = 2c_1 \zeta_1^2 - \zeta_1^3,$$

where the integration constant is taken to be zero to ensure ζ falling off exponentially toward infinity for the solitary wave (or a different constant for analyzing cnoidal waves). In addition, we have the boundary condition, under the assumed symmetry of $\zeta(-\theta) = \zeta(\theta)$ that at the wave crest, $\zeta_1 = a$, $\zeta_1' = 0$, by which c_1 is determined. Therefore,

$$\zeta_1 = a \operatorname{sech}^2(\theta), \quad c_1 = a/2, \quad k_0^2 = 3a/4. \tag{110}$$

This is the Korteweg–de Vries (1895) solitary wave. And we also have, by (106)–(108),

$$u_2 = \zeta_2 + \zeta_1(\zeta_1 - a)/2 \tag{111}$$

from the compatibility. This is needed for the solution in the next step.

$O(\alpha^3)$ — for the third order, we find from the original expansions the equations as

$$c_0 \zeta_3 - u_3 = -c_1 \zeta_2 - c_2 \zeta_1 + \zeta_1 u_2 + \zeta_2 u_1$$
$$+ (k_0^2/3)[u_2'' + (3\zeta_1 + 2k_1)u_1''] + (2/15)k_0^6 u_1^{(6)}, \tag{112}$$

$$\zeta_3 - c_0 u_3 = c_1 u_2 + c_2 u_1 - u_1 u_2 - (k_0^2/2)(\zeta_1')^2. \tag{113}$$

Substituting the lower-order equations and their solutions in (112) and (113) and after some straightforward, albeit lengthy algebra, we obtain, for (u_3, ζ_3) to be solvable, the equation

$$(k_0^2/3)\zeta_2'' + (3\zeta_1 - a)\zeta_2 = Z_2(\zeta_1; k_1, c_2), \tag{114a}$$

$$Z_2(\zeta_1; k_1, c_2) = -\tfrac{3}{2}\zeta_1^3 + 3(a + k_1)\zeta_1^2 + (2c_2 - 2ak_1 - \tfrac{19}{20}a^2)\zeta_1. \tag{114b}$$

We notice that by (109), ζ_1' is a solution of the homogeneous equation of (114a), i.e.,

$$(k_0^2/3)\zeta_1''' + (3\zeta_1 - a)\zeta_1' = 0.$$

Hence (114a–b) can be integrated once by Lagrange's scheme to give

$$\frac{k_0^2}{3}(\zeta_1'\zeta_2' - \zeta_2\zeta_1'') = -\frac{3}{8}\zeta_1^4 + (a + k_1)\zeta_1^3 + \left(c_2 - ak_1 - \frac{19}{40}a^2\right)\zeta_1^2. \tag{115}$$

There are two boundary conditions. First, if we invoke that ζ_2 decays toward infinity no slower than ζ_1, then the term with ζ_1^2 on the right-hand side of (115) is lower in order than the left-hand side and must therefore have its coefficient vanish, thus providing one relation between c_2 and k_1 In addition, at the wave crest, $\zeta_1 = a$ (with $\zeta_1' = 0$), we have $\zeta_2 = 0$ (by definition of $\zeta = a$), by which k_1 is determined. Hence we have for (u_2, ζ_2) the solution

$$\zeta_2 = -\tfrac{3}{4}a^2 \mathrm{sech}^2(\theta)\tanh^2(\theta), \quad k_1 = -\tfrac{5}{8}a, \quad c_2 = -\tfrac{3}{20}a^2, \tag{116}$$

with u_2 given by (111). This is Laitone's (1960) solitary wave of second order. In addition, we have, by (113) and the lower-order relations, the new relation

$$u_3 = \zeta_3 + \left(\zeta_1 - \frac{a}{2}\right)\zeta_2 - \zeta_1^3 + \frac{3}{4}\zeta_1^2 + \frac{2}{5}\zeta_1, \tag{117}$$

which is needed for the next order analysis.

Solution can be pursued to higher orders by following the same procedure as just developed. At the fourth order next, we obtain a solvability condition for (u_4, ζ_4) in terms of an equation for ζ_3 that is just like (114a–b) for ζ_2, except with a new inhomogeneous term $Z_3 = Z_3(\zeta_1, \zeta_2, k_2, c_3)$, and can again be integrated by the same argument as for the lower-order case. When all this is carried through, with rather lengthy algebra, the final result is

$$\zeta = aS^2 - a^2 \tfrac{3}{4}S^2 T^2 + a^3(\tfrac{5}{8}S^2 - \tfrac{101}{80}S^4)T^2 \quad (S = \text{sech}(\theta), \ T = \tanh(\theta))$$

$$(118a)$$

$$c = c_0(1 + \tfrac{1}{2}\alpha - \tfrac{3}{20}\alpha^2 + \tfrac{3}{56}\alpha^3 + \cdots), \qquad (c_0 = \sqrt{gh} = 1), \qquad (118b)$$

$$k = k_0(1 - \tfrac{5}{8}\alpha + \tfrac{71}{128}\alpha^2 + \cdots), \qquad (k_0^2 = \tfrac{3}{4}a), \qquad (118c)$$

in agreement with that given by Chappelear (1962) and Grimshaw (1971).

To sum up, several remarks are in order. First, we note that exactly the same solution results for the wave elevation ζ, regardless of which accompanying velocity base is adopted (our $\hat{u}$ base used here is different from those used in the previous contributions). While the velocity will differ for different bases, the final flow field [recovered from the field solution of (23) and (24)] will nevertheless be identical on the grounds that the series solution is absolutely and uniformly convergent. This shows the degree of uniformity of the theory.

Further, as viewed from perturbation theory, this constructive method of solution has a salient feature in requiring solvability conditions in every step, from the very first. We note that the dispersion effects (bearing derivatives of higher order) enter the expansion at a level higher than the leading linear order in keeping balance with the nonlinear effects as required for the Boussinesq family. Consequently, the nth-order solution is completed only until the $(n + 1)$st-order equations are considered for solvability. By this pattern, the solvability equations, which assume the same form as (114a b) for all the higher orders, are expected to be integrable on the same ground as argued for (114a–b), except for the algebra soon becoming formidable. The present constructive method illustrates how the so-called "asymptotic reductive perturbation method" is developed by ascertaining a central pattern that the same solvability condition as that found in the first fundamental round continues to hold for solving all the successive higher orders, thereby suggesting its countable infinite times of validity of the methodology. (For this problem, for example, starting from the $(n + 1)$st-order perturbation equations for (u_{n+1}, ζ_{n+1}), we derive under solvability condition a differential equation for ζ_n just like (114a–b) except with a new inhomogeneous term $Z_n(\zeta_1, \ldots, \zeta_{n-1}; k_{n-1}, c_n)$, which can similarly be solved. The ζ_n solution then relates u_{n+1} to ζ_{n+1} and the lower-order solutions for the next $(n + 2)$nd-order problem, $\forall n \geqslant 3$.) This solution procedure also illustrates a theoretical proof of existence of a solitary wave in asymptote to

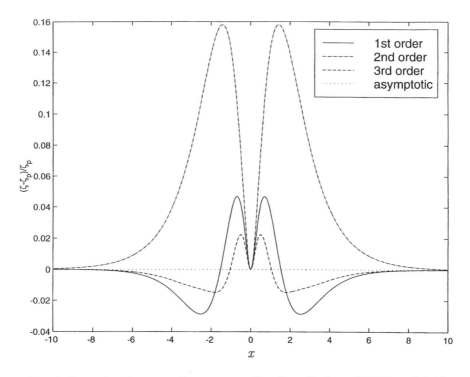

FIG. 5. Comparison between solitary wave profiles of amplitude $a = 0.75823$ predicted by different models, shown with distribution of relative errors with reference to the permanent wave exact solution ζ_p (with maximum local error less than 10^{-5}): asymptotic solution (the present FNFD wave theory) — indistinguishable from the zero line; first order (KdV, in solid line); second order (Laitone, in dash-dot line); third order (Chappelear/Grimshaw, in dashed line).

the full Euler model. On this basis we can assert that the solitary wave solution supported asymptotically by the Euler model forms a one-parameter family, e.g., in wave amplitude, just like the first-order KdV theory.

Finally, we point out that the above illustration of existence of solitary waves, by asymptotic construction, is strongly supported by numerical results obtained by Qu *et al.* (2000) in studies pursued in parallel to this analytical approach, with preliminary results presented in Section IV.A. As shown in Figure 5, the third-order solution for solitary waves shows improvement over the lower-order results as compared based on the numerical results of the present exact theory up to quite high amplitude such as that shown for $\alpha = 0.75823$, with a slight slack in accuracy for greater values of α, before approaching the fastest solitary wave at

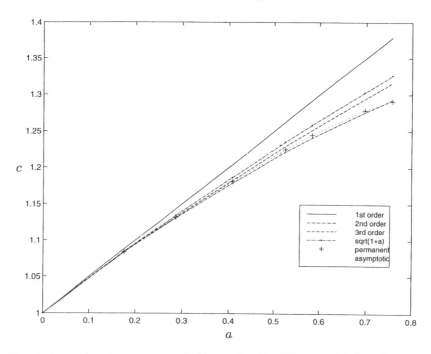

FIG. 6. Comparison between wave velocities predicted by different models for solitary wave up to amplitude $a = 0.75823$ (with the same notation as for Figure 5).

$\alpha = 0.79587$, with $c = 1.2942$ and the highest solitary wave (with a 120° crest corner) at $\alpha = 0.83320$ with $c = 1.2909$, as found by Qu *et al.* (2000). The improvement is quite noticeable from the figure, both in wave profile and wave velocity, as shown in Figure 6 for the latter. A conspicuous feature noted here is that the errors of the first three orders of results by expansion appear to be oscillatory in approaching the exact solution of the permanent wave theory, and that the second-order theory is noticeably poorer than the first-order result in predicting wave profile, but is excellent (best of the three) in representing the wave velocity to amplitudes as high as $a = 0.76$.

IX. Nonlinear Wave–Wave Interactions

Nonlinear wave interaction theory is of great importance not only in basic science but also in technological development and applications to many disciplines. A simple class of problems is concerned with two solitary waves propagating in one or opposite directions along a straight channel of uniform rectangular cross-sectional shape and engaging in binary encounters, or collisions. The salient features of nonlinear wave–wave interactions, such as wave modulations in shape, amplitude and phase, mass-energy transfer, and possible wave reflection and transmission that may occur in more general premises (such as in nonuniform channels of arbitrary shape), are quite well exhibited in the Boussinesq class of binary wave collisions. We present a brief recounting of the theoretical and historical background for this simple class.

A. BIDIRECTIONAL LONG-WAVE MODEL

We adopt the Boussinesq model (71) and (72), now extended to include an external pressure disturbance:

$$\zeta_t + [(1 + \zeta)u]_x = 0, \tag{119}$$

$$u_t + uu_x + \zeta_x - \tfrac{1}{3}u_{xxt} = -(p_a)_x, \tag{120}$$

where u stands for $\bar{u}(x, t)$, with the rest water depth $h = 1$ and the linear wave velocity $c = \sqrt{h} = 1$ being understood.

We introduce a "velocity potential" $\varphi(x, t)$ such that

$$u(x, t) = \varphi_x. \tag{121}$$

(Notice the difference between this φ and the ϕ of (15).) In terms of φ, (120) can be integrated once, yielding

$$\zeta + \varphi_t + \tfrac{1}{2}(\varphi_x)^2 - \tfrac{1}{3}\varphi_{xxt} = -p_a, \tag{122}$$

with the rest condition implied at infinity. Substituting (121) and (122) into (119) to eliminate ζ, we obtain for φ the equation

$$\varphi_{tt} - \varphi_{xx} = \tfrac{1}{3}\varphi_{xxtt} - [(\varphi_x)^2 + \tfrac{1}{2}(\varphi_t)^2]_t - (p_a)_t, \tag{123}$$

in which the terms on the left-hand side are of $O(\varepsilon^{3/2})$, those on the right are of $O(\varepsilon^{5/2})$, and the terms of order $O(\varepsilon^{7/2})$ are left with the error term, as explained before on scaling. In deriving (123), the term $\partial_t(\varphi_t)^2$ follows from using the lowest order approximation of (123), i.e., $\varphi_{tt} = \varphi_{xx}$, to convert the quadratic term $(\varphi_t\varphi_x)_x$ into $\varphi_x\varphi_{xt} + \varphi_t\varphi_{tt}$, with the error term intact, which remains at two orders above the lowest one therein. From the derivation, this equation clearly holds valid for bidirectional waves (i.e., when both right-going and left-going waves coexist). In addition, we also have from Boussinesq (1872) the equation

$$\varphi_{tt} - \varphi_{xx} = \tfrac{1}{3}\varphi_{xxtt} - [\tfrac{3}{2}(\varphi_x)^2]_t,$$

known as *Boussinesq's one-equation model*, which is valid only for unidirectional waves [in which case, $(\varphi_x)^2 = (\varphi_t)^2$]. With respect to attaining a single-equation model, the fundamentals have been elucidated by Lin and Clark (1959) and Benney and Luke (1964).

We next apply the multiple scale expansion in terms of the new variables:

$$\xi_\pm = \varepsilon^{1/2}(t \mp x), \qquad \tau = \varepsilon^{3/2}t, \tag{124}$$

$$\varphi(x, t) = \varepsilon^{1/2}[\varphi_0(\xi_+, \xi_-; \tau) + \varepsilon\varphi_1(\xi_+, \xi_-; \tau) + \cdots], \tag{125a}$$

$$p_a = \varepsilon^2[P_+(\xi_+) + P_-(\xi_-)]. \tag{125b}$$

Clearly, this scaling restores the original set for WNWD wave motions of the Boussinesq class with $\alpha = O(\varepsilon) \ll 1$ as delineated in the foregoing, while the slow time τ is one order higher than the fast time t in accordance with the physical rate of the second-order wave velocity c_1 of (101) for admitting unsteady phenomena that arise slowly as viewed in the wave frame. Corresponding to this scale, the differential operators are related by

$$\partial_t = \varepsilon^{1/2}[(\partial_- + \partial_+) + \varepsilon\partial_\tau], \qquad \partial_x = \varepsilon^{1/2}(\partial_- - \partial_+), \tag{126}$$

where

$$\partial_+ = \partial/\partial\xi_+, \qquad \partial_- = \partial/\partial\xi_-, \qquad \partial_\tau = \partial/\partial\tau.$$

Here, forcing p_a is assumed to consist of a right-going and a left-going component, or R wave and L wave for brevity, both of $O(\varepsilon^2)$ (anticipating incompatibity if p_a is lower in order). With this asymptotic expansion, the leading term of (123), of order $O(\varepsilon^{3/2})$ for φ_0 yields

$$4\partial_+\partial_-\varphi_0 = 0. \tag{127}$$

This equation has the general solution

$$\varphi_0 = \varphi_+(\xi_+; \tau) + \varphi_-(\xi_-; \tau), \tag{128}$$

comprising an R wave and an L wave, which are so far indeterminate (since their dependence on slow time τ is still unknown), but will be fully specified in the next order analysis (as is generic for the reductive perturbation method).

The next order terms of $O(\varepsilon^{5/2})$ in (123) are found to give for φ_1 the equation

$$-4\partial_-\partial_+\varphi_1 = 2\partial_\tau\partial_+\varphi_+ + (3/2)\partial_+(\partial_+\varphi_+)^2 - (1/3)\partial_+^4\varphi_+ + \partial_+P_+$$
$$+ 2\partial_\tau\partial_-\varphi_- + (3/2)\partial_-(\partial_-\varphi_-)^2 - (1/3)\partial_-^4\varphi_- + \partial_-P_-$$
$$- (\partial_+ + \partial_-)(\partial_+\varphi_+)(\partial_-\varphi_-). \tag{129}$$

The solvability condition to eliminate a linear growth in φ_1 with increasing ξ_+ and ξ_- requires the secular terms in the first two lines on the right-hand side of (129) to vanish separately with respect to ξ_+ and ξ_-, thus yielding the following set of equations:

$$\partial_\tau\partial_\pm\varphi_\pm + \tfrac{3}{4}\partial_\pm(\partial_\pm\varphi_\pm)^2 - \tfrac{1}{6}\partial_\pm^4\varphi_\pm = -\tfrac{1}{2}\partial_\pm P_\pm, \tag{130a}$$

$$4\partial_+\partial_-\varphi_1 = (\partial_+ + \partial_-)(\partial_+\varphi_+)(\partial_-\varphi_-),$$

where the $\pm$ subscripts are vertically ordered. These two equations have the integral

$$\partial_\tau\varphi_\pm + \tfrac{3}{4}(\partial_\pm\varphi_\pm)^2 - \tfrac{1}{6}\partial_\pm^3\varphi_\pm = -\tfrac{1}{2}P_\pm, \tag{130b}$$

$$\varphi_1 = (\varphi_-\partial_+\varphi_+ + \varphi_+\partial_-\varphi_-)/4. \tag{131}$$

For the wave elevation, we have the corresponding expansion

$$\zeta(x, t) = \varepsilon[\zeta_+(\xi_+; \tau) + \zeta_-(\xi_-; \tau) + \varepsilon\zeta_1(\xi_+, \xi_-, \tau) + \cdots]. \tag{132}$$

Substituting (130)–(132) in (122) and applying the above results, we obtain

$$\zeta_\pm = -\partial_\pm\varphi_\pm, \tag{133}$$

$$-\partial_\tau\zeta_\pm + \tfrac{3}{4}\partial_\pm(\zeta_\pm)^2 + \tfrac{1}{6}\partial_\pm^3\zeta_\pm = -\tfrac{1}{2}\partial_\pm P_\pm. \tag{134}$$

We thus have the final model equations by collecting (131)–(134) expressed in terms of (x, t), and finally with $\varepsilon = 1$ by rescaling, as follows:

$$\zeta(x, t) = \zeta_+(x, t) + \zeta_-(x, t) + \zeta_1(x, t), \tag{135}$$

$$\left(\pm\frac{1}{c}\partial_t + \partial_x\right)\zeta_\pm + \frac{3}{4}\partial_x\zeta_\pm^2 + \frac{1}{6}\partial_x^3\zeta_\pm = -\frac{1}{2}\partial_x P_\pm. \tag{136}$$

$$\zeta_1 = \tfrac{1}{2}\zeta_+\zeta_- + \tfrac{1}{4}(\varphi_+\partial_x\zeta_- - \varphi_-\partial_x\zeta_+) - \tfrac{1}{2}(P_+ + P_-). \tag{137}$$

This set of equations constitutes a model for evaluating the generation and evolution of nonlinear dispersive waves moving bidirectionally in uniform rectangular channels.

In the absence of external forcing, (136) becomes the classical KdV equation named after Korteweg and de Vries (1895). This equation is shown by Zabusky and Kruskal (1965) to possess a number of remarkable properties including the phenomenon of reappearance of initial flow configurations assigned to a spatially periodic motion, giving rise to a temporally periodic response, on which phenomenon (and additional wave–wave interaction behaviors) the term *soliton* is coined. The KdV equation has since furnished a rich field in which an enormous amount of intensive research has blossomed into what may be called a distinguished colorful modern history of applied and pure mathematics, physics, biology, and engineering and applied science. Close collaborations between various disciplines also emerged as a distinct mark of the time as similar phenomena are found to be commonplace in nonlinear systems.

In the presence of external forcing, eq. (136), known as "the forced KdV (or fKdV) equation," has been adopted as an alternative to the other WNWD wave models for studies of various phenomena of generation of waves propagating and evolving in soliton-bearing systems through exchanges of mass, momentum, and energy with exterior systems in critical state of resonance, as will be discussed in Section X.

This simple version of bidirectional long-wave model has been extended (Wu, 1993a) to cover more general cases of nonuniform channels of arbitrary geometric shape.

B. BINARY HEAD-ON COLLISIONS OF BIDIRECTIONAL SOLITONS

For the class of head-on collisions of solitons, the special case of total reflection of a soliton by an inviscid vertical wall has been investigated by Chan and Street (1970) using the SUMMAC integration of the Navier–Stokes equations, by Byatt-Smith (1971) using a conformal mapped integral equation, by Maxworthy (1976) and Weidman and Maxworthy (1978) experimentally, by Power and Chwang (1984) using perturbation methods for a comparative study, and by Yih and Wu (1995) from a different approach. For end-wall reflection, the theory has been extended to higher orders by Su and Mirie (1980). In another direction, the theory has been generalized by Wu (1995) for bidirectional interactions between multiple solitons of the Boussinesq class and by Cooker *et al.* (1997) for the Euler equations.

For analyzing binary head-on collisions of solitons, (135)–(137) can be recombined as

$$\zeta(x, t) = \zeta_+(\xi_+ + \tfrac{1}{4}\varphi_-, \tau) + \zeta_-(\xi_- + \tfrac{1}{4}\varphi_+, \tau) + \tfrac{1}{2}\zeta_+(\xi_+, \tau)\zeta_-(\xi_-, \tau), \quad (138)$$

since the two-term expansion of Taylor's series for $\zeta_\pm$ in (138) (about their original phaseline) recovers (135) and (137), leaving the error term unaltered. Specifically, we begin with prescribing two head-on approaching solitons, ζ_+ and ζ_-,

$$\zeta_\pm(\xi_\pm, \tau) = \alpha_\pm \operatorname{sech}^2\theta_\pm, \qquad \theta_\pm = (3\alpha_\pm/4)^{1/2}(\xi_\pm + \alpha_\pm t/2 + s_\pm), \quad (139)$$

$$\varphi_\pm = -(3\alpha_\pm/4)^{1/2}\tanh\theta_\pm, \quad\quad\quad\quad\quad (140)$$

which satisfy (136) (with $P_\pm = 0$), with arbitrary amplitude, α_+ and α_-, and arbitrary phase constants, s_+ and s_- (for prescribing the initial position of each soliton), respectively. With $s_\pm = 0$, this solution has the "time-reversal" symmetry, $\zeta(x, -t) = \zeta(-x, t)$, as warranted by differential equations (136). Then (138) indicates that both solitons recover their original entity through the collision except that each bears a terminal phase shift *toward* their center of mass, by a total time retardation $\Delta t_\pm = [\varphi_\mp(t=\infty) - \varphi_\mp(t=-\infty)]/4 = (\alpha_\mp)^{1/2}$. The phase shifts are the only mark left, permanently, by the nonlinear encounter.

In a wall reflection encounter, an incoming soliton of amplitude α has its crest first asymptotically accelerating to reach the wall at $t = -t_0$, remaining at the wall with its amplitude first increasing, as the trailing tail continues coming inward and converting kinetic energy to potential energy, to reach the maximum amplitude of $(2\alpha + \alpha^2/2)$ at $t = 0$, and then varying with the time-reversal symmetry and leave the wall at $t = t_0$. For $\alpha \ll 1$, a good estimate of the "phase-locking" duration is given by Temperville (1979) as $2t_0 = (2/\sqrt{3})[\ln(\sqrt{3} + 1) - \ln(\sqrt{3} - 1)]\alpha^{-1/2}$, and is independently evaluated by Power and Chwang (1984) to give $2t_0 = 1.52(\alpha)^{-1/2}$.

For asymmetrical head-on collisions, the encounter scenario is similar; however, the symmetry is somewhat modified from the wall reflection case. The crests of two incoming solitons do not actually merge; instead, the smaller one evanesces at a station $P_m(x = x_0, t = -t_0)$ (at which $\zeta_x = \zeta_{xx} = 0$), with the taller gaining in height to a maximum at $t = 0$ and reversing with time and finally with the smaller soliton reappearing to part at station $P_p(x = -x_0, t = t_0)$. The duration in $(-t_0 < t < t_0)$ can still be defined as the phase-locking period. A typical result given by Wu (1995) is shown in Figure 7 for $\alpha_+ = 0.6$, $\alpha_- = 0.3$ with time steps of

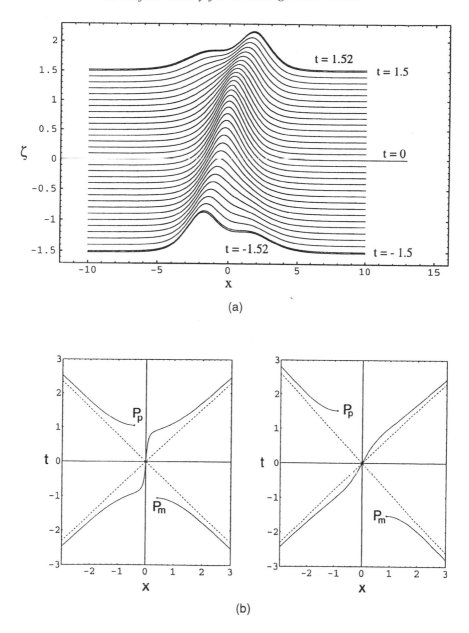

FIG. 7. (a) Wave evolution through head-on collision of two solitary waves of amplitude $\alpha_+ = 0.6$ and $\alpha_- = 0.3$, shown throughout the phase-locking period ($-t_0 < t < t_0$, $t_0 = 1.522$) with time interval $\Delta t = 0.1$. (b) Trajectories of wave crests of two head-on colliding solitons with $\alpha_+ > \alpha_-$, the crest of the left-going weaker wave evanesces during the period between P_m and P_p (at which $\zeta_x = \zeta_{xx} = 0$): (left figure) $\alpha_+ = 0.6$, $\alpha_- = 0.55$ ($x_0 = 0.418$, $t_0 = -1.055$ at P_m); (right) $\alpha_+ = 0.6$, $\alpha_- = 0.3$ ($x_0 = 0.893$, $t_0 = -1.522$ at P_m).

variation in wave profile from $t = -t_0$ to $t = t_0$, where $t_0 = 1.522$, $x_0 = 0.893$. For $|\alpha_+ - \alpha_-| \ll 1$, we have (Wu, 1995), approximately, $t_0 = 0.38[(\alpha_+)^{-1/2} + (\alpha_-)^{-1/2}]$, $x_0 = 0.38|(\alpha_+)^{-1/2} - (\alpha_-)^{-1/2}|$.

In higher-order theory, the time-reversal symmetry becomes further modified, as shown by Su and Mirie (1980) and Cooker *et al.* (1997).

C. Binary Overtaking Collisions of Unidirectional Solitons

The process of overtaking collisions between two right-going solitons, with a soliton of height α_1 overtaking a weaker one of height α_2, can be described by the R-wave KdV equation (134) alone [since (135)–(137) for bidirectional waves are now uncoupled], namely,

$$\partial_t \zeta - (3/4)\partial_\xi \zeta^2 - (1/6)\partial_\xi^3 \zeta = 0 \qquad (\xi = t - x). \tag{141}$$

For overtaking collisions, the asymptotic phase shifts imparted to each participating soliton are well known (Hirota, 1973; Whitham, 1974), while the overtaking processes have been evaluated more in detail by Yih and Wu (1995) and Wu (1995). Adopting Hirota's (1973) exact solution for the binary system satisfying (141) and the transformation

$$\zeta = (4/3)(\log f)_{\xi\xi}, \tag{142}$$

we see that a solution of (141) implies that f satisfies the following quadratic equation:

$$(f_\xi - f\partial_\xi)(f_\tau - \tfrac{1}{6}f_{\xi\xi\xi}) + \tfrac{1}{2}(f_{\xi\xi}^2 - f_\xi f_{\xi\xi\xi}) = 0. \tag{143}$$

For a single soliton, we have

$$f = 1 + f_j = 1 + \exp(\theta_j), \qquad \theta_j = \beta_j(\xi + s_j + \beta_j^2 t/6), \quad (j = 1), \tag{144a}$$

where β_j is a real constant and s_j an arbitrary phase constant, so that

$$\zeta = \tfrac{1}{3}\beta_1^2 \,\mathrm{sech}^2[\tfrac{1}{2}\beta_1(\xi + s_1 + \tfrac{1}{6}\beta_1^2 t)], \tag{144b}$$

which is a soliton of amplitude $\alpha_1 = \beta_1^2/3$.

For two right-going solitons, we apply Hirota's (1973) exact solution of (143):

$$f(x, t) = 1 + f_1(\xi, t) + f_2(\xi, t) + a_{12}f_1(\xi, t)f_2(\xi, t), \tag{145a}$$

$$f_j = \exp(\theta_j), \qquad a_{12} = (\beta_1 - \beta_2)^2/(\beta_1 + \beta_2)^2. \tag{145b}$$

where θ_j is given by (144a). For two solitons of amplitude $\alpha_j = \beta_j^2/3$,

($j = 1, 2; \alpha_1 > \alpha_2$), we choose s_j such that at $t = 0$, $\zeta(-x, 0) = \zeta(x, 0)$, which is achieved with

$$f_j(x, 0) = b_j \exp(-\beta_j x) \qquad b_j = \exp(\beta_j s_j) \quad (j = 1, 2), \qquad (146a)$$

$$b_1 = b_2 = a_{12}^{-1/2} = (\beta_1 + \beta_2)/(\beta_1 - \beta_2). \qquad (146b)$$

To verify this, we write (145a) for $t = 0$ in the form

$$f(x, 0) = 1 + b_1 e^{-\beta_1 x} + b_2 e^{-\beta_2 x} + a_{12} b_1 b_2 e^{-(\beta_1 + \beta_2)x} = e^{-(\beta_1 + \beta_2)x} \psi(x),$$

$$\psi(x) = e^{(\beta_1 + \beta_2)x} + b_1 e^{\beta_2 x} + b_2 e^{\beta_1 x} + a_{12} b_1 b_2,$$

The factor $e^{-(\beta_1 + \beta_2)x}$ alone has no contribution to ζ because by (142), the second derivative of its logarithm is zero. Therefore, the required symmetry is achieved by equating $f(-x, 0)$ with $\psi(x)$, which yields $b_1 = b_2 = b$ and $a_{12} b_1 b_2 = 1$, just as was to be shown. From this result we see that a greater soliton overtaking a smaller one gains momentum, having finally accelerated forward with a forward phase shift by $\Delta x_1 = 2s_1$, at the cost of the smaller one, which is pushed back, eventually suffering a backward phase shift by $\Delta x_2 = 2s_2$, while keeping the net momentum intact.

This symmetry also facilitates the analysis of the wave properties near the center of the encounter. In fact, Wu (1995) obtained from this solution the following results:

$$\zeta(0, 0) = (\alpha_1 - \alpha_2), \qquad (147)$$

$$\zeta_x(0, 0) = 0, \qquad (148)$$

$$\zeta_{xx}(0, 0) = -(3/2)(\alpha_1 - \alpha_2)(\alpha_1 - 3\alpha_2). \qquad (149)$$

To verify these results, we use $\zeta(x, 0)$ from (142), now with $f(x, 0)$ reduced to a form symmetric about $x = 0$ as

$$f(x, 0) = A_- \cosh A_+ x + A_+ \cosh A_- x \qquad (A_\pm = (\beta_1 \pm \beta_2)/2), \qquad (150)$$

from which the results readily follow. Here, (148) is a consequence to the symmetry identified, (147) gives the resultant wave elevation at the center of encounter, equal to $(\alpha_1 - \alpha_2)$, in contrast to $\zeta(0, 0)$ being greater than the sum of the wave heights of two head-on colliding solitons. Finally, (149) provides the criticality criterion in terms of the amplitude ratio of $\alpha_1/\alpha_2 = 3$ which differentiates the wave profile curvature at ($x = 0$, $t = 0$) into three distinct regimes,

$$\zeta_{xx}(0, 0) <, =, \text{ or } > 0 \quad \text{according to} \quad \alpha_1/\alpha_2 >, =, \text{ or } < 3. \qquad (151)$$

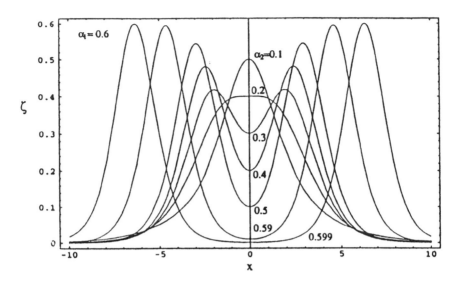

Fɪɢ. 8. In overtaking collisions of two unidirectional solitons, all wave profiles attain a fore-and-aft symmetry at time instant $t = 0$, about which the system has the time-reversal symmetry, $\zeta(x, -t) = \zeta(-x, t)$. Symmetrical wave profiles (at $t = 0$) have the single-peak, double-peak, and the critical flattened peak patterns according to the amplitude ratio $\alpha_1/\alpha_2 >$, $<$, or $= 3$; here $\alpha_1 = 0.6$, versus a list of values for α_2.

Thus, the two soliton crests either pass through each other or remain separated throughout the encounter according to whether $\alpha_1/\alpha_2 > 3$ or $1 < \alpha_1/\alpha_2 < 3$, respectively. At the critical condition, $\alpha_1/\alpha_2 = 3$, the single peak becomes instantaneously flattened to zero curvature, at which $\zeta_x = \zeta_{xx} = 0$. With no phase locking during the overtaking, the two solitons reemerge afterward recovering their initial forms, with the stronger wave being advanced and the weaker one retarded in phase from their original path lines.

The three distinct regimes of peak merging are clearly exhibited in Figure 8 in which the fore-and-aft symmetric wave profiles at $t = 0$ are shown for various values of the amplitude ratio α_1/α_2, with $\alpha_1 = 0.6$ and a set of designated α_2's. The resultant wave height at the center of symmetry, $\zeta(0, 0) = \alpha_1 - \alpha_2$, reflects in totality a reduction in potential energy as required for offsetting the net gain in kinetic energy as the stronger soliton accelerates and the weaker retards during the overtaking collision. In the two-peak regime, the twin peaks of the two interacting solitons reach at

$t = 0$ the shortest distance apart and this minimum distance increases as the soliton pair becomes increasingly closer in height.

Figure 9 illustrates these main features of overtaking collisions referred to the center-of-mass frame. Figure 9a shows the single-peak evolution of a merged peak of two overtaking solitons of amplitude $\alpha_1 = 0.6$, and $\alpha_1 / \alpha_2 = 6$, with its amplitude dipping slightly to a minimum of $(\alpha_1 - \alpha_2) = 0.5$ at $x = 0$ and $t = 0$ while the wave broadens in proportion. The critical case of peak separation is delineated in Figure 9b for $\alpha_1 = 0.6$ and $\alpha_1/\alpha_2 = 3$, in which case the wave spreads the widest of all the cases at the point of symmetry, with the curvature vanishing at the crest. Figure 9c traces out the double-peak evolution for $\alpha_1 = 0.6$ and $\alpha_1/\alpha_2 = 12/11$, in which the fore-aft symmetry of the profile takes place at $t = 0$ when the two peaks are the least far apart. The critical criterion separating the single-peak and double-peak regimes for overtaking soliton encounters was first noted by Zabusky (1967), proved for its existence and numerically analyzed by Lax (1968), experimentally measured by Weidman and Maxworthy (1978), and analytically determined by Wu (1995). Finally, we mention that the time-reversing symmetry is shown by Su and Mirie (1980) and Cooker *et al.* (1997) to be further modifed under the full nonlinear effects.

D. Mass and Energy Transfer between Overtaking Solitons

A remarkable feature of overtaking collisions between two comparable solitons of amplitude α_1 and α_2, especially in the limit of vanishingly small values of $(\alpha_1/\alpha_2 - 1)$, is that excess mass and energy will be continually transferred from the greater soliton behind to the slightly smaller one ahead until the smaller one keeps growing and gaining in speed to finally outrun the weakening soliton behind in accomplishing the overtaking collision without ever coming close between their crests. Indeed, for $0 < (\alpha_1/\alpha_2 - 1) \ll 1$, the shortest distance between the two crests can be shown from the above solutions given by (142) and (150) to be $2x_c$, where, asymptotically,

$$x_c = \frac{2}{\sqrt{3}}(\sqrt{\alpha_1} + \sqrt{\alpha_2})^{-1} \ln\left(2\frac{\sqrt{\alpha_1} + \sqrt{\alpha_2}}{\sqrt{\alpha_1} - \sqrt{\alpha_2}}\right), \tag{152}$$

so the shortest approach (equal to $2x_c$) between two comparable solitons increases logrithmically beyond all bounds as $(\alpha_1/\alpha_2 - 1) \to 0$. It thus seems perplexing whether unidirectional solitons of nearly equal amplitudes would ever stop interacting when separated sufficiently far. The answer, affirmative

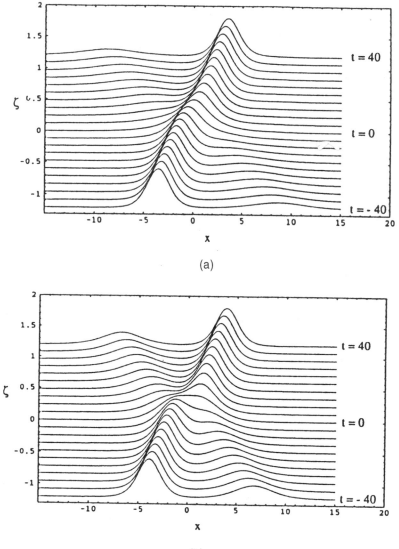

(a)

(b)

FIG. 9. The three regimes of wave patterns during overtaking collisions of two right-going solitons of amplitude ratio α_1/α_2: (a) A single-peak pattern, with $\alpha_1 = 0.6$ and $\alpha_1/\alpha_2 = 6$, >3 [the merged single peak reaching its minimum height of $(\alpha_1 - \alpha_2) = 0.5$ at $t = 0$]. (b) A critical overtaking encounter with $\alpha_1 = 0.6$ and $\alpha_1/\alpha_2 = 3$ [the merged single peak has its crest of vanishing curvature].

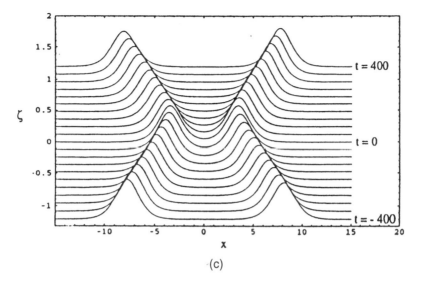

(c)

FIG. 9. Continued. (c) A double-peak pattern with $\alpha_1 = 0.6$ and $\alpha_1/\alpha_2 = 12/11$, <3 [the two peaks remain separated throughout, with their gap valley bottom reaching the elevation of $(\alpha_1 - \alpha_2) = 0.05$ at $t = 0$]. The time sequences of all wave profiles are referred to the center-of-mass (CM) frame, which is fixed at $x = 0$.

in accord with the above scenario, is in fact based on the *exact solution* of the KdV equation, regardless of how far apart the shortest distance is between the crests of the interacting solitons.

Even more remarkable is the exceedingly slow rate of transferring mass and energy between two such solitons. The total net transfer of mass is simply

$$\Delta m = m_1 - m_2 = 4\beta_1/3 - 4\beta_2/3 = 4\sqrt{\alpha_1/3} - 4\sqrt{\alpha_2/3}, \qquad (153)$$

as $m_j = 4\beta_j/3$ ($j = 1, 2$) is their excess mass. To find the rate of transfer, the perspective is best viewed from the center-of-mass (CM) frame, which translates relative to the absolute frame with a constant velocity

$$c_m = 1 + (\beta_1^2 + \beta_2^2 - \beta_1\beta_2). \qquad (154)$$

With respect to the new coordinates (x', t') fixed in the CM frame defined by

$$x = x' + c_m t, \qquad t' = t, \qquad (155)$$

the wave functions f_1 and f_2 in (145a–b) become,

$$f_j = \exp[\beta_j(c'_j t - x')], \qquad c'_1\beta_1 = -c'_2\beta_2 = \beta_1\beta_2(\beta_1 - \beta_2).$$

Hence the net momentum of the system vanishes in the CM frame since $m_1 c'_1 + m_2 c'_2 = 0$.

In the CM frame, continuity equation (119) becomes

$$\frac{\partial \zeta}{\partial t} + \frac{\partial}{\partial x'}[(1 + \zeta)u - c_m \zeta] = 0.$$

Now integrating this equation from $x' = -\infty$ to $x' = 0$, with t fixed, yields for the rate of mass transfer from the left to the right side across the CM plane the equation

$$-\frac{d}{dt}m_-(t) \equiv -\frac{d}{dt}\int_{-\infty}^{0} \zeta(x', t)\,dx' = (1 + \zeta_c)u_c - c_m\zeta_c, \qquad (156)$$

where ζ_c and u_c refer to ζ, u at $(x' = 0, t)$. Finally, we define the mass flux transfer velocity, $v_T(t)$, by

$$-\frac{d}{dt}m_-(t) = (1 + \zeta_c)v_T. \qquad (157)$$

This rate has been calculated by Wu and Zhang (1996b) using the exact KdV solution for the binary encounter between a soliton of amplitude $\alpha_1 = 0.6$ and a weaker one of amplitude $\alpha_2 = 0.55$. In the CM frame, the flow velocity and wave elevation at the origin, $u_c(t)$ and $\zeta_c(t)$, are shown in Figure 10a to reach their temporal maximum at $t = 0$ (when the two wave crests are at the shortest distance apart) equal to $u_c(0) = 0.062$ and $\zeta_c(0) = 0.05$, which are an order of magnitude smaller than their values at the crest. In addition, the velocity of mass transfer across the CM plane, $v_T(t)$, computed using the exact KdV solution, is found to reach at $t = 0$ its maximum $v_T(0) = 0.00058$, which, as shown in Figure 10b, is three orders smaller than the flow velocity at the crest, yet these numerical results are found, on integration of (157) in t, to be in perfect agreement with the first principle (153) on the final total mass transfer. The precise significance of this finding is worthy of further in-depth examination to study concerns such as the philosophical question of whether such small rates of mass-energy transfer between waves also hold true in general for the Euler model.

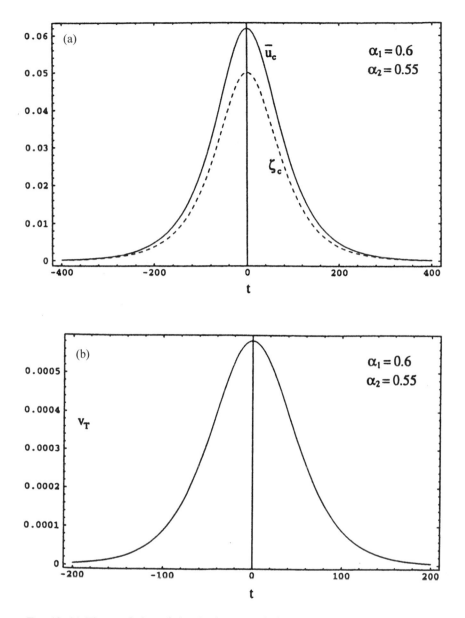

FIG. 10. (a) Time variation of the depth-mean velocity $\bar{u}_c$ and wave elevation ζ_c at CM section during an overtaking collision between two comparable solitons, $\alpha_1 = 0.6$, $\alpha_2 = 0.55$. (b) Time variation of mass transfer velocity, v_T, of transporting wave excess mass across the CM section.

X. Soliton Generation by Resonant Forcing

Forced generation of water waves is a common experience, from the ripples produced by throwing a pebble into a placid pond, watching fascinating ship waves, to water waves excited by winds of all strengths. Naval architecture is one of the oldest highly skillful areas of engineering in history; its modern advances and applications to ocean engineering have attracted broad interest. However, to witness, with apprehension, any manifestation of generation of nonlinear waves by forcing sustained at resonance of solitary-wave-supporting systems is an exceedingly rare experience. The first reported observation appears to be the legendary "chance encounter" by John Scott Russell (1844) who observed an impressive wave self-organized around the prow of a channel boat that was suddenly stalled from being towed by a pair of horses. That wave emerged quickly to stretch straight across the channel and move forward from the stalled boat as a free wave with an impressive constant velocity in the solitary form of a rounded smooth heap of water he called "a solitary wave of translation." Afterwards, Russell carefully conducted a series of experiments, finding out the somewhat narrow parametric window of a transcritical speed range in which the phenomena were manifested and discovering various salient properties of the solitary wave of translation that have stimulated much interest in this subject.

A. Numerical and Experimental Discoveries

In a program devoted to studies on mitigating tsunami and storm hazards, Wu (1979, 1981) introduced the so-called "generalized Boussinesq" (or gB) model:

$$\zeta_t + \nabla \cdot [(h + \zeta)\bar{u}] = -h_t, \tag{158}$$

$$\bar{u}_t + \bar{u} \cdot \nabla \bar{u} + \nabla \zeta = -\nabla p_a + \frac{h}{2} \frac{\partial}{\partial t} \nabla [h_t + \nabla \cdot (h\bar{u})] - \frac{h^2}{6} \nabla^2 \bar{u}_t. \tag{159}$$

This model extends the classical Boussinesq model [originally for physically closed wave systems in $(1 + 1)$ dimensions] to a physically open system with variable bathymetry in $(2 + 1)$ dimensions together with admitting external forcing functions in surface pressure $p_a(x, y, t)$ and moving ocean floor at depth $h(x, y, t)$. This set of basic equations, derived by a method based on the $(\bar{u}, \zeta)$ system of variables, is in agreement with the present theory, in particular (96) and (97), to the same order of accuracy as Boussinesq's.

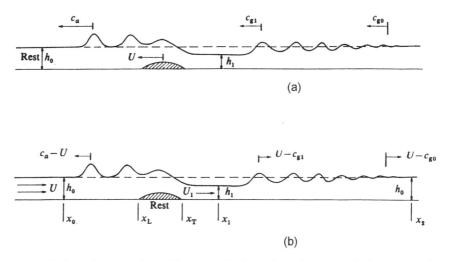

(a)

(b)

FIG. 11. Periodic generation of forward-radiating solitons by transcritical stationary disturbances, here a humped topography moving at constant velocity U in water of uniform rest depth h_0: (a) In laboratory frame. (b) In body frame.

This gB model was first applied to predict the $(1 + 1)$-dimensional forced flow field produced, from an initial state at rest, by a localized presssure distribution,

$$p_a(x, t) = P_0 \cos^2(x + Ut)\pi/L \qquad (-L/2 < x + Ut < L/2), \qquad (160)$$

exerted at the surface of a water layer of uniform rest depth $(h = 1)$ over a range of $L/h(>1)$ and Froude number $F = U/\sqrt{gh} = U$ in a transcritical regime $(0.8 < F < 1.2)$. The solution was computed numerically by Wu and Wu (1982) using a predictor-corrector finite-difference scheme applied to a sufficiently large computation region, which is allowed to move with the left-progressing forcing function and supplemented by an adequate "open-boundary condition" as reported. Refreshingly interesting numerical results were obtained bringing forth the discovery of a remarkable phenomenon that can be summarized as follows.

The new phenomenon so discovered exhibits, as shown in Figure 11, that a forcing distribution of surface pressure or a submerged topography moving steadily with a transcritical velocity in a shallow rectangular water channel can generate, periodically and indefinitely, a succession of solitary waves, radiating upstream of the disturbance, one after another, to form a regular procession continually advancing forward of the disturbance. This

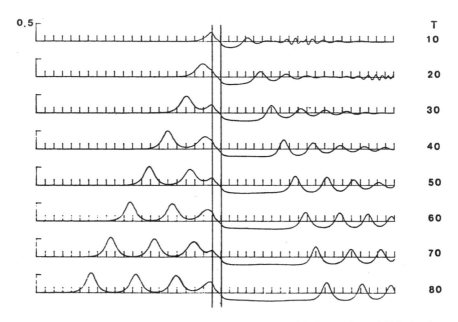

FIG. 12. One of the very first few numerical results obtained by Wu and Wu (1982) that lent to the working team a clear physical concept and results of significance leading to the discovery of this remarkable phenomenon. Wave profiles are shown in dimensionless time sequence as generated by p_a of (160) (with $P_0 = 0.2$, $U = 1$) located between the two bars fixed in the body frame.

steadily moving disturbance is immediately followed by a prolonging region of a uniformly depressed water surface while a train of very weakly nonlinear and weakly dispersive waves develops to trail further behind, appearing like a train of cnoidal waves and falling off into the quiet water beyond. This wave pattern indicates that the excess mass carried away by the upstream advancing solitary waves is extracted from the depressed region, which is uniform in depth (h_1, say, $h_1 < h$) and prolonging its length in time. This is depicted in Figure 12 by a plot of wave profile, this being one of the very first few numerical outputs obtained by Wu and Wu (1982). Presented in Figure 13 is a sequence of perspective photographic images taken of this phenomenon by George Yates while conducting laboratory experiments together with the corresponding numerical results based on the gB model.

Suppose the new solitary waves are generated periodically in this slow-time process with a period T_s, a quantity that is a functional determined as a part of the final solution. With the trailing region depressed uniformly to

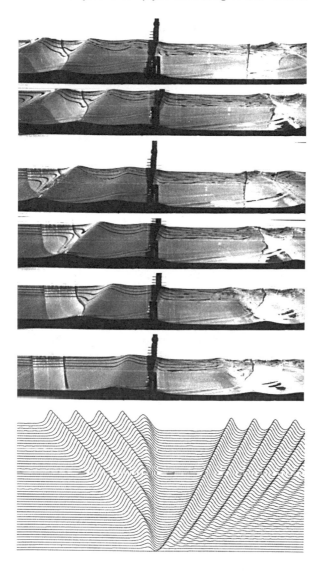

FIG. 13. Perspective views of laboratory experiment (top) exhibit the phenomenon of periodic generation of upstream emitting solitary waves by a bottom topography which is under tow on vertical struts at critical speed of Foude number $F = 1$. Resulting waves are shown in snapshots at (dimensionless) times of 33.1, 45.0, 55.8, 65.2, and 78.3. By the final picture (top), two waves have been radiated forward free of the disturbance while the third is forming at the bump. A corresponding numerical result (bottom) is shown for comparison. Close agreement can be seen between the surface profile plots and the longitudinal cross sections of water layer darkened by optical refraction (along the frontal water channel glass wall). (Adapted from George T. Yates (1990) by permission of World Scientific.)

depth h_1, we then have, by the principle of mass conservation, $Uh - U_1 h_1 = m_s/T_s$, where m_s is the excess mass of each new solitary wave so produced, and U_1 is the uniform horizontal velocity (relative to the moving disturbance) of the fluid within the depressed region. Based on the first principle of mass-momentum conservation, more averaged relationships between the key flow quantities have been found independently by Grimshaw and Smyth (1986), Wu (1987), Lee (1985), and Lee *et al.* (1989). In addition to the Froude number $F = U/\sqrt{gh} = U$ as a *de facto* free parameter, the water depth h_1 of the depressed layer is another parameter that plays a key role both in transporting fluid mass forward and in producing the terminal train of cnoidal-like waves. When the phenomenon of periodic emission of solitary waves is manifest, h_1 is determined as a part of the solution, giving $h_1 < h$. However, it is possible that for a certain special group of characteristic forcing, the resulting wave produced in the same range of resonance of the system is not only stationary but also very stable with respect to finite perturbations, as has been found from numerical experimentation by Camassa (1990) and Camassa and Wu (1991a,b) (e.g., the one known as ζ_{ss} given in Section X.C), in which case, $h_1 = h$. In view of this possibility, Wu (1987) stressed the point that the first-principle approach should not resolve completely a problem of periodic solitary wave generation but only lead to a solution that still has one additional free parameter left open, whether it is the depth of the uniformly depressed water layer, or whatever, that has to be closed empirically (e.g., by experiment), because stability analysis is beyond the first principle.

Experimentally, this remarkable phenomenon was first discovered in ship model experiments in a shallow-water towing tank by Sun and coworkers (1982) at Zhongsan University in Guangzhou and independently by J. V. Wehausen and coworkers (Huang *et al.*, 1982) at the University of California at Berkeley. Both teams discovered periodic generation of upstream-radiating waves by a ship model towed at transcritical speeds in a water tank filled to shallow depth, with each wave fanning outward to stretch straight across the tank and growing briskly to outrace the ship model as a free solitary wave, periodically one after another, as reported by Huang *et al.* (1982). Also independently, Wu and Wu (1982) reported their discovery made in 1981 by numerical methods for simpler two-dimensinal wave generation by resonant forcing. These numerical results have been attested in laboratory by Wu and coworkers towing a bent metal sheet, a humped yard measuring stick, fastened on struts towed in a water tank filled to 5–8 cm deep, immediately verifying the periodic production of processions of solitary waves for the two-dimensional case in autumn 1982. This experimental discovery lent to the working team and the author a first-hand appreciation, in resonance, of the expression by John Scott Russell (1865) acclaiming that "it was the happiest day of my life" for witnessing his first solitary wave. This led to a systematic experimental study by Lee (1985) and Lee *et al.* (1989). As shown in Figure 14, broad agreement has been found between experiment and theory based on the gB and KdV models. Noticeable departures do exist, such as for the trailing waves in a high subcritical range that have large amplitude on theory but are experimentally observed to be breaking; these aspects can be ascribed to the viscous effects that are neglected here but are of interest for further studies. Separate investigations have been pursued by Ertekin *et al.* (1984) and further developed by Ertekin *et al.* (1986) using both the Boussinesq model and Green–Naghdi's equations. Computational studies have been extended to three-dimensional forcing distributions by Schember (1982), Ertekin *et al.* (1986), Mei and Choi (1987), Wu and Wu (1988), and Lee and Grimshaw (1989, 1990), and to ship model studies by Chen and Sharma (1995).

In addition to the generalized Boussinesq model, the forced Korteweg–de Vries (fKdV) equation (136) has been adopted by Akylas (1984), Cole (1985), and Grimshaw and Smyth (1986) among many others, to investigate this general problem. Solutions to this interesting phenomenon have been sought and reported by Casciola and Landrini (1996) and Tang and Chang (1997) based on the Euler equations, and by Tang *et al.* (1990), Zhang and Chwang (1996), and Chang (1997) using the Navier–Stokes model. In addition, a good number of analogous phenomena have been observed in physical, biological, geophysical, and various other disciplines with contributions similar to that made to fluid mechanics. Notably, we may mention the studies on forced generation of internal solitary waves in stratified and rotating fluids, including Zhu (1986), Grimshaw and Smyth (1986), Zhu *et al.* (1986, 1987), Hanazaki (1991), Wu and Lin (1994), Choi (1993), Choi and Wu (1996), and Grue *et al.* (1997) among others. Some of the advances in these contexts have been discussed by Keller (1985).

As for the window in the parametric domain for manifestation of this category of phenomena, there seems to have no generally accepted specifics. A set of upper/lower limits has been given by Grimshaw and Smyth (1986). Physically, one can assume (beyond the Boussinesq level) an upper limit at the Froude number $F = 1.3$, since this is the speed of the fastest solitary wave ($F = 1.294$; Qu *et al.*, 2000) so no solitary wave can outrace such a fast moving disturbance to radiate forward to become a free wave. For greater Froude numbers, we could have either a stationary wave accompanying,

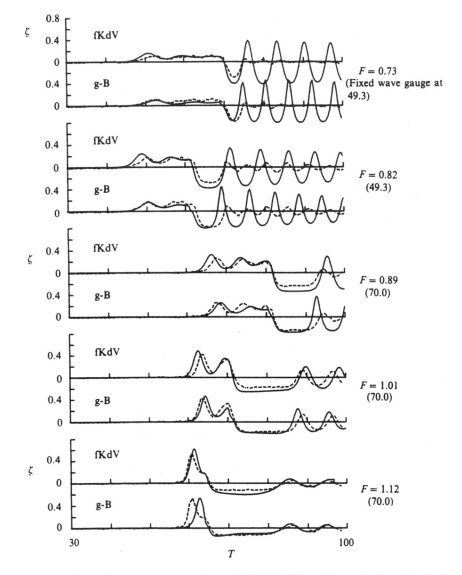

FIG. 14. Comparison between experimental (dashed line) and numerical results of the forced KdV equation (136) and generalized Boussinesq model (158) and (159) (solid lines) on waves generated by (left-going) transcritical (Froude number $F \simeq 1$) bottom bump $P_a = b_m \cos^2(\pi x/2L)$, $-L < x < L$, $b_m = 0.2$, $L/h = 1.23$ (in body frame). Data were taken (Lee *et al.*, 1989) from wave gauges fixed at locations (in parentheses, in units of h) upstream of the disturbance starting line.

possibly smooth in shape, or a hydraulic jump leading ahead of such a highly supercritical forcing. On the low side, Lee *et al.* (1989) reported that such a phenomenon may persist down to Froude numbers as low as $F = 0.25$ if based on accurate numerical experimentation or on linear theory arguments. In sharp contrast, however, we point out emphatically the numerical results that stationary waves are generated by transcritical stationary forcing that appear numerically robust against arbitrary perturbations (intended to excite bifurcation to new solutions with periodical emission of solitary waves) but remain stably stationary, as reported by Camassa (1990) and Camassa and Wu (1991a), and designated as the ζ_{ss} wave in Section X.C.

As a comment on this very important subject, we note that the mathematical problems discussed in this section are initial value problems on the entire real line with inhomogeneous terms in the equations representing external forcing that open systems may admit. Unlike for the initial value problem for corresponding closed systems, there are no analytical methods, such as the inverse scattering transform (IST), Backlund transform, etc., to reduce the nonlinear equations to integrable forms for constructing exact solutions, and obtaining them, so far solely by numerical means, is of interest. In this respect, various numerical methods of finite-difference, finite-element, and pseudo-spectral schemes have been developed to furnish solutions for examining the physical properties of the problems in question. Still, efficient algorithms are always useful, especially for complicated cases.

B. Soliton Generation by Boundary Forcing

The initial-boundary problem for the KdV equation on half real line:

$$u_t + uu_x + \varepsilon^2 u_{xxx} = 0 \qquad (x \geqslant 0, t \geqslant 0) \tag{161a}$$

$$u(x, 0) = 0, \quad x \geqslant 0, \tag{161b}$$

$$u(0, t) = u_0(t), \quad t \geqslant 0, \tag{161c}$$

also known as the "quarter-plane $(x \geqslant 0, t \geqslant 0)$ problem," is another category that lacks analytical methods for constructing solutions. Here, eqs. (161a–c) are written in terms of u, with the coefficients scaled as shown for convenience, and $u_0(t)$ is specified as a boundary forcing function. This problem was first solved numerically by Chu *et al.* (1983) using finite differences with $\varepsilon = 1$ and $u_0(t)$ given as a trapezoidal pulse, i.e., with it rising linearly and quickly to a value ($u_0 = $ const.) at which it remains for a given duration before falling off linearly to its initial free state. Their results show that solitons are produced, and the number of solitons increases with increasing duration of the constancy of forcing, with their amplitudes quickly approaching the value $2u_0$, and their speeds reaching $2u_0/3$ [which is the speed of a free soliton of amplitude $2u_0$, according to eq. (161a)], except for the final one, which invariably is slightly weaker. The results thus indicate that the periodic process will continue indefinitely as long as steady forcing is maintained, with each new soliton quickly becoming free after leaving the boundary. After the forcing of a finite duration is terminated, the initial state of rest is quickly restored everywhere except in the region occupied by the progressing free solitons so produced.

This investigation of Chu *et al.* (1983) has been revisited by Camassa and Wu (1989) with two objectives, the first being to broaden the forcing function pattern to examine variations in the responses and further to attain necessary and sufficient numerical details for pursuing the second objective, which is an attempt at adapting the IST (inverse scattering transform, for $-\infty < x < \infty$) method, with some appropriate assumptions, to achieve approximate solutions analytically. In the first aspect, two types of forcing distributions are used, one being the trapezoidal form with a different rise-up time $(0.1, 0.01)$ and the other a Gaussian function $(u(0, t) = U_0 \exp[-\tau(t - t_0)^2]$, $\tau = 60$, $t_0 = 0.4$ with $U_0 = 2.5, 4)$. From these numerical simulations using (161) with $\varepsilon = 0.022$ [for directly adopting the numerical scheme of Zabusky and Kruskal (1965) and with rescaling for making comparisons], the principal features are verified to agree with Chu *et al.* (1983). Further, in all the cases of the quarter-plane problem numerically evaluated, both $u_x(0, t)$ and $u_{xx}(0, t)$ are found to have an average invariably almost zero. Since these two quantities are of crucial importance for the IST of the KdV equation (on which values the initial scattering data depend), the assumptions of

$$u_x(0, t) = 0 \quad \text{and} \quad u_{xx}(0, t) = 0 \qquad (162)$$

are introduced as a first approximation. In addition, a boundary value of $u(0, t) = Q = $ const. for $(0 < t < 1)$ and $u(0, t) = 0$ for $(t > 1)$ is specified to accomplish constructing an IST scheme for this quarter-plane problem. This approximate IST method is found to provide qualitative predictions in good agreement with the numerical results, within 10–20% uniformly from the onset of generation of solitons to their large-time asymptote. This study thus exemplifies that when IST or other analytical methods are not directly applicable to a problem, it may still be open ended for approximate schemes to be devised.

C. STABILITY AND BIFURCATION OF SOLITONS RESONANTLY GENERATED

Returning to the phenomenon of periodic upstream-radiating solitons produced by stationary disturbances sustained at resonance with the wave system, we now present some studies on the basic mechanism underlying this remarkable phenomenon by examining the instability and bifurcation of typical stationary solutions produced by critical forcing. Such studies have been made by Camassa (1990), Camassa and Wu (1991a, b), Yates (1990), Yates and Wu (1991), and Gong and Shen (1994) to investigate this general problem modeled by the forced KdV (fKdV) equation

$$\zeta_t + (\mu - \zeta)\zeta_x - \zeta_{xxx} = P_x. \tag{163}$$

This equation can be deduced from (136) for the stationary forcing frame, with $\mu = 6(F - 1)/k^2$, a detuning parameter as a measure of departure of the Froude number F from the critical value of $F = 1$, and with the scaling of x by k^{-1}, t by k^{-3}, ζ by k^2, P by k^4, and $k - 2\pi h/\lambda$ so that k is explicitly scaled out in (163). The problem is to investigate the stability and bifurcation of the following steady solution of (163) [as a typical representation of those found by Wu (1987) and Yates (1990) and adopted for stability studies]:

$$\zeta_s = 12 \operatorname{sech}^2(x) \quad \text{with} \quad P = 12(\mu - 4) \operatorname{sech}^2(x), \tag{164}$$

where μ, as defined above, is a constant parameter.

To explore the stability properties of the stationary solution (164), we consider an arbitrary perturbation $\eta(x, t)$ imposed on the steady solution, $\zeta(x, t) = \zeta_s(x) + \eta(x, t)$, which, with ζ satisfying (163), yields for η the nonlinear evolution equation (with variable coefficient):

$$\eta_t + \partial_x[(\mu - \zeta_s)\eta - \eta^2/2 - \eta_{xx}] = 0. \tag{165}$$

To resolve this nonlinear eigenvalue problem, we first consider its linearized version by assuming η to be small, then by separation of variables, $\eta =$

$\exp(\sigma t)f(x)$, we obtain for $f(x)$ the linear eigenvalue problem defined by

$$\mathscr{L}f(x) = \sigma f(x), \tag{166a}$$

$$\mathscr{L} = DK, \quad D = \partial_x, \quad K = D^2 + \zeta_s(x) - \mu, \tag{166b}$$

$$f^{(n)}(x) \to 0 \quad \text{sufficiently fast as } |x| \to \infty \quad (n = 0, 1, 2), \tag{166c}$$

with eigenvalue σ and eigenfunction $f(x)$. This eigenvalue problem is distinctive in having the differential equation (166a) in third order and not being self-adjoint. The operator $\mathscr{L}$ has the symmetry property

$$\mathscr{L}(-x) = -\mathscr{L}(x) \quad \text{with} \quad \zeta_s(-x) = \zeta_s(x), \tag{167}$$

and is related to its adjoint system by

$$\mathscr{L}^\dagger g(x) = \sigma_* g(x), \qquad \mathscr{L}^\dagger = -KD, \tag{168}$$

where σ_* is the complex conjugate of σ, and operator $\mathscr{L}$ and its adjoint $\mathscr{L}^\dagger$ are related by

$$D\mathscr{L}^\dagger = -DKD = -\mathscr{L}D.$$

Furthermore, the two systems have the following properties:

1. Equation (166a) has, under condition (166c), the integral invariant $\sigma \int f(x)\,dx = 0$, hence

$$m_e \equiv \int_{-\infty}^{\infty} f(x)\,dx = 0 \qquad \text{unless } \sigma = 0. \tag{169}$$

This shows that the "perturbed excess mass" m_e must vanish when $\sigma \neq 0$, but *need not* vanish when $\sigma = 0$. This singular behavior of the eigenvalue problem marks the source of challenging difficulties in calculating the eigenfunctions.

2. If $[\sigma, f(x)]$ is an eigenpair, so are $[-\sigma, f(-x)]$ and $[\pm\sigma_*, f_*(\pm x)]$. This follows immediately from the symmetry property given above. Therefore ζ_s is unstable whenever the eigenvalue σ has a nonzero real part, whichever its sense.

3. If $[\sigma, f(x)]$ is an eigenpair, its adjoint pair is $[\sigma_*, g(x)]$ where

$$Dg(x) = f_*(-x) \quad \text{or} \quad g(x) = \int_{-\infty}^{x} f_*(-x)\,dx. \tag{170}$$

In addition, there are more specific properties associated with this eigenvalue system, as found by Camassa (1990) and Wu (1993b) that may be useful for numerical purposes.

In pursuing our stability analysis, we first explore the fixed-point solutions for stationary perturbations, i.e., $\sigma = 0$, along the boundary of neutral stability on linear theory. For $\sigma = 0$, (166a) has a first integral, which is the well-known Schrödinger equation:

$$Kf \equiv (D^2 + \zeta_s(x) - \mu)f = 0. \tag{171}$$

With the given ζ_s it follows that (171) has exactly three fixed-point eigenvalues $\mu_n = n^2$, $(n = 1, 2, 3)$ with the eigenfunctions

$$f_0(x, \mu_1) = S(5T^2 - 1), \quad f_0(x, \mu_2) = S^2 T, \quad f_0(x, \mu_3) = S^3, \tag{172}$$

where $S = \mathrm{sech}(x)$, $T = \tanh(x)$. These fixed-point solutions furnish a key foundation for deveoping sophisticated singular perturbation techniques to calculate the eigenvalue pair $[\sigma, f(x)]$, first in a neighborhood of $\mu = \mu_n$, and to carry out analytical continuation of the solution $f(x, \mu_n)$ [with expansion about $f_0(x, \mu_n)$] to all their ranges as desired. This is a tedious task requiring skill and physical insight that is invariably dictated by the solvability condition for the solution of (166a) to exist under the regularity condition (166c) and the integral invariance condition (169). The last condition alone differentiates the fixed points into two kinds, (FPi) characterized by $f_0(x, \mu_1)$ and $f_0(x, \mu_3)$ for not satisfying the invariance condition (169), and (FPii) with $f_0(x, \mu_2)$ for satisfying (169), but not for its next higher order term of $f(x, \mu_2)$. The results of this eigenvalue problem can be summed up as follows.

Fa: The eigenvalue integral curve in the parametric plane emanating from a fixed point of type (FPi) exists only on one side of (FPi), here for $\mu < 1$ about μ_1, and $\mu < 9$ about μ_3, along which σ is purely real, as shown in the lower part of Figure 15.

Fb: The eigenvalue integral curve emanating from a fixed point of type (FPii) exists on both sides of (FPii), being purely real on one side and complex on the other, with the real component at least one order smaller than the imaginary component. In this case, σ of $f_0(x, \mu_2)$ is purely real for $\mu > 4$ and complex for $\mu < 4$ about $\mu_2 = 4$.

Fc: The fixed points of type (FPi) and (FPii) occur interlaced.

The final result of the local expansion calculation complemented by numerical computations for analytical continuation of the eigenvalue is shown in Figure 15 for this case. Numerical calculations of eigenpairs require high-precision schemes primarily owing to the small ratios of the

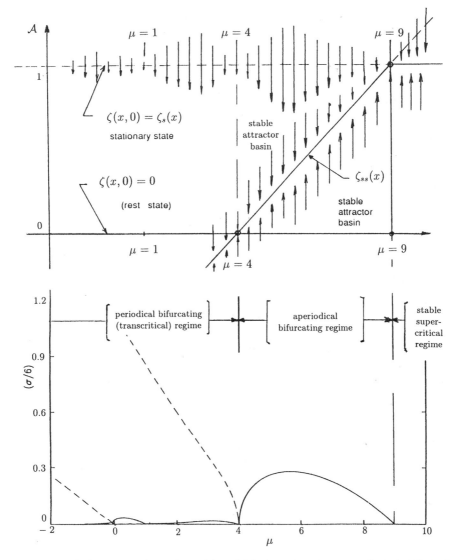

FIG. 15. The lower part shows the real (in solid line) and imaginary part (in dashed line) of the eigenvalue σ corresponding to $\zeta_s(x)$ of (164) as the basic steady flow. The upper part is the bifurcation diagram indicating bifurcation of the basic flow $\zeta_s(x)$ and the stable attractor basin for the second stationary solution $\zeta_{ss}(x)$, due to forcing $P(x)$, as found by Camassa (1990) and Camassa and Wu (1991) in terms of the initial amplitude $\mathscr{A} = \zeta(x, 0)/\zeta_s(x)$ and μ: instability (dashed line); stability (solid line), as indicated with arrows.

real-to-imaginary parts whenever σ is complex; this requirement renders some mode-expansion methods inadequate, as found by Wu (1988) and Camassa (1990).

A bifurcation diagram of the stationary solution with its characteristic critical forcing has been constructed based on the eigenvalue distribution and numerical simulation, as presented in Figure 15 in terms of the dimensionless amplitude $\mathscr{A} = \zeta(x, 0)/\zeta_s$ as a constant parameter versus μ. Thus, the line of $\mathscr{A} = 1$ is initially the stationary state, which should remain stationary if stable. But for $\mu < 4$, $\sigma = \sigma_r + i\omega$ is complex, forming a regime we call the *periodical bifurcating regime*, in which the phenomenon of periodic generation of forward-radiating solitons is invariably observed. The rate of growth of the first soliton to maturity and that of the successive ones is related to the value of σ_r and the period of radiation is locked in with ω. Variations of these rates with varying $\mathscr{A}$ and μ have been found by detailed numerical simulations to be dictated by the pertinent eigenvalue attained.

For the range $4 < \mu < 9$, σ is purely real, forming a regime called the *aperiodical bifurcating regime*, in which an initially established steady basic flow bifurcates into a new solution $\zeta_{ss}(x)$, which is found by Camassa (1990) and Camassa and Wu (1991a) to be globally stable and numerically robust. Existence of the stable solution $\zeta_{ss}(x)$ is determined numerically within a broad basin of attraction centered about a slightly curved line passing through the points $(\mu = 4, \mathscr{A} = 0)$ and $(\mu = 9, \mathscr{A} = 1)$ as the valley bottom.

Finally, for $\mu > 9$, the eigenvalue does not exist on linear stability analysis, rendering the theory inconclusive. Fortunately, a proof is given by Camassa (1990) for this regime showing that the perturbation flow of $\eta = \zeta - \zeta_s$ possesses a Hamiltonian function with its positivity sufficient to ensure nonlinear global stability in the Lyapunov sense. Mathematically, there is an exchange of stability between line $\mathscr{A} = 1$ and the extended line of ζ_{ss}. Numerically, extensive simulations have shown strong evidence of robust stability for ζ_{ss}. Physically, the Froude number of the forcing, with $F > 1 + 9a/8$ (for $\mu > 9$ and for wave amplitude $a = 4k^2/3$ of KdV solitons), is too fast to be outraced by any free wave in this regime, which for this reason is called the *stable supercritical regime*.

Additional studies on another steady forced wave with symmetry also even in x have been found by Camassa and Wu (1991b) to possess the same qualitative features as the present case. This may suggest, on first sight, that the general properties of eigensolutions for the class of symmetric forced waves may have much in common. Even more, for skew-symmetric waves pertaining to skew-symmetric forcings, the studies by Yates (1990) and

Yates and Wu (1991) have brought forth evidence that a broad similarity exists between the stationary waves produced by symmetrically and skew-symmetrically forcings and that they share a common basis on their general properties.

D. Standing Solitons Generated in Faraday Resonance

A new kind of forced soliton generation is the standing solitary waves generated within an elongated horizontal water tank resonantly excited under vertical oscillation. Pioneering on this study, Faraday (1831) observed subharmonic excitation of a liquid layer contained in a basin that is forced to oscillate vertically with frequency twice the natural frequency of the excited waves in the basin. The Faraday waves have been studied by Rayleigh (1883) explaining on linear theory why waves excited are parametrically subharmonic, by Benjamin and Ursell (1954) finding that the excited waves can be described by Mathieu's equation, by Ockendon and Ockendon (1973) extending the view to nonlinear self-interaction, by Miles (1984a,b) who evaluated the nonlinear interaction coefficients, and by others. For a review, see Miles and Henderson (1990) who further considered the effects of dissipation and surface tension together with comparison between theory and experiment.

A remarkable family of the parametrically excited waves has been discovered by Wu *et al.* (1984) who first showed experimentally the existence of standing cross waves sloshing to and fro across the narrow width of an elongated rectangular tank, subject to vertical oscillation, with the wave profile modulated by a hyperbolic-secant envelope along the tank's length, at a forcing frequency shifted slightly *below* twice the natural frequency ω_0 of the (0, 1)-mode of the water waves within the tank. For this mode Larraza and Putterman (1984) and Miles (1984b) showed that the standing soliton does not exist for values of $kh \leqslant 1.022$ where h is the rest water depth and $k = \pi/b$, b being the width of the tank.

Subsequently, it has been discovered independently by Denardo *et al.* (1990), and Guthart and Wu (1991) that for $kh \leqslant 1.022$ and for a forcing frequency shifted slightly *above* twice the natural frequency ω_0 of the (0, 1) mode, a new family of cross waves is found to exist that sloshes across the width of the tank and is modulated by the hyperbolic-tangent envelope, or kink wave that by nature is standing along the length of the tank, as depicted in Figure 16.

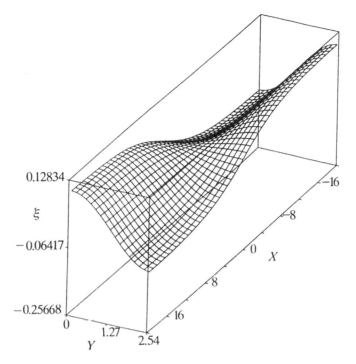

Fig. 16. Computer-generated plot of the surface elevation for the sloshing kink wave as presented by Guthart and Wu (1991).

For analysis, the present FNFD theory can be readily extended to include the more general phenomenon of the Faraday wave in a water tank under oscillation [e.g., $h(r, t) = h(r) \exp(i\omega t)$] simply by replacing the constant gravity acceleration g in (12) and (13) by a time-dependent value with reference to the tank frame. It has been known that linear waves and nonlinear solitons are found to exist in their specific regions but cannot be observed in all regions of the parameter space for which theoretical solutions formally exist. This leads to the question of hydrodynamic stability as discussed by Craik (1993), Guthart (1992), and Guthart and Wu (1993). For nonlinear Faraday-type solitons, Guthart and Wu have performed linear stability analysis, its extension including nonlinear effects as well as numerical simulations of the forced and damped NLS equations, together with broad experimental investigations of evolution of the solitons in the laboratory along the boundary of linear and nonlinear analyses. These observations qualitatively confirm all of the predicted behavior for the evolution of solitons in the stable and unstable parameter regions when the forcing and frequency detuning are sufficiently small in amplitude. Some new phenomena of transition of standing solitons (under certain variable modulated oscillations) to bifurcations and chaos have been discovered and studied by Wei *et al.* (1990) and Chen and Wei (1994).

XI. Channel Shape Effects on Wave Propagation and Generation

For channel flow applications, the Boussinesq-class theory dealing with nonlinear waves in gradually varying channels of arbitrary cross-sectional shape is of pratical importance. Long waves evolving in such (straight) arbitrary channels have been investigated by Peregrine (1967, 1968), Shen (1969), Shuto (1974), Miles (1979), Chang *et al.* (1979), Kirby and Vengayil (1988), Teng (1990), and Teng and Wu (1992, 1994, 1997) among others. This general problem of long waves can be described by the generalized channel Boussinesq (cgB) model equations (Teng and Wu, 1997):

$$(2b_0\tilde{\zeta} + S\tilde{\zeta}^2)_t + [2b_0(\tilde{h} + \tilde{\zeta})\bar{u}]_x = (A_d)_t, \tag{173}$$

$$\bar{u}_t + \bar{u}\bar{u}_x + \tilde{\zeta}_x - \tfrac{1}{3}\kappa^2\tilde{h}^2\bar{u}_{xxt} = -(\tilde{p}_a)_x, \tag{174}$$

where $2b_0(x)$ is the unperturbed channel breadth (at the rest water surface level $z = 0$), S is the sectional channel sidewall slope with respect to the vertical (ranging over water surface sloshing stretch), the symbol $\tilde{f}(x, t)$ denotes the sectional surface mean value of f averaged across the channel at section x at time t, $\bar{u}(x, t)$ is the sectional mean velocity along the channel axis, averaged over the rest cross-sectional area $A_0(x)$, and A_d is the *reduction* in cross-sectional area due to a submerged object moving below the rest water surface level. The factor $\kappa(x)$ is defined by

$$\kappa^2(x) = \frac{3}{\tilde{h}^2}(\tilde{\tilde{\psi}} - \bar{\psi}), \tag{175}$$

where $\psi(y, z)$ is related to $\phi_2(r, z, t)$ of the series expansion (14) for the velocity potential ϕ such that $\phi_2 = -\psi(y, z)\bar{u}_{xxt}$, with ψ satisfying the following equations:

$$\psi_{yy} + \psi_{zz} = 1 \qquad ((y, z) \in A_0(x)),$$

$$\psi_z(y, z = 0) = \tilde{h}(x), \qquad (-b < y < b),$$

$$\mathbf{n} \cdot \nabla\psi = 0 \qquad \text{(on channel wall)},$$

where $\mathbf{n}$ is the normal vector at the channel boundary. From the above system of equations defining ψ, $\kappa(x)$ is seen to depend solely on the geometry of channel shape and hence is called the *channel shape factor*. It can further be shown (Teng and Wu, 1997) that κ^2 as defined above is real and positive. Moreover, $\kappa = 1$ for channels of rectangular cross section, even with varying size, which can therefore serve as the standard reference. A set of common geometric shapes can be found in Teng and Wu (1992) with their shape factor κ values.

For *uniform* channels of arbitrary shape, the shape factor κ becomes a constant. In this case it is clear that by the similarity transformation:

$$x = \kappa x', \quad t = \kappa t', \quad \zeta(x, t) = \zeta'(x', t'), \quad u(x, t) = u'(x', t'), \quad p_a(x, t) = p'_a(x', t'),$$

$$(176)$$

(173) and (174) are reduced to one for an analogous rectangular channel (with $\kappa' = 1$) provided the two channels have the same mean water depth serving as the common length scale. From this we have the following theorem (Teng and Wu, 1992).

Uniform channel analogy theorem: To a long wave of wavenumber k, period T, phase velocity c, and amplitude α evolving in a κ-shaped uniform channel, there corresponds an analogous wave of wavenumber k', period T', phase velocity c', and amplitude α', evolving in an analogous rectangular channel according to

$$k' = \kappa k, \quad T' = T/\kappa, \quad c' = c, \quad \text{and} \quad \alpha' = \alpha, \tag{177}$$

in addition to having the waves satisfy the similarity relations in (176).

The term $\tilde{p}_a$ in channel equation (174) is significant in representing external pressure disturbances and also qualitatively any outside forcing that renders the system physically open to having exchanges of mass, momentum, and energy with other systems. In the presence of such open exchanges, the remarkable thing is to realize that all space- and time-related quantities must in principle obey the similarity law. It may be somewhat obvious in relating the corresponding wave patterns in analogous channels by the similarity law, as shown by Teng and Wu (1994). But it is of great interest in applying this theorem for correlating such slow-time nonlinear phenomena as that of periodic production of upstream-radiating solitary waves in analogous channels by resonant forcing, e.g., surface pressure and/or submerged topography moving with transcritical velocities. If a given resonant forcing generates solitary waves at period T' in a rectangular channel, then corresponding solitary waves will be produced in a κ-shaped channel by a similarly corresponding forcing at a period T which is κ times

T'. This theoretical prediction has recently been established experimentally by Teng and Wu (1997). And this agreement is significant in thoroughly reflecting the similarity law on functionals such as the period $T = \kappa T'$ of soliton generation as a part of the whole solution.

Finally, we mention that concerning nolinear waves with a primary component in one spatial dimension and one weak transverse component, there is the Kadomtsev and Petviashivili equation (1970), which is a two-dimensional version of the KdV equation. With regard to nonlinear wave propagation in channels with significantly curved and sharp-cornered bends, Shi *et al.* (1998) have investigated the transmission and reflection of a solitary type of waves for such channels with vertical sidewalls and with water of uniform rest depth. It is found that the wave transmission and reflection coefficients depend on one single dimensionless parameter, namely, the ratio of the channel width b to the effective wavelength λ_e. Quantitative results for predicting these coefficients are obtained that may facilitate further studies for more general problems.

XII. Conclusion

With the intent of facilitating scientific and engineering research with applications to water wave problems under various conditions approaching that in nature, a new theoretical model has been developed for evaluating unsteady, fully nonlinear, fully dispersive, three-dimensional gravity-capillary waves on a single layer of water of variable depth. It is formulated in terms of four alternative sets of basic variables for establishing the model equations. These systems of model equations are reduced in spatial dimensions by one from the underlying Euler equations, but otherwise retain the full capacity of the primary system in every respect. The model equations are closed either in differential form, for convenience in theoretical studies and reductive analysis, or in integral form, for ready computation of exact solutions. In exactness, this theory is fully established by the preliminary numerical results obtained for the large-time asymptotic wave profile and wave velocity of terminally evolving solitary waves in comparison with the exact solutions on the permanent solitary wave formalism. In unification, this theory recovers all the well-known linear, weakly and stongly nonlinear, weakly and strongly dispersive, two- and three-dimensional water wave models as special cases pertaining to specific limiting regimes of the key parameters. These reductions include the weakly nonlinear fully dispersive

(WNFD) Stokes wave theory, Boussinesq's WNWD two-equation model and its high-order versions, the KdV-type model for unidirectional waves with or without forcing, the family of FNWD wave models including Airy's and Green–Naghdi's versions, the bidirectional wave models of various orders for studying wave–wave interactions, waves in variable channels, and the related processes of mass-energy transfer. These approximate versions of the underlying Euler model have various degrees of validity to suit different premises for developmental studies and applications of pertinent interest.

Comprehensive coverage by the present theory of the aspects of non-linearity, dispersion, and bathymetry variation (as the three primary physical effects pertaining to water wave modeling) has been illustrated by applications. In adopting the present theory to determine the Stokes wave in water of arbitrary depth, by perturbation expansion to third order, the results elucidate that this theory not only commands the dispersive effect over its entire range, but also finds the contributions from the nonlinear effects to it. In the task of calculating the solitary wave to third order based on using the $(\hat{u}, \zeta)$ system of equations (not being used previously) by the reductive perturbation method, the procedure illuminates the entire course showing the existence of the solution to all higher orders by construction (not presented), thus implying the existence of solitary waves, at least asymptotically as an exact solution supported by the Euler equations. The trend of convergence is clearly suggested by the leading three orders of the solution. In analytical execution, the present theory further exhibits, to some degree, its unifying facility and simplicity.

The recent studies on generation of nonlinear waves by resonant forcing of the wave-supporting system have opened a vast new field of rich resources. The early results have caught workers in the field by refreshing surprise. The salient facts appear to lay in sharp contrast to our previous intuition that to stationary disturbances, however strong, one would not expect to see any responses (from deterministic systems) but the stationary. Regarding the basic mechanism underlying the remarkable phenomena of solitary waves periodically produced by stationary critical forcing, the results from the stability and bifurcation study presented in Section X.C have furnished enlightening exposition. These new findings may provide stimuli to encourage us to look deeper at nonlinear phenomena of wave–wave and wave–body interactions at critical states. One such instance concerns the classical hydraulic analogy, especially for the case when the pertinent processes may leave a long enough time for slow nonlinear events to manifest. In another instance, the results given in Section IX.D on mass-energy transfer between overtaking solitons may invite deeper study. A philosophical question here is whether such vanishingly weak transport of mass and energy may also hold true, by nature, for the underlying Euler model.

In conclusion, it is hoped that this new theoretical model may offer opportunities for further studies of these very interesting problems. It may also provide a sound foundation for making new advances in areas requiring considerations of other physical effects that are of importance under differing premises.

Acknowledgments

The author is indebted to Gerald Whitham for his enlightening views, appreciates having interesting discussions with Jin Zhang, George Yates, Wooyoung Choi, Roberto Camassa, Michelle Teng, and Wendong Qu, and would like to thank Wendong Qu and John Kao for achieving the preliminary numerical results quoted here. This study has been supported by NSF through the Hazard Mitigation Program, to which subject further applications of theory are being studied under grants CMS-9503620 and CMS-9615897.

References

Agnon, Y., Madsen, P. A., and Shäffer, H. A. (1999). A new approach to higher order Boussinesq models. *J. Fluid Mech.* **399**, 319–333.
Akylas, T. R. (1984). On the excitation of long nonlinear water waves by a moving pressure distribution. *J. Fluid Mech.* **141**, 455–466.
Beale, J. T., Hou, T. Y., and Lowengrub, J. (1996). Convergence of a boundary integral method for water waves. *SIAM J. Numer. Anal.* **33**, 1797–1843.
Benjamin, T. B., and Feir, J. E. (1967). The disintegration of wave trains in deep water. Part 1. Theory, *J. Fluid Mech.* **27**, 417–430.
Benjamin, T. B., and Ursell, F. (1954). The stability of the plane free surface of a liquid in vertical periodic motion. *Proc. R. Soc. London Ser. A* **225**, 505–515.
Benney, D. J., and Luke, J. C. (1964). On the interactions of permanent waves of finite amplitude. *J. Math. Phys.* **43**, 309–313.
Boussinesq, J. (1871). Théorie de l'intumescence liquide appelee onde solitaire ou de translation se propageant dans un canal rectangulaire. *C.R. Acad. Sci. Paris* **72**, 755–759.
Boussinesq, J. (1872). Théorie des ondes et des remous qui se propagent le long d'un canal rectangulaire horizontal, en communiquant au liquide contenu dans ce canal des vitesses sensiblement pareilles de la surface au fond. *J. Math. Pures Appl. Ser. 2* **17**, 55–108.
Byatt-Smith, J. G. B. (1971). An integral equation for unsteady surface waves and a comment on the Boussinesq equation. *J. Fluid Mech.* **49**, 625–633.

Camassa, R. (1990). Part 1. Forced generation and stability of nonlinear waves; Part 2. Chaotic advection in a Rayleigh-Benard flow. Ph.D. thesis, California Institute of Technology, Pasadena.

Camassa, R., and Wu, T. Y. (1989). The Kortweg–de Vries model with boundary forcing. *Wave Motion* 11, 495–506.

Camassa, R., and Wu, T. Y. (1991a). Stability of forced steady solitary waves. *Phil. Trans. R. Soc. Lond. A* 337, 429–466.

Camassa, R., and Wu, T. Y. (1991b). Stability of some stationary solutions for the forced KdV equation. *Physica D* 51, 295–307.

Casciola, C. M., and Landrini, M. (1996). Nonlinear long waves generated by a moving pressure disturbance. *J. Fluid Mech.* 325, 399–418.

Chan, R. K.-C., and Street, R. L. (1970). A computer study of finite-amplitude water waves. *J. Comput. Phys.* 16, 68–94.

Chang, J.-H. (1997). Interaction of solitary waves with structures in viscous fluid. Ph.D. dissertation, National Cheng-Kung University, Tainan, Taiwan.

Chang, P., Melville, W. K., and Miles, J. W. (1979). On the evolution of a solitary wave in a gradually varying channel. *J. Fluid Mech.* 95, 401–414.

Chappelear, J. E. (1962). Shallow water waves. *J. Geophys. Res.* 67, 4693–4704.

Chen, X., and Sharma, S. D. (1995). A slender ship moving at a near critical speed in a shallow channel. *J. Fluid Mech.* 291, 263–285.

Chen, X., and Wei, R. (1994). Dynamic behavior of a non propagating soliton under a periodically modulated oscillation. *J. Fluid Mech.* 259, 291–303.

Choi, W. (1993). Forced generation of solitary waves in a rotating fluid and their stability. Ph.D. thesis, California Institute of Technology, Pasadena.

Choi, W. (1995). Nonlinear evolution equations for two-dimensional surface waves in a fluid of finite depth. *J. Fluid Mech.* 295, 381–394.

Choi, W., and Camassa, R. (1996). Weakly nonlinear internal waves in a two-fluid system. *J. Fluid Mech.* 313, 83–103.

Choi, W. Y., and Wu, T. Y. (1996). Vortex solitons in a rotating fluid within a non-uniform tube. *Wave Motion* 24, 243–262.

Chu, C. K., Xiang, L. W., and Baransky, Y. (1983). Solitary waves induced by boundary motion. *Phys. Fluids* 16, 1565–1572.

Cole, J. D. (1968). *Perturbation Methods in Applied Mathematics.* Blaisdell Publ. Co., Waltham, MA.

Cole, S. L. (1985). Transient waves produced by flow past a bump. *Wave Motion* 7, 579–587.

Cooker, M. J., Weidman, P. D., and Bale, D. S. (1997). Reflection of a high-amplitude solitary wave at a vertical wall. *J. Fluid Mech.* 342, 141–158.

Craik, A. D. D. (1993). The stability of some three-dimensional and time-dependent flows. In *Proc. IUTAM Symp. on Nonlinear Instability of Nonparallel Flows* (S. P. Lin, W. R. C. Phillips, and D. T. Valentine, eds.), Clarkson University, July 26–31, 1993, pp. 382–396. Springer-Verlag, New York.

Daily, J. W., and S. C. Stephan (1952). The solitary wave. *Proc. Third Conf. on Coastal Eng.*, pp. 13–30. ASCE, Reston, VA.

Denardo, B., Wright, W., Putterman, S., and Larraza, A. (1990). Observation of a kink soliton on the surface of a liquid. *Phys. Rev. Lett.* 64–13, 1518–1521.

Ertekin, R. C., Webster, W. C., and Wehausen, J. V. (1984). Ship-generated solitons. In *Proc. 15th Symp. Naval Hydrodynamics*, pp. 347–364. National Academy of Sciences, Washington, DC.

Ertekin, R. C., Webster, W. C., and Wehausen, J. V. (1986). Waves caused by a moving disturbance in a shallow channel of finite width. *J. Fluid Mech.* 169, 275–292.

Faltinsen, O. M. (1977). Numerical solution of transient nonlinear free-surface motion outside or inside moving bodies. In *Proc. 2nd Int. Conf. Numer. Ship Hydrodyn.*, pp. 347–357. University of California, Berkeley, CA.

Faraday, M. (1831). On a peculiar class of acoustical figures; and on certain forms assumed by groups of particles upon vibrating elastic surfaces. *Phil. Trans. R. Soc. London* **121**, 299–340.

Friedrichs, K. O., and Hyers, D. H. (1954). The existence of solitary waves. *Comm. Pure Appl. Math.* **7**, 517–550.

Gong, L., and Shen, S. S. (1994). Multiple supercritical solitary wave solutions of the stationary forced Korteweg–de Vries equation and their stability. *SIAM J. Appl. Math.* **54**, 1268–1290.

Green, A. E., and Naghdi, P. M. (1976). A derivation of equations for wave propagation in water of variable depth. *J. Fluid Mech.* **78**, 237–246.

Grimshaw, R. H. J. (1971). The solitary wave in water of variable depth. *J. Fluid Mech.* **46**, 611–622.

Grimshaw, R. H. J., and Smyth, N. F. (1986). Resonant flow of a stratified fluid over topography. *J. Fluid Mech.* **169**, 429–464.

Grue, J., Friis, H. A., Palm, E., and Rusas, P. O. (1997). A method for computing unsteady fully nonlinear interfacial waves. *J. Fluid Mech.* **351**, 233–252.

Guthart, G. (1992). On the existence and stability of standing solitary waves in Faraday resonance. Ph.D. thesis, California Institute of Technology, Pasadena.

Guthart, G., and Wu, T. Y. (1991). Observation of standing kink cross wave parametrically excited. *Proc. R. Soc. Lond. A* **434**, 435–440.

Guthart, G., and Wu, T. Y. (1993). On the stability of standing solitons in Faraday resonance. In *Proc. IUTAM Symp. on Nonlinear Instability of Nonparallel Flows* (S. P. Lin, W. R. C. Phillips, and D. T. Valentine, eds.), Clarkson University, July 26–31, 1993, pp. 397–406. Springer-Verlag, New York.

Hanazaki, H. (1991). Upstreaming-advancing nonlinear waves in an axisymmetric resonant flow of rotating fluid past an obstacle. *Phys. Fluids A* **3**, 3117–3120.

Hirota, R. (1973). Exact N-soliton solutions of the wave equations of long waves in shallow-water and in nonlinear lattices. *J. Math. Phys.* **14**(7), 810–814.

Hou, T. Y., Teng, Z., and Zhang, P. (1996). Well-posedness of linearized motion for 3-D water waves far from equilibrium. *Comm. Partial Diff. Equat.* **21** (9/10) 1551–1585.

Huang, D. B., Sibul, O. J., Webster, W. C., Wehausen, J. V., Wu, D. M., and Wu, T. Y. (1982). Ships moving in the transcritical range. In *Proc. Conf. Behavior of Ship in Restricted Waters*, Vol. 2, pp. 26/1–26/10. Bulgarian Ship Hydrodynamics Center, Varna.

Kadomtsev, B. B., and Petviashvili, V. I. (1970). The stability of solitary waves in weakly dispersing media. *Dokl. Akad. Nauk SSR* **192**, 753–756.

Kármán, Th. von (1938). Eine praktische Anwendung der Analogie zwischen Überschall-strömung in Gasen und überkritischer Strömung in offenen Gerinnen. *Zeit. Angew. Math. Mech.* **18**, 49–56.

Kaup, D. J. (1975). A higher-order water-wave equation and the method for solving it. *Prog. Theor. Phys.* **54**, 396–408.

Keller, J. B. (1985). Soliton generation and nonlinear wave propagation. *Phil. Trans. R. Soc. Lond. A* **315**, 367–377.

Kirby, J. T., and Vengayil, P. (1988). Nonresonant and resonant reflection of long waves in varying channels. *J. Geophy. Res.* **93**, 10782–10796.

Korteweg, D. J., and de Vries, G. (1895). On the change of form of long waves advancing in a rectangular channel, and on a new type of long stationary waves. *Philos. Mag.* **39**, 422–443.

Kupershmidt, B. A. (1985). Mathematics of dispersive water waves. *Comm. Math. Phys.* **99**, 51–73.

Laitone, E. V. (1960). The second approximation to cnoidal and solitary waves. *J. Fluid Mech.* **9**, 430–444.

Larraza, A., and Putterman, S. (1984). Theory of non-propagating surface-wave solitons. *J. Fluid Mech.* **148**, 443–449.

Lax, P. D. (1968). Integrals of nonlinear equations of evolution and solitary waves. *Comm. Pure Appl. Math.* **21**, 467–490.

Lee, S.-J. (1985). Generation of long water waves by moving disturbances. Ph.D. thesis, California Institute of Technology, Pasadena.

Lee, S.-J., and Grimshaw, R. H. J. (1989). Precursor waves generated by three-dimensional moving diatrubances. In *Engineering Science, Fluid Dynamics* (G. T. Yates, ed.), pp. 59–74. World Scientific, Singapore.

Lee, S.-J., and Grimshaw, R. H. J. (1990). Upstream-advancing waves generated by three-dimensional moving disturbances. *Phys. Fluids A.* **2**, 194–201.

Lee, S.-J., Yates, G. T., and Wu, T. Y. (1989). Experiments and analysis of upstream-advancing solitary waves generated by moving disturbances. *J. Fluid Mech.* **199**, 569–593.

Levi-Civita, T. (1925). Détermination rigoureuse des ondes permanentes d'ampleur finie. *Math. Ann.* **93**, 264–314.

Lighthill, J. M. (1978). *Waves in Fluids.* Cambridge Univ. Press, Cambridge.

Lin, C. C., and Clark, A., Jr. (1959). On the theory of shallow water waves. *Tsing Hua J. of Chinese Studies Special* **1**, 54–62; also in *Selected Papers of C.C. Lin*, 1987, pp. 352–360. World Scientific, Singapore.

Liu, H., and Wu, T. Y. (1998). Some solitary wave solutions of fully nonlinear weakly dispersive wave models. In *Postprint 3rd Intern. Conf. on Nonlinear Mechanics*, Shanghai University, China, Aug. 17–20, 1998.

Longuet-Higgins, M. S., and Cokelet, E. D. (1976). The deformation of steep surface waves on water I. A numerical method of computation. *Proc. R. Soc. Lond. A* **350**, 1–26.

Madsen, P. A., and Schäffer, H. A. (1998). Higher-order Boussinesq-type equations for surface gravity waves: Derivation and analysis. *Phil. Trans. R. Soc. Lond. A* **356**, 3123–3184.

Madsen, P. A., and Sorensen, O. R. (1992). A new form of the Boussinesq equations with improved linear dispersion characteristics. Part 2. A slowly varying bathymetry. *Coastal Eng.* **18**, 183–204.

Madsen, P. A., Murray, R., and Sorensen, O. R. (1991). A new form of the Boussinesq equations with improved linear dispersion characteristics. Part 1. *Coastal Eng.* **15**, 371–388.

Madsen, P. A., Banijamali, B., Schäffer, H. A., and Sorensen, O. R. (1996). Boussinesq type equations with high accuracy in dispersion and nonlinearity. In *Proc. 25th Int. Coastal Engineering Conf.*, pp. 95–108. ICCE.

Maxworthy, T. (1976). Experiments on collisions between solitary waves. *J. Fluid Mech.* **76**, 177–185.

Mei, C. C. (1983). *The Applied Dynamics of Ocean Surface Waves.* John Wiley, New York.

Mei, C. C., and Choi, H. S. (1987). Forces on a slender ship advancing near the critical speed in a wide canal. *J. Fluid Mech.* **179**, 59–76.

Miles, J. W. (1979). On the Korteweg–de Vries equation for a gradually varying channel. *J. Fluid Mech.* **91**, 181–190.

Miles, J. W. (1984a). Nonlinear Faraday resonance. *J. Fluid Mech.* **146**, 285–302.

Miles, J. W. (1984b). Parametrically excited solitary waves. *J. Fluid Mech.* **148**, 451–460; **154**, 535.

Miles, J., and Henderson, D. (1990). Parametrically forced surface waves. *Ann. Rev. Fluid Mech.* **22**, 143–165.

Nekrasov, A. I. (1921). On waves of permanent type. I. *Izv. Ivanovo-Voznesensk. Politekhn. Inst.* **3**, 52–65.

Nwogu, O. (1993). Alternative form of Boussinesq equations for nearshore wave propagation. *J. Wat. Ways, Port, Coastal Ocean Eng.* **119**(6), 618–638. ASCE.

Ockendon, J. R., and Ockendon, H. (1973). Resonant surface waves. *J. Fluid Mech.* **59**, 397–413.

Olver, P. J. (1986). *Applications of Lie Groups to Differential Equations.* Springer-Verlag, New York.

Peregrine, D. H. (1967). Long waves on a beach. *J. Fluid Mech.* **27**, 815–827.

Peregrine, D. H. (1968). Long waves in a uniform channel of arbitrary cross-section. *J. Fluid Mech.* **32**, 353–365.

Power, H., and Chwang, A. T. (1984). On the reflection of a planar solitary wave at a vertical wall. *Wave Motion* **6**, 183–195.

Prandtl, L. (1931). *Abriss der Strömungslehre.* Vieweg, Braunschweig.

Qu, W., Wu, T. Y., and Kao, J. (2000). Computation of strongly and fully nonlinear water waves (to appear).

Rayleigh, Lord (1876). On waves. *Phil. Mag.* **1**, 257–279; reprinted in *Scientific Papers*, vol. 1, pp. 251–271. Cambridge Univ. Press, Cambridge.

Rayleigh, Lord (1883). On maintained vibrations. *Phil. Mag.* **15**, 229–235; On the crispations of fluid resting upon a vibrating support. *Phil. Mag.* **16**, 50–58. Reprinted, 1990, in *Scientific Papers*, **2**, 188–193; 212–219. Cambridge Univ. Press, Cambridge.

Russell, John Scott (1844). Report on waves. In *Rept. 14th Meeting of the British Association for the Advancement of Science,* pp. 311–390. John Murray, London.

Russell, John Scott (1865). *The Modern System of Naval Architecture*, Vol. 1, p. 208. Day and Son, London.

Schäffer, H. A., and Madsen, P. A. (1995). Further enhancements of Boussinesq-type equations. *Coastal Eng.* **26**, 1–14.

Schember, H. R. (1982). A new model for three-dimensional nonlinear dispersive long waves. Ph.D. thesis, California Institute of Technology, Pasadena.

Shen, M. C. (1969). Asymptotic theory of unsteady three-dimensional waves in a channel of arbitrary cross section. *SIAM J. Appl. Math.* **17**, 260–271.

Shi, A. M., Teng, M. H., and Wu, T. Y. (1998). Propagation of solitary wave through significantly curved shallow water channels. *J. Fluid Mech.* **362**, 157–176.

Shuto, N. (1974). Nonlinear long waves in a channel of variable section. *Coastal Eng. Japan* **17**, 1–12.

Stokes, G. G. (1847). On the theory of oscillatory waves. *Trans. Cambridge Phil. Soc.* **8**, 441–455. In *Math. Phys. Papers*, Vol. 1, pp. 197–229. Cambridge Univ. Press, Cambridge, 1880.

Su, C. H., and Mirie, R. M. (1980). On head-on collisions between two solitary waves. *J. Fluid Mech.* **98**, 509–525.

Sun, M. G. (1982). Experimental photographs. (Private communication.)

Tang, C. J., and Chang, J.-H. (1997). Local grid refinement for nonlinear waves. *J. Chinese Inst. Engineers* **20**, 285–293.

Tang, C. J., Patel, V. C., and Landweber, L. (1990). Viscous effects on propagation and reflection of solitary waves in shallow water. *J. Comput. Phys.* **88**, 86–113.

Temperville, A. (1979). Interactions of solitary waves in shallow water theory. *Arch. Mech.* **31**, 177–184.

Teng, M. H. (1990). Forced emissions of nonlinear water waves in channels of arbitrary shape. Ph.D. thesis, California Institute of Technology, Pasadena.

Teng, M. H., and Wu, T. Y. (1992). Nonlinear water waves in channels of arbitrary shape. *J. Fluid Mech.* **242**, 211–233.

Teng, M. H., and Wu, T. Y. (1994). Evolutioon of long water waves in variable channels. *J. Fluid Mech.* **266**, 303–317.

Teng, M. H., and Wu, T. Y. (1997). Effects of channel cross-sectional geometry on long wave generation and propagation. *Phys. Fluids* **9**(11), 3368–3377.

Tsai, W., and Yue, D. K. P. (1996). Computation of nonlinear free-surface flows. *Ann. Rev. Fluid Mech.* **28**, 249–278.

Wehausen, J. V., and Laitone, E. V. (1960). Surface waves. In *Handbuch der Physik*, Vol. IX (S. Flügge and C. Truesdell, eds.). Springer-Verlag, New York.

Wei, G., Kirby, J. T., Grilli, S. T., and Subramanya, R. (1995). A fully nonlinear Boussinesq model for surface waves. Part 1. *J. Fluid Mech.* **294**, 71–92.

Wei, R., Wang, B., Mao, Y., Zheng, X., and Miao, G. (1990). Further investigation of nonpropagating solitons and their transition to chaos. *J. Acoust. Soc. Am.* **88**, 469–472.

Weidman, P. D., and Maxworthy, T. (1978). Experiments on strong interactions between solitary waves. *J. Fluid Mech.* **85**, 417–431.

Whitham, G. B. (1974). *Linear and Nonlinear Waves.* Wiley, New York.

Wu, D. M., and Wu, T. Y. (1982). Three-dimensional nonlinear long waves due to moving surface pressure. In *Proc. 14th Symp. on Naval Hydrodynamics*, pp. 103–125. National Academy Press, Washington, DC.

Wu, D. M., and Wu, T. Y. (1988). Precursor solitons generated by three-dimensional disturbances moving in a channel. In *IUTAM Symp. on Nonlinear Water Waves* (K. Horikawa and H. Maruo, eds.), Tokyo, Japan, Aug. 25–28, 1987, pp. 69–75. Springer-Verlag, New York.

Wu, J., Keolian, R., and Rudnick, I. (1984). Observation of a nonpropagating hydrodynamic soliton. *Phys. Rev. Lett.* **52**, 1421–1424.

Wu, T. Y. (1979). On tsunami propagation: Evaluation of existing models. In *Tsunamis — Proc. National Science Foundation Workshop*, May 7–9, 1979, pp. 110–149. Tetra Tech. Inc., Pasadena, CA.

Wu, T. Y. (1981). Long waves in ocean and coastal waters. *J. Engng Mech. Div. ASCE* **107**, 501–522.

Wu, T. Y. (1987). Generation of upstream advancing solitons by moving disturbances. *J. Fluid Mech.* **184**, 75–99.

Wu, T. Y. (1988). Forced generation of solitary waves. In *Applied Mathematics, Astrophysics, A Symposium to Honor C. C. Lin*, (D. J. Benney, F. H. Shu, and C. Yuan, eds.), Massachusetts Institute of Technology, June 22–24, 1987, pp. 198–212. World Scientific, Singapore.

Wu, T. Y. (1993a). A bidirectional long-wave model. *Meth. Appl. Anal.* **1**(1), 108–117.

Wu, T. Y. (1993b) Stability of nonlinear waves resonantly sustained. In *Proc. IUTAM Symp. on Nonlinear Instability of Nonparallel Flows* (S. P. Lin, W. R. C. Phillips, and D. T. Valentine, eds.), Clarkson University, July 26–31, 1993, pp. 367–381. Springer-Verlag, New York.

Wu, T. Y. (1995). Bidirectional soliton street — The Inaugural Pei-Yuan Chou Memorial Lecture. In *6th Asian Congress of Fluid Mechanics*, Singapore, May 21–26, 1995. *Acta Mech. Sinica* **11**, 289–306.

Wu, T. Y. (1997). On modeling nonlinear water waves. In *Proc. 12th Int'l Workshop on Water Waves and Floating Bodies* (Centennial Celebration of Georg Weinblum), Marseilles, France, March 16–20, 1997, pp. 321–324.

Wu, T. Y. (1998a). Nonlinear waves and solitons in water. *Physica D* **123**, 48–63.

Wu, T. Y. (1998b). On fully nonlinear water waves. In *Proc. 3rd Int. Conf. on Nonlinear Mechanics* (Chien Wei-zang, ed.), Aug. 17–20, 1998, pp. 119–124. Shanghai Univ. Press.

Wu, T. Y. (1998c). Modeling and computing nonlinear dispersive water waves. In *Proc. 3rd Int. Conf. on Hydrodynamics* (H. Kim, H. S. Lee, and S. J. Lee, eds.), Seoul, Korea, Oct. 12–15, 1998, pp. 3–9. UIAM Publ., Seoul.

Wu, T. Y. (1999a). Modeling nonlinear dispersive water waves. *J. Eng. Mech. Div. ASCE* **125**, 747–755.

Wu, T. Y. (1999b). On modeling unsteady fully nonlinear dispersive interfacial waves. In *Fluid Dynamics at Interface, 13th U.S. National Congress of Appl. Mech. Yih Memorial Symp.*, (Wei Shyy and R. Narayanan, eds.), Gainesville, FL, June 21–26, 1998, pp. 171–178. Cambridge Univ. Press, Cambridge.

Wu, T. Y., and Lin, D. M. (1994). Oceanic internal waves—their run-up on a sloping seabed. *Physica D* **77**, 97–107.

Wu, T. Y., and Zhang, J. E. (1996a). On modeling nonlinear long waves. In *Mathematics Is for Solving Problems: A Volume in Honor of Julian Cole on His 70th Birthday*, pp. 233–247. SIAM.

Wu, T. Y., and Zhang, J. E. (1996b). Mass and energy transfer between unidirectional interacting solitons (A tribute to Prof. C. C. Yu in honor of his 80th anniversary). *Chinese J. Mechanics* **12**(1), 79–84.

Yates, G. T. (1990). Some antisymmetric solutions with permanent form of the forced KdV equation. In *Engineering Science, Fluid Dynamics*, (G. T. Yates, ed.) pp. 119–134. World Scientific, Singapore.

Yates, G. T., and Wu, T. Y. (1991). Stability of solitary waves under skewed forcing. Volume Festschrift in honor of Marshall P. Tulin. In *Mathematical Approaches in Hydrodynamics*, (T. Miloh, ed.) pp. 193–206. SIAM.

Yeung, R. W. (1982). Numerical methods in free-surface flows. *Ann. Rev. Fluid Mech.* **14**, 395–442.

Yih, C. S., and Wu, T. Y. (1995). General solution for interaction of solitary waves including head-on collisions. *Acta Mech. Sinica* **11**(3), 193–199.

Zabusky, N. J. (1967). A synergetic approach to problems of nonlinear dispersive wave propagation and interaction. In *Nonlinear Partial Differential Equations*. Academic Press, New York.

Zabusky, N. J., and Kruskal, M. D. (1965). Interaction of "solitons" in a collisionless plasma and the recurence of initial states. *Phys. Rev. Lett.* **15**, 240–243.

Zakharov, V. E. (1968). Stability of periodic waves of finite amplitude on the surface of a deep fluid. *Z. Prik. Mek. Tekh. Fiziki* **9**, 86–94.

Zhang, D. H., and Chwang, A. T. (1996). Numerical study of nonlinear shallow water waves produced by a submerged moving disturbance in viscous flow. *Phys. Fluids*, **8**, 147–155.

Zhu, J. L. (1986). Internal solitons generated by moving disturbances. Ph.D. thesis, California Institute of Technology, Pasadena.

Zhu, J. L., Wu, T. Y., and Yates, G. T. (1986). Generation of internal runaway solitons by moving disturbances. In *Proc. 16th Symp. on Naval Hydrodyn.*, pp. 186–197. National Academy Press, Washington, DC.

Zhu, J. L., Wu, T. Y., and Yates, G. T. (1987). Internal solitary waves generated by moving disturbances. In *Proc. Third Intl. Symp. Stratified Flows* (E. J. List and G. H. Jirka, eds.), pp. 74–83. ASCE.

Coastal Hydrodynamics of Ocean Waves on Beach

JIN E. ZHANG

Economics and Finance, City University of Hong Kong, Kowloon, Hong Kong
and
Engineering Science, California Institute of Technology, Pasadena, California

THEODORE Y. WU

Engineering Science, California Institute of Technology, Pasadena, California

and

THOMAS Y. HOU

Applied Mathematics, California Institute of Technology, Pasadena, California

ADVANCES IN APPLIED MECHANICS, VOL. 37
ISBN 0-12-002037-8
ISSN 0065-2165/01 $35.00

I. Introduction

Coastal regions around the globe are becoming increasingly precious resources that need to be sustained. To maintain and enhance their environmental qualities, it is essential to understand well the basic mechanisms underlying the physical processes occuring in great varieties and to acquire expertise at predicting them under specific situations. Mastering this knowledge is a prerequisite for monitoring the health status of these regions and, more importantly, for developing theory and new technology for controlling and modifying natural processes to suppress ill effects and enhance desirable ones.

With respect to coastal hydrodynamics, theoretical modeling has to take into account the physical parameters that play key roles in the processes ranging from daily routines to severe resonantly excited extremes, such as hurricane and tsunami events. Over this wide scope, the primary roles include three-dimensional topography, short to very long timescales, non-linear and dispersive effects, influences of fluid viscosity and dissipative losses, factors of fluid density variation and earth rotation for geophysical motions, etc. Different phenomena are found manifest in specifically differing parametric regimes. For instance, the three-dimensional aspect cannot be avoided for field applications. The nonlinear and dispersive effects become very essential and must be fully accounted for around a moving (sometimes vastly moving) waterline or when waves become very strong. These three factors may be regarded as the three primary parameters for coastal hydrodynamics. The other parameters may have their pertinent regime for becoming important. Necessarily, therefore, a comprehensive theoretical model will have to cover the three primary ones. For example, the field observations of devastating damages caused by tsunamis indicate convincingly that it would be infeasible to predict them theoretically without strong three-dimensional wave aspects together with the nonlinear and dispersive effects playing their full exact roles. The challenges in developing a comprehensive theoretical model for ocean waves on continental shelves and beaches stem from the great diversity and range of the parametric domain that needs to be covered. It is in this context that coastal hydrodynamic modeling is constituted.

Historically, studies on run-up of ocean waves on beach have been ongoing for several decades. In theoretical research, analytical solutions to problems of nonlinear wave run-up on beaches exist only for the special case of two-dimensional waves on plane beaches (of uniform slope), while disregarding dispersion and other effects. The pioneering paper by Carrier and Greenspan (1958) presents an elegant exact solution to the nonlinear shallow-water equations that has kindled strong interest in this important problem. Subsequently, there are the welcome contributions by Tuck and Hwang (1972) introducing an ingenious transformation for mapping the nonlinear shallow-water equations into a linear form, and by Spielvogel (1975) making applications of Carrier–Greenspan's theory. Along this line, Synolakis (1987) gives an approximate formula for linear run-up of a solitary wave on plane beach and finds that his nonlinear run-up is equal to the linear result based on an approximate matching condition at the junction point between the plane beach and the ocean beyond, assuming the wave is small in amplitude. Recently, Li and Raichlen (1999) have found a nonlinear correction to Synolakis's formula. For the three-dimensional case, Carrier and Noiseux (1983) have analyzed the reflection of a tsunami obliquely incident on sloping plane beach based on the linear shallow-water wave theory. More recently, Brocchini and Peregrine (1996) presented a weakly nonlinear theory in weakly two-plus-one formulation for calculating run-ups of tsunamis at small angles of incidence. Kanoglu and Synolakis (1998) studied long-wave evolution and run-up on piecewise linear one- and two-dimensional bathymetries with a model of linear shallow-water equations, which we use in Section II.

In numerical studies, the difficult problem of capturing moving waterlines on sloping beaches has been tackled via various approaches, popularly by moving the wet-to-dry boundary from grid to grid in time steps, or by obtaining a local solution around the waterline by extrapolation (Hibberd and Peregrine, 1979), and so on. The results, however, are approximate in nature, requiring iterative computations and bearing cumulated errors. Another approach is to avoid the issue by adopting Lagrange's variables, as reported by Pedersen and Gjevik (1983), Zelt (1986, 1991), and Zelt and Raichlen (1990). However, the Lagrangian form gives rise to several new nonlinear terms in the basic equations that render numerical computations considerably more complicated. While the Lagrangian method is still amenable for one-space-plus-one-time $(1 + 1)$-dimensional problems, its complications become excessive for $(2 + 1)$-dimensional cases. Another alternative is the boundary integral equation method (BIEM), evaluated by finite-difference or finite-element schemes, as was used by Kim *et al.* (1983), Grilli (1997), and Grilli *et al.* (1997) to study two-dimensional wave run-up problems. In this respect, the state of the art has been brought up to date by the timely expository surveys and developments of Yeh *et al.* (1997). The numerical BIEMs are noted to have been applied almost exclusively to only two-dimensional problems. The achievement in grappling with nonlinear effects is noticeable. For instance, the horizontal velocities under a shoaling wave crest are found by Grilli *et al.* (1994) on fully nonlinear 2-D BIEM to become strongly nonuniform over depth (in some cases by more than 200%), an effect that first-order shallow-water wave theory cannot describe. However, the methodology for resolving strongly and fully nonlinear wave problems is still nearly confined to the two-dimensional case, and the primary difficulties involved with extending them to three dimensions are thought to be mainly computational in nature. Further, the direct numerical approach generally cannot lend us clear insights into the solutions to elucidate the basic mechanisms underlying the phenomenon investigated. They are so computational in nature and so well covered in literature that they are regarded as beyond the present scope. In surmounting some of these restrictions, Tuba Ozkan-Haller and Kirby (1997) present a Fourier–Chebyshev collocation method for the shallow-water wave equations in resolving a problem of periodic three-dimensional shoreline run-up for normal incidence of ocean waves on a periodically curved beach, yielding results of significance.

In experimental studies, Hall and Watts (1953) investigated the vertical rise of solitary waves on impermeable slopes. Their maximum run-up data have been updated by Synolakis (1987) with more experimental results. Yeh *et al.* (1989) studied bore propagation near the shoreline, the transition from bore to wave run-up, and the ensuing run-up motion on a uniform sloping beach. Zelt (1991) studied the run-up of nonbreaking and breaking solitary waves on plane impermeable beaches. His measurement of wave elevation can serve well for comparisons with theoretical and numerical studies. Liu *et al.* (1995) presented a series of large-scale laboratory experiments on the interactions of solitary waves climbing up a circular island. Li and Raichlen (1999) presented their laboratory measurement of wave evolution with advanced technologies well applied.

In summary, the literature seems to have focused on the nonlinear shallow-water theory for modeling two-dimensional ocean waves on shelves and beaches (see, e.g., Liu *et al.*, 1995), while leaving the dispersive effects and the geometry of three dimensionality almost unattended. Although some of the analytical and numerical results have been validated by comparisons with experiment, the scope may be still too small for generalization and the need for comprehensive models still exists. In pursuing the ultimate validity, we note that the nonlinear shallow-water wave model is known for its failure to support solitary waves within the transcritical speed regime, which is in contrast with the experimental findings pioneered by John Scott Russell. With the dispersive effects entirely neglected, some of the remarkable phenomena of soliton generation by resonant forcing on soliton-bearing systems and their bifurcation as found by experiment would not even manifest. For ocean waves on beaches in particular, the dispersive effects are noted here to possess both a long-time cumulative and local instantaneous scales. The former is known to grow with time algebraically (e.g., over long travels along long shelves), whereas the latter arises in regions having large gradients of the surface elevation, such as at the waterline receding in the wave run-down limit, as the dispersion effects are represented by higher derivatives of the surface elevation. It is therefore of interest to have the dispersion effects examined critically.

This article is intended to develop a comprehensive study on modeling three-dimensional ocean waves coming from an open ocean of uniform depth and obliquely incident on beach with arbitrary offshore slope distribution, while evolving under balanced effects of nonlinearity and dispersion. To simplify this challenging task, we consider here a family of beach configurations that is uniform in the longshore direction as a first approximation for beaches with negligible longshore curvature. In addition, the beach slope variation will be assumed to have such distributions that the ocean waves will evolve on beach without breaking. The overall approach adopted here begins with development of a three-dimensional linear shallow-water wave theory, followed by taking, step by step, the nonlinear and dispersive effects into account. In Section II, the linear theory is shown to provide a fundamental solution involving a central function, called the *beach-wave function*, that delineates the evolution of the incoming train of simple (sinusoidal) waves during interaction with any beach belonging to this broad family of beach configurations. This linear theory can easily afford to cover such factors as oblique wave incidence, arbitrary distribution of offshore beach slope, and wavelength variations with respect to beach breadth. In this respect, it would be very expensive — even infeasible — to obtain by direct numerical approaches. This achievement of the linear theory is followed in Section III by a comparative study using the nonlinear shallow-water wave equations for the case of normal wave incidence. For this task, a hybrid Lagrangian–Eulerian method is developed by incorporating the local Lagrangian determination of moving waterline with the Eulerian field computation of the interior flow domain to establish an efficient exact method. The numerical scheme is further simplified by mapping the time-dependent flow domain in physical space into a fixed region for computation. From comparison with the exact analytical solutions available, this method is shown to be uniformly accurate and highly efficient, and is used as such to correlate the nonlinear shallow-water wave model with its linear version, thereby giving the linear theory a certain established accuracy. In Section IV, the Fourier spectra assembled on linear theory provide a shortcut to finding the run-up of a solitary wave obliquely incident from ocean on a parabolic beach. The results exhibit salient features of the wave impact and reflection that vary with changes in the geometric and physical effects. The linear solution is used in turn for comparison with the corresponding results obtained on the nonlinear shallow-water wave model by using the hybrid Lagrangian–Eulerian method.

Following this, the dispersion effect is then first estimated for its correction and can be amended to the nonlinear solution of Carrier–Greenspan's. Finally, the effects of the two primary parameters, namely, the nonlinear and dispersive effects in two-dimensional beach-wave motion, are integrated into a generalized Boussinesq model to obtain results in Section V for an overall comparison and discussion. In addition, the linear results are further applied to develop a ray theory that can evaluate both the incident and reflected waves. The related mass and energy transport is also evaluated in terms of the Lagrangian material drift of the water particles and the longshore current so induced over the beach region by obliquely incident waves.

II. Linear Nondispersive Theory of Oblique Wave Run-Up on Uniform Beach

We begin by recounting linear theory of three-dimensional run-up of long ocean waves obliquely incident on a uniform beach of variable offshore slope, which is connected to an open ocean of uniform depth h_0 (Figure 1), with the bathymetry

$$h = \begin{cases} h(x) & (0 \leqslant x \leqslant l, \ -\infty < y < \infty) \\ h_0 & (x > l, \ h_0 = \text{const.}), \end{cases} \tag{2.1a}$$

$$h(0) = 0, \quad h'(0) \neq 0, \quad \text{and} \quad h(l - 0) = h_0, \tag{2.1b}$$

the depth function $z = -h(x)$ being assumed smooth, otherwise arbitrary. A train of simple waves is incoming from the ocean, obliquely incident on the beach, and reflected back to the ocean. The waves are assumed to be sufficiently small and long compared to h_0, and the beach slope $h'(x)$ not exceedingly gradual so that the process proceeds without wave breaking; the effects of dissipation are left for further consideration. The problem is to determine accordingly the wave motion in the flow field and the run-up on the beach.

For this class of problems, an appropriate theoretical model for the assumed inviscid and irrotational flow is the generalized Boussinesq equations (Wu, 1979, 1981):

$$\zeta_t + \nabla \cdot [(h + \zeta) \, \boldsymbol{u}] = 0, \tag{2.2}$$

$$\boldsymbol{u}_t + \boldsymbol{u} \cdot \nabla \boldsymbol{u} + \nabla \zeta = \frac{h}{2} \nabla [\nabla \cdot (h \boldsymbol{u}_t)] - \frac{h^2}{6} \nabla^2 \boldsymbol{u}_t, \tag{2.3}$$

where ζ is the wave elevation above the rest water surface at $z = 0$, the time t in subscript denotes differentiation, $\boldsymbol{u} = (u, v, 0)$ is the depth-averaged horizontal flow velocity, and $\nabla = (\partial_x, \partial_y, 0)$ is the horizontal gradient operator. Here, the length is scaled by h_0 and the time by $(h_0/g)^{1/2}$, g being the gravitational acceleration. This theoretical model is based on the assumption that

$$\tilde{a} = a/h_0 \ll 1, \qquad \varepsilon = (h_0/\lambda)^2 = O(\tilde{a}), \tag{2.4}$$

for waves of typical amplitude a and length λ on water of depth $h(x)$. On these scales, the terms in (2.2) and (2.3) are of $O(\tilde{a}\varepsilon^{1/2})$ for the linear terms,

Jin E. Zhang et al.

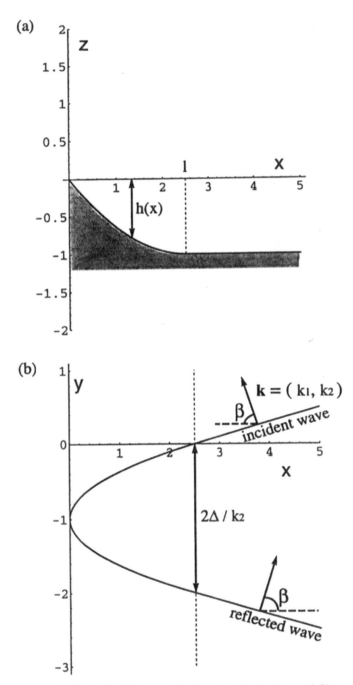

FIG. 1. A sketch of the ocean bathymetry: (a) side view and (b) top view.

of $O(\tilde{a}^2\varepsilon^{1/2})$ for the nonlinear terms, and of $O(\tilde{a}\varepsilon^{3/2})$ for the dispersive terms (the third derivatives). In the beach region, especially in the neighborhood of the moving waterline, all of these effects are comparable in magnitude and in significance.

Before we explore the nonlinear and dispersive effects step by step in later sections, it is useful first to discuss the linear theory, because it provides analytical convenience for producing useful results and furnishes such a basic tool as to lend us directly qualitative insights into the salient characteristics of the physical mechanism underlying the often intriguing phenomena. Thus we assume that the two parameters $\tilde{a}$ and ε are so small that (2.2) and (2.3) may be reduced to give the linear nondispersive long (LNDL) wave model as follows:

$$\zeta_t + \nabla \cdot (h\boldsymbol{u}) = 0, \tag{2.5}$$

$$\boldsymbol{u}_t + \nabla\zeta = 0. \tag{2.6}$$

This classical model is based on the assumptions that the waves are typically long relative to water depth, the pressure distribution is hydrostatic, and the vertical acceleration of the fluid particles is negligible (e.g., Lamb, 1932). Under this premise, eliminating $\boldsymbol{u}$ between (2.5) and (2.6) gives

$$(h(x)\,\zeta_x)_x + h(x)\,\zeta_{yy} - \zeta_{tt} = 0 \qquad (0 < x < l), \tag{2.7a}$$

$$\zeta_{xx} + \zeta_{yy} - \zeta_{tt} = 0 \qquad (x > l), \tag{2.7b}$$

with conditions

$$[\zeta] = 0 \quad \text{and} \quad [\zeta_x] = 0 \qquad (x = l), \tag{2.8a,b}$$

$$|\zeta(0, y, t)| < \infty. \tag{2.8c}$$

where $[f]$ denotes the jump of f across $x = l$; conditions (2.8a–c) are imposed for the smoothness and boundedness of solutions within the flow domain, including the waterline.

In the open ocean, we prescribe the incident and reflected waves as

$$\zeta = A(k)\,e^{i(k_2 y + k_1(x-l) - kt)} + B(k)\,e^{i(k_2 y - k_1(x-l) - kt)}, \tag{2.9a}$$

$$k_1 = -k\cos\beta, \qquad k_2 = k\sin\beta, \tag{2.9b}$$

where β is the wave incidence angle between the incoming wave vector $k = (k_1, k_2)$ and the $-x$ axis (see Figure 1), A is the given amplitude of the incident wave, and B is the unknown (complex) amplitude of the reflected wave, with the real component of complex quantities implied for physical interpretation.

Within the sloping beach region ($0 < x < l$), the wave field may assume the form

$$\zeta = \zeta(x', s) = C(k) F(x'; \kappa, \beta) e^{i\kappa s} \quad (s = y' \sin \beta - t', \quad \kappa = kl, \quad 0 < x' < 1),$$
(2.10)

where $x' = x/l$, $y' = y/l$, $t' = t/l$, with the prime omitted in the sequel for brevity, but with k converted into κ, and $C(k)$ is an undetermined (complex) constant. The function $F(x)$, to be called the *beach-wave function*, satisfies the following equation:

$$(h(x) F_x)_x + \kappa^2 (1 - h(x) \sin^2 \beta) F = 0 \quad (\kappa = kl) \qquad (2.11a)$$

$$F(0; \kappa, \beta) = 1. \qquad (2.11b)$$

This beach-wave function, normalized under (2.11b), represents the wave field evolving from incident simple waves on the beach of a broad category of configurations that is uniform in the longshore direction and otherwise arbitrary in seaward depth distribution. In addition, the depth function $h(x)$ is transformed by this new scaling into

$$h(x) = h(h_0 x'/\alpha) = h(x'/\alpha) \quad \text{with} \quad \alpha \equiv h_0/l, \qquad (2.11c)$$

and $h_0 = 1$ by the first scaling. Physically, the parameter α is the slope of a plane beach, or the "overall mean slope" of arbitrary beach of category (2.1). Thus, the beach-wave function depends in general on three parameters, $F = F(x; \kappa, \beta, \alpha)$; however, α can be omitted as understood. In view of its broad scope and utility, it is of value to devote a general study on the beach-wave function to realize its properties as a class of new transcendental functions.

For arbitrary $h(x)$ in general, with $h'(0) \neq 0$, (2.11a) can be written as

$$F'' + \frac{1}{x} p(x) F' + \frac{1}{x^2} q(x) F = 0, \qquad (2.12a)$$

$$p(x) = x\frac{h'(x)}{h(x)}, \qquad q(x) = \kappa^2 \left(\frac{1}{h(x)} - \sin^2\beta\right)x^2, \qquad \text{(2.12b)}$$

where the prime denotes differentiation with respect to x. We assume $p(x)$ and $q(x)$ to be analytic and regular in $0 \leqslant x \leqslant 1$, so that

$$p(x) = \sum_{n=0}^{\infty} p_n x^n, \qquad q(x) = \sum_{n=0}^{\infty} q_n x^n \qquad \text{(2.13)}$$

exist, and $x = 0$ is a regular singular point. By Frobenius's theory, (2.12) has a series solution

$$F(x) = \sum_{n=0}^{\infty} b_n x^{n+\nu}, \qquad \text{(2.14a)}$$

where ν satisfies the indicial equation

$$\nu^2 + (p_0 - 1)\nu + q_0 = 0, \qquad \text{(2.14b)}$$

which has two roots, ν_1 and ν_2, say, with $\nu_1 + \nu_2 = 1 - p_0$, $\nu_1\nu_2 = q_0$. For the run-up problem at hand, we are interested in those beach slopes that make $\nu_1 = 0$; otherwise, either the run-up of (2.10) and (2.14a) is singular at $x = 0$ if $\nu_1 < 0$, or the run-up is identically zero if $\nu_1 > 0$, neither case being of any physical interest. The root of $\nu_1 = 0$ requires that

$$q_0 = 0, \qquad \text{(2.15)}$$

which we shall assume to hold, as is true for beaches of constant slope. Under this condition, the other root of (2.14b) is $\nu_2 = 1 - p_0$. For beaches with $h'(0) \neq 0$ [by condition (2.1b)], $p_0 = 1$ by (2.12b), so the indicial equation (2.14b) has a double root

$$\nu_1 = \nu_2 = 0 \qquad (p_0 = 1). \qquad \text{(2.16)}$$

This implies that one solution of $F(x)$ has a regular series expansion (2.14a) and the other solution is logarithmically singular at $x = 0$, which must be discarded under condition (2.8c). This general property of $\nu_1 = \nu_2 = 0$ therefore holds for all the beaches whose slope does not vanish at waterline, as assumed. Whence the solution for the case of regular run-up is of the form

$$F(x; \kappa, \beta) = \sum_{n=0}^{\infty} b_n x^n, \tag{2.17}$$

where $b_0 = 1$ by normalization (2.11b), and b_n values are determined by the recurrence formula,

$$n^2 b_n = - \sum_{r=0}^{n-1} (r p_{n-r} + q_{n-r}) b_r \qquad (n = 1, 2, \ldots), \tag{2.18}$$

with $p_0 = 1$ and $q_0 = 0$ being understood.

Application of matching conditions (2.8a,b) determines $B(k)$ and $C(k)$ in (2.9) and (2.10) as

$$B(k) = A(k) \exp\{2i\Delta(\kappa, \beta)\}, \tag{2.19}$$

$$C(k) = A(k) R(\kappa, \beta) \exp\{i\Delta(\kappa, \beta)\}, \tag{2.20}$$

where

$$R(\kappa, \beta) = 2(M^2 + N^2)^{-1/2}, \qquad \Delta(\kappa, \beta) = \arg(M + iN), \tag{2.21}$$

$$M(\kappa, \beta) = F(1; \kappa, \beta) = \sum_{n=0}^{\infty} b_n, \qquad N(\kappa, \beta) = \frac{1}{\kappa_1} F_x(1; \kappa, \beta) = \frac{1}{\kappa_1} \sum_{n=0}^{\infty} n b_n,$$
$$\tag{2.22}$$

and where $\kappa_1 = -\kappa \cos \beta$, $\kappa = kl$. This result shows that the incident waves are totally reflected (since $|B| = |A|$, as should be expected in the absence of dissipation), bearing with them a terminal phase lag of $2\Delta(\kappa, \beta)$ as a mark of the wave–beach interaction. In the beach region, the evolving waves propagate alongshore with a phase lag $\Delta(\kappa, \beta)$ (which is the algebraic mean of the incident and reflected wave phases) and deliver a *relative run-up* $R(\kappa, \beta)$ [relative to $A(\kappa)$ of the incoming wave] on the beach. Here, the run-up is taken, after Keller (1961), as that given by $R(\kappa, \beta)$, which is actually the wave height at the rest waterline on linear theory, if this height is extrapolated horizontally to intercept the sloping beach beyond. It is on this empirical basis that the three-dimensional linear theory is extended to provide a nonlinear interpretation.

With respect to the mathematical properties of the series solution (2.17) and (2.18) for the general case of arbitrary depth distribution, it is known (e.g., Copson, 1948) that if both p_n and q_n series are convergent, within a radius of convergence $|x| = X_o \geqslant 1$, as is assumed to hold in general, then the series solution (2.17) and (2.18) converges absolutely and uniformly within $|x| = X_o$. We note in addition that the option of discarding the other solution of (2.17) and (2.18), because of its logarithmic singularity at the waterline, leaves exactly one initial condition to specify at waterline, which, with $b_0 = 1$ by (2.11b), gives solution (2.9) and (2.10) two (complex) constants, B and C, and they are completely determined by the two (complex) conditions in (2.8a,b) for arbitrary amplitude of the incoming wave. This makes the solution complete and physically compatible with the wave run-up problem in general. Consequently, we can regard the beach-wave function thus defined and determined as forming a special family of transcendental functions. In general, these functions can be numerically computed. In some particular cases of geometry, they become known functions already tabulated, as will be exemplified below.

A. OBLIQUE RUN-UP ON SLOPING PLANE BEACH

A special case of basic interest is the sloping plane beach,

$$h(x) = \alpha x \qquad (0 < x < l, \; \alpha l = h_0 = 1). \qquad (2.23a)$$

Because α can be rescaled to unity (by $x' - x/l$, with $h_0 = 1$), we can set $\alpha = 1$ and omit the prime for simplicity. Then (2.11a) becomes

$$xF_{xx} + F_x + \kappa^2(1 - x \sin^2 \beta)F = 0 \qquad (0 < x < 1, \; \kappa = kl). \quad (2.23b)$$

This is the model equation adopted by Eckart (1951) and Carrier and Noiseux (1983). For this case, (2.23b) has for the beach region $(0 < x < 1)$ the solution

$$\zeta(x,s) = C(k) \exp(-\kappa_2 x + i\kappa s) G(a; 1; 2\kappa_2 x) \qquad (a = (1 - \kappa \csc \beta)/2,$$

$$\kappa_2 = \kappa \sin \beta), \qquad (2.24)$$

where $G(a; b; z)$ is the confluent hypergeometric function (Abramowitz and Stegun, 1964):

$$G(a; b; z) = \sum_{n=0}^{\infty} \frac{a_n}{b_n} \frac{z^n}{n!} - 1 + \frac{a}{b} z + \frac{a_2}{b_2} \frac{z^2}{2!} + \cdots, \qquad (2.25a)$$

$$a_n = \prod_{j=1}^{n} (a + j - 1) = a(a + 1)(a + 2) \cdots (a + n - 1), \qquad (2.25b)$$

and likewise for b_n. [We recall that the other (logarithmically singular) solution of (2.23b) has been dismissed.] By matching conditions (2.8a, b), $B(k)$ in (2.19) and $C(k)$ in (2.20) are found (Zhang and Wu, 1999) as

$$B(k) = A(k)\, e^{2i\Delta}, \qquad C(k) = A(k)\, R(\kappa, \beta)\, e^{i\Delta}, \qquad (2.26a)$$

$$R(\kappa, \beta) = 2(M_0^2 + N_0^2)^{-1/2}, \qquad \Delta(\kappa, \beta) = \arg(M_0 + iN_0), \qquad (2.26b)$$

$$M_0(\kappa, \beta) = e^{-\kappa_2} G(a; 1; 2\kappa_2) \qquad (\kappa_2 = \kappa \sin \beta), \qquad (2.26c)$$

$$N_0(\kappa, \beta) = e^{-\kappa_2}[(1 + \kappa \csc \beta) G(a; 2; 2\kappa_2) - G(a; 1; 2\kappa_2)] \tan \beta. \qquad (2.26d)$$

To illustrate the coastal dynamics for this simple beach–ocean configuration, we take $\beta = 60°$ for the wave incidence angle, and $\boldsymbol{k} = (-1/l, \sqrt{3}/l)$, $(\kappa = 2)$ to obtain $a = -0.6547$, $M_0 = -0.4622$, $N_0 = 0.2889$, and hence the run-up and the phase lag as

$$R = 3.670 \quad \text{and} \quad \Delta = 2.583\, \text{rad}. \qquad (2.27)$$

As shown in Figure 2, the perspective wave profile exhibits a superposed pattern of the incident and reflected waves, which induce a quite pronounced run-up on the sloping beach, with an amplitude about 3.670 times that of the incident wave.

B. Normal Incident Wave Run-Up on Sloping Plane Beach

For normal incident waves sloshing on a uniform plane beach of slope $\alpha = 1$, the solution can be readily obtained by direct analysis of (2.23) with $\beta = 0$ or by deducing from the above solution (2.24) in the limit of $G(a, b, -z/a)$ as $a \to \infty$. In either way, we attain the following classical

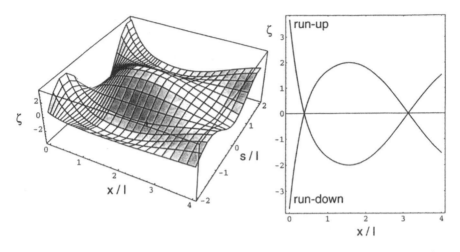

FIG. 2. Wave profile on the beach $(0 < x/l < 1)$ and ocean $(x/l > 1)$ for $\beta = 60°$, $\kappa = 2$.

result:

$$\zeta(x,t) = \begin{cases} A\left[\exp(-i\kappa(x-1+t)) + \exp(i\kappa(x-1-t) + i2\Delta_0)\right] & (x > 1) \\ AR_0(\kappa)\exp(-i\kappa t + i\Delta_0)J_0(2\kappa\sqrt{x}) & (0 < x < 1) \end{cases}$$

$$(2.28a,b)$$

$$R_0(\kappa) = 2[J_0^2(2\kappa) + J_1^2(2\kappa)]^{-1/2}, \tag{2.29a}$$

$$\Delta_0(\kappa) = \arg(J_0(2\kappa) + iJ_1(2\kappa)), \tag{2.29b}$$

where $J_n(x)$ is the Bessel function of the first kind.

As a numerical example, we take $\kappa = 2$, then

$$R_0(\kappa) = 4.968, \qquad \Delta_0(\kappa) = 3.306 \, \text{rad.} \tag{2.30}$$

The corresponding wave profile, as shown in Figure 3, displays a pattern with the specified phase lag and run-up that is somewhat greater than that in the previous case of $\beta = 60°$, by a margin of about 35%.

C. EDGE WAVES AT GRAZING INCIDENCE ON SLOPING PLANE BEACH

With the incidence angle β approaching $\pm\pi/2$, the incoming waves become nearly along shore. This limit turns out to be of particular

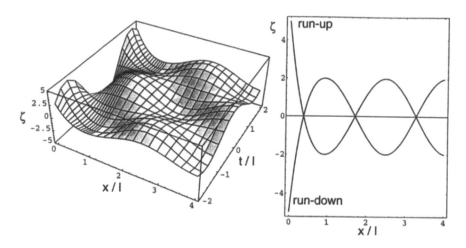

FIG. 3. Perspective wave profile on the beach and in the ocean for normal incidence at $\beta = 0°$ and wave number $\kappa = 2$.

significance in view of the eigenmodes of beach waves arising within the grazing incidence range (Zhang and Wu, 1999). In this limit, (2.26d) indicates that $\tan \beta \to \infty$, with N_0 either becoming unbounded for arbitrary κ (in which case $\zeta \to 0$ for all $x > 0$, a case of no physical interest) or with N_0 vanishing like $(\pi/2 - \beta)$ to yield nontrivial solutions for certain *eigenvalues* of κ, $\kappa = \kappa_n = \kappa_n l$, such that

$$N_0(\kappa_n, \beta \to \pi/2) = 0 \qquad (n = 1, 2, \ldots), \qquad (2.31\text{a})$$

of which the first several κ_n's are found (Zhang and Wu, 1999) as

$$\kappa_n = k_n l = 2.5337, 4.5788, 6.5986, 8.6101, 10.6177, 12.6232, \ldots \quad (2.31\text{b})$$

which suggests that $(\kappa_{n+1} - \kappa_n) = 2$ for $n \gg 1$, as is the case. Condition (2.31a,b) further implies, by (2.26b), that $\Delta(\kappa_n, \beta = \pm \pi/2) = 0$, signifying that the reflected wave and the edge wave on beach are both in phase with the incident wave. Thus in this case, the wave pattern in the open ocean is that of simple harmonic waves propagating along the shore and is connected to a *longshore edge wave* along the inclined beach with no phase lag, while the run-up of the κ_n-mode wave is $R_g(\kappa_n)$ given by

$$R_g(\kappa_n) = 2e^{\kappa_n}/|M_0((1 - \kappa_n)/2; 1; 2\kappa_n)|, \qquad (2.32)$$

which is finite. As an example, for $\kappa_1 = 2.5337$, there is a quite pronounced run-up of $R_g(\kappa_1) = 4.061$.

Within a range of incidence close to grazing, $85° < \beta < 90°$, we find the remarkable feature that R remains small except in a narrow band of κ centered about the eigenvalues $\kappa = \kappa_n$, $(n = 1, 2, \ldots)$ as illustrated in Figure 4 where R is shown versus κ for a set of values of β as listed. For β not too close to the grazing angle, say, $\beta < 60°$, R increases with increasing κ from the value of $R(0, \beta) = 2$ at $\kappa = 0$, with some slight undulations as shown in Figure 4. These eigenvalues also mark the narrow band in which the phase angle $\Delta(\kappa, \beta)$ jumps by π as κ increases across each κ_n, as shown in Figure 5.

The mechanism underlying the eigenmode waves arising within the grazing incidence range is discovered by observing that from (2.24) and (2.26d), the condition of $N_0(\kappa_n) = 0$ with $\beta = \pi/2$ implies $\zeta_x = 0$ at $x = 1$, which further infers, on account of (2.6), that the normal velocity $u = 0$ at $x = 1$ for all t. Physically, this signifies that the beach–ocean interface at $x = 1$ will remain geometrically a fixed vertical plane (with $\kappa = \kappa_n$, $\beta = \pi/2$), which therefore has zero rate of working and hence causes no energy transfer across this vertical plane although this plane is transparent to

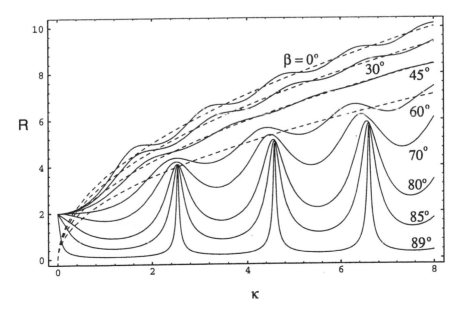

FIG. 4. Variation of the run-up function $R(\kappa, \beta)$ with $\kappa = kl = kh_0/\alpha$. Solid line, for a list of β indicated; dashed line, asymptotic formula [eq. (2.45e)].

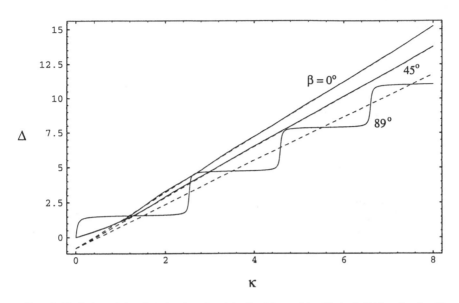

FIG. 5. Variation of the phase lag function $\Delta(\kappa, \beta)$ with $\kappa = kl = kh_0/\alpha$. Solid line, for $\beta = 0°$, 45°, and 89°; dashed line, asymptotic formula [eq. (2.45b)].

pressure continuity. In this light, the eigenmode waves can be regarded as being trapped within the beach region bounded by $x = 0$ and $x = 1$ planes and being sustained by the exterior pressure field of the waves at grazing incidence. This mechanism is thought to hold nearly true for incidence angles close to grazing.

The edge waves on a sloping plane beach may have analogy with some special wave patterns in a river with trapezoidal cross section, or the propagating waves along a trapezoidal channel studied by Teng and Wu (1992) when the eigenmodes of beach waves correspond to the modes of river waves. The longshore current of the edge wave is of fundamental importance in studying sand transport along a river. An interesting point here is that the eigenmode is not complete, therefore one can not build an arbitrary wave, e.g., a KdV type of solitary wave, with these eigenmodes. This means that a solitary wave cannot propagate for a long distance in a channel with trapezoidal shape based on the LNDL model. The nonlinear and dispersive effects play crucial roles in supporting a solitary wave in such a channel.

D. Normal Run-Up on Parabolic Beach

For a train of simple waves at normal incidence on a parabolic beach, we take $\beta = 0$ and

$$h(x) = mx + (1 - m)x^2 \qquad (0 < x < 1, 0 < m < 2), \qquad (2.33)$$

with $h(0) = 0$, $h'(x) = m + 2(1 - m)x$, and $h''(x) = 2(1 - m)$, so that $h'(x) > 0$ for $(0 < x < 1)$ and the beach is convex (or concave) up for $0 < m < 1$ (or $1 < m < 2$). In this case, eq. (2.11) for the beach-wave function has a unique solution:

$$F(x; \kappa, m) = F(a, b; c; z) = \sum_{n=0}^{\infty} \frac{a_n b_n}{c_n} \frac{z^n}{n!} = 1 + \frac{ab}{c} z + \frac{a_2 b_2}{c_2} \frac{z^2}{2!} + \cdots, \qquad (2.34a)$$

which is a hypergeometric function (Abramowitz and Stegun, 1964) with

$$a = \frac{1}{2} + \frac{1}{2}\left(1 + \frac{4\kappa^2}{m - 1}\right)^{1/2}, \quad b = 1 - a, \quad c = 1, \quad z = \frac{m - 1}{m} x, \qquad (2.34b)$$

and a_n, b_n, c_n are defined by (2.25b). This series converges well in our computations throughout $(0 \leqslant x < 1)$ for $(0.5 < m < 2)$, and the limited range of convergence for $(0 < m < 0.5)$ can be fully overcome by using known transformations for analytic continuation to give

$$F(a, b; c; z) = (1 - z)^{-a} F(a, c - b; c; z/(z - 1)) \qquad (z = (m - 1)x/m)$$

$$= \left(\frac{\mu}{x + \mu}\right)^a F\left(a, a; 1; \frac{x}{x + \mu}\right) \quad \left(\mu = \frac{m}{1 - m} > 0, \text{ for } 0 < m < 1\right),$$

$$(2.34c)$$

in which the first step is a known formula (Abramowitz and Stegun, 1964, 15.3.4) and the result is for the case at hand, with the new series converging for $(0 \leqslant x \leqslant 1)$, and for $(0 < m < 1)$.

Figure 6a shows the variation of the run-up function, $R(\kappa, \beta)$, versus $\kappa = kl$ for waves incident on the parabolic beach $h(x) = 2x - x^2$ and also for the plane beach $h(x) = x$ over a set of different incidence angles β. Figure 6b shows the variation of the run-up function, $R(\kappa, \beta = 0)$, at normal

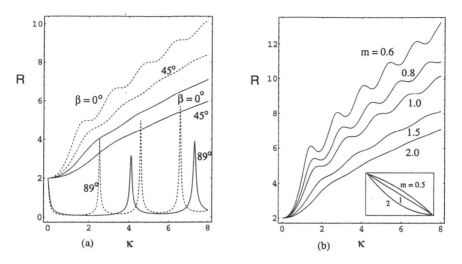

FIG. 6. (a) Variation of run-up function, $R(\kappa, \beta)$, with increasing $\kappa = kl$. Solid line, for parabolic beach $h(x) = 2x - x^2$; dashed line, for place beach $h(x) = x$; at incidence angle $\beta = 0°$, $45°$, $89°$. (b) Variation of run-up function, $R(\kappa, \beta = 0)$, with increasing $\kappa = kl$ for parabolic beaches, $h(x) = mx + (1 - m)x^2$, $m = 0.6$, 0.8, 1.0, 1.5, 2.0 (see the inset).

incidence on different parabolic beaches, $h(x) = mx + (1 - m)x^2$, $m = 0.6$, 0.8, 1.0, 1.5, 2.0. The beach becomes a plane for $m = 1.0$. The results of Figure 6a are obtained with eqs. (2.21) and (2.22) and the series solution of the beach-wave function (2.17), while the results of Figure 6b were obtained with eqs. (2.21) and (2.22) and the hypergeometric beach-wave function in eq. (2.34). These numerical results show that for the incidence angle up to about $\beta = 60°$, the run-up on parabolic beach is increased, or decreased, relative to the plane beach, as m is decreased ($m < 1$, more concave downward), or increased ($m > 1$, more concave upward) from $m = 1$. For greater values of β, especially near the grazing angle of $\beta = 90°$, the run-up function R remains small except near a set of "eigenvalues of κ," as exhibited for the case of $\beta = 89°$ shown here and as has been discussed in detail in Section II.C. These eigenvalues are shown to depend on the beach slope and curvature.

For incidences not close to the grazing, an important result obtained here indicates that the run-up on parabolic beaches is heavily weighted by the beach slope *near* the waterline, such that the run-ups in the two cases of plane and parabolic beaches are very nearly equal for equal beach slope at waterline. This important finding is thought to hold for gradually varying beach bathymetry in general.

E. Normal Run-Up on a Family of x^α Beaches

To make beach configuration representation more comprehensive near the waterline, we let the depth function have a branch-point behavior such that

$$h(x) = x^\alpha \qquad (0 < \alpha < 2). \qquad (2.35)$$

For normal incidence of a train of simple waves, eq. (2.11) has a unique solution:

$$F(x;\kappa,\alpha) = \left(\frac{\kappa}{2-\alpha}\right)^\nu \Gamma\left(\frac{1}{2-\alpha}\right) x^{(1-\alpha)/2} J_\nu\left(\frac{2\kappa}{2-\alpha} x^{(2-\alpha)/2}\right), \qquad (2.36)$$

where $\kappa = kl$, $\nu = (\alpha - 1)/(2 - \alpha)$, and $J_\nu(z)$ is the Bessel function of the first kind (Abramowitz and Stegun, 1964). We observe that for $\alpha > 2$ and $\alpha = 2$, the above $F(x)$ is singular and is therefore of no physical interest.

For $0 < \alpha < 2$, we find that

$$F(x) = 1 + O(x^{2-\alpha}). \qquad (2.37)$$

The corresponding run-up and phase lag functions can be obtained accordingly. For $\alpha = 1$ in particular,

$$F(x) = J_0(2\kappa\sqrt{x}), \qquad (2.38)$$

the solution reduces to known result on plane beach. Since

$$h'(0) = \begin{cases} 0 & (1 < \alpha < 2) \\ 1 & (\alpha = 1) \\ \infty & (\alpha < 1) \end{cases}, \qquad (2.39)$$

the beach behaves near the waterline $x = 0$ like a vertical wall for $\alpha < 1$ and like a horizontal plane for $1 < \alpha < 2$. This is quite different from the beach discussed in the last section, for which the beach always has a finite slope m/l at the waterline.

F. Asymptotic Expansion of the Beach-Wave Function and the Ray Theory

We now study the asymptotic behavior of the beach-wave function for the class of waves of length very small compared with the beach breath l, but still large relative to the ocean depth ($h_0 = 1$), i.e.,

$$\kappa = kl \gg 1 \quad \text{and} \quad kh_0 \ll 1. \tag{2.40}$$

For this case, the WKB (Wentzel, Kramers, and Brillouin) method is useful, because the entire family of arbitrary beach geometry can be handled by this asymptotic theory. In this asymptotic limit for κ large, the beach-wave function may assume the asymptotic expansion:

$$F(x) = C_1 \exp\left[\kappa \sum_{n=0}^{\infty} \frac{1}{\kappa^n} S_n(x)\right] = C_1 \exp\left[\kappa S_0(x) + S_1(x) + \frac{1}{\kappa} S_2(x) + \cdots\right], \tag{2.41a}$$

where C_1 is an undetermined (complex) constant. Substituting (2.41a) into (2.11a), and collecting the terms for the various orders of κ^{-1} yields the leading order equation:

$$S_0(x) = i \int_0^x \left(\frac{1}{h(x)} - \sin^2 \beta\right)^{1/2} dx \equiv i\theta(x), \tag{2.41b}$$

and the second-order equation

$$S_1(x) = -\frac{1}{4} \ln\left[h(x)(1 - h(x) \sin^2 \beta)\right] - i\frac{\pi}{4}. \tag{2.41c}$$

Consequently, the asymptotic expansion of the beach-wave function is given by

$$F(x; \kappa, \beta, h(x)) = C_1[h(x)(1 - h(x) \sin^2 \beta)]^{-1/4} \cos(\kappa\theta(x) - \pi/4) + O(\kappa^{-1}). \tag{2.42}$$

Note that the beach-wave function of (2.42) and (2.17) are two different things. The latter obeys (2.1b) and gives $F(0; \kappa, \beta, h(x)) = 1$, whereas the former is singular at a waterline like $F \sim h^{-1/4}$ and $F_x \sim h(x)^{-5/4}$ as $x \to 0$. But the former is a good approximation of the latter when x is about one wavelength or so off $x = 0$.

From (2.21) and (2.22), the run-up and phase lag function is only sensitive to the value and the first-order derivative of the beach-wave function at $x = 1$. Therefore it is acceptable to use (2.42) to derive them. Given the asymptotic formula (2.42) of the beach-wave function, we can apply it to compute $M(\kappa, \beta)$ and $N(\kappa, \beta)$ by using (2.22) and obtain for oblique incidence of simple waves on arbitrary beach the result

$$M(\kappa, \beta) = F(1; \kappa, \beta) = C_1(\cos \beta)^{-1/2} \cos[\kappa\theta(1) - \pi/4] + O(\kappa^{-1}), \qquad (2.43)$$

$$N(\kappa, \beta) = \frac{1}{\kappa_1} F_x(1; \kappa, \beta) = C_1(\cos \beta)^{-1/2} \sin[\kappa\theta(1) - \pi/4] + O(\kappa^{-1}). \qquad (2.44)$$

The corresponding run-up and phase lag functions, computed by (2.21), are given by

$$R(\kappa, \beta, h(x)) = 2(M^2 + N^2)^{-1/2} = 2C_1^{-1}(\cos \beta)^{1/2} + O(\kappa^{-1}), \qquad (2.45a)$$

$$\Delta(\kappa, \beta, h(x)) = \kappa\theta(1) - \frac{\pi}{4} = \kappa \int_0^1 \left(\frac{1}{h(x)} - \sin^2 \beta\right)^{1/2} dx - \frac{\pi}{4} + O(\kappa^{-1}),$$

$$(2.45b)$$

leaving a constant C_1 to be determined. Equation (2.45a) says that the run-up is not sensitive to the *overall* shape of $h(x)$. And we have just seen from the case of parabolic beaches that the run-up is sensitive to the local slope at waterline, which for a sloping plane beach is

$$R(\kappa, \beta, x) = 2(\pi k l \cos \beta)^{1/2} = 2(\pi k \cos \beta/h'(0))^{1/2}. \qquad (2.45c)$$

Notice that $l = 1/h'(0)$. We therefore have determined C_1 as follows:

$$C_1 = \left(\frac{\pi k}{h'(0)}\right)^{-1/2}, \qquad (2.45d)$$

and the run-up function on beach of arbitrary bathymetry as

$$R(\kappa, \beta, h(x)) = 2\left(\frac{\pi k \cos \beta}{h'(0)}\right)^{1/2} + O(\kappa^{-1}). \qquad (2.45e)$$

This general result states that the run-up appears to be heavily weighted by the beach slope near the waterline such that, roughly, as a rule, the arbitrary beach run-up can be estimated by that on a plane beach tangential to the arbitrary beach at the waterline. The statement was first numerically verified for parabolic beach by Zhang and Wu (1999).

Because eq. (2.42) is singular at the waterline, it cannot be used to characterize the local behavior of the beach-wave function near the waterline. But because the beach can be approximated with a sloping plane beach locally near the waterline, we can use the confluent hypergeometric beach-wave function (2.24) as the inner solution matching with the outer solution (2.42) to approximate the beach-wave function for arbitrary beach.

With asymptotic formulas (2.42) and (2.45) of oblique run-up and phase lag for sinusoidal incoming waves obliquely incident on an arbitrary beach with mild slope, the wave elevation becomes

$$\zeta(x, s) = \eta(x) \cos(ks + \Delta), \qquad s = y \sin \beta - t, \qquad (2.46a)$$

$$\eta(x) = 2 A_0 \cos[\kappa x \cos \beta - \kappa \cos \beta + \Delta(\kappa, \beta)]$$

$$= 2 A_0 \cos[\kappa(x - 1) \cos \beta + \kappa\theta(1) - \pi/4] + O(\kappa^{-1}) \qquad (x > 1); \qquad (2.46b)$$

$$\eta(x) = A_0 R(\kappa, \beta) F(x; \kappa, \beta, h(x)) = 2 A(x) \cos(\kappa\theta(x) - \pi/4) + O(\kappa^{-1})$$

$$(0 < x < 1), \qquad (2.46c)$$

$$A(x) = A_0(\cos^{1/2} \beta)[h(x)(1 - h(x) \sin^2 \beta)]^{-1/4}, \qquad (2.46d)$$

$$\theta(x) = \int_0^x \left(\frac{1}{h(x)} - \sin^2 \beta\right)^{1/2} dx \qquad (2.46e)$$

On the beach, $0 < x < 1$, $\zeta(x, y, t)$ can be written as

$$\zeta(x, y, t) = A(x) \cos(S_i(x, y, t)) + A(x) \cos(S_r(x, y, t)), \qquad (2.47a)$$

$$S_i(x, y, t) = -\kappa\theta(x) + \kappa y \sin \beta - \kappa t + \kappa\theta(1), \qquad (2.47b)$$

$$S_r(x, y, t) = \kappa\theta(x) + \kappa y \sin \beta - \kappa t + \kappa\theta(1) - \frac{\pi}{2}, \qquad (2.47c)$$

where S_i and S_r are the phase functions of the incident and reflected waves, respectively.

From the *classical ray theory* (e.g., Whitham, 1974), the incident wave assumes on the beach the form $\zeta_i(x, y, t) = A(x) \exp[iS(x, y, t)]$, with

$$k_1 = \frac{\partial S}{\partial x} = -\sqrt{k_b^2 - k_2^2}, \quad k_2 = \frac{\partial S}{\partial y} = \text{const.}, \quad \omega = -\frac{\partial S}{\partial t}, \quad (2.48a)$$

where the circular frequency $\omega = k_b\sqrt{h(x)} = \text{const.}\ (= k)$, $k_b = k\sqrt{1/h(x)} = \sqrt{k_1^2 + k_2^2}$ is the wave number on the beach, and $k_2 = k \sin \beta = \text{const.}$ by virtue of $\partial \omega/\partial y = 0$. After integrating the equations in (2.48a), we obtain the phase function $S = S(x, y, t)$ to be exactly the same as $S_i(x, y, t)$ in eq. (2.47b). After integrating the equation $dy/dx = k_2/k_1$, we obtain the ray trajectory of the incident waves as

$$y - y_0 = \int_1^x -\frac{\sin \beta}{\sqrt{1/h(x) - \sin^2 \beta}}\, dx. \quad (2.48b)$$

For sloping plane beach in particular, the integral can be carried out in closed form, which we use to exhibit the key features of ray behavior as a standard case. For $h(x) = x$, we have (Zhang and Wu, 1999)

$$\theta(x) = \csc \beta \arcsin(\sqrt{x} \sin \beta) + \sqrt{x(1 - x \sin^2 \beta)}, \quad (2.49a)$$

$$y - y_0 = \sin^{-2} \beta [\sin \beta \sqrt{x(1 - x \sin^2 \beta)} - \arcsin(\sqrt{x} \sin \beta)$$
$$- \sin \beta \cos \beta + \beta], \quad (2.49b)$$

where $y = y_0$ at $x = 1$. Following the ray, we can compute the wave amplitude by the formula:

$$\frac{A(x)}{A_0} = \left[\frac{(c_g)_0}{c_g}\frac{d\sigma_0}{d\sigma}\right]^{1/2} = \left[\frac{1}{\sqrt{x}}\frac{\cos \beta}{k_1/k_b(x)}\right]^{1/2}, \quad (2.50)$$

where $d\sigma$ is the width of a ray filament and $d\sigma_0$ its value at $x = 1$. For our linear nondispersive model, $c_g = c = \sqrt{x}$ on the beach. One can easily find that $A(x)$ given by (2.50) is the same as that of equation (2.46d). Thus the ray theory gives the incident wave on the beach as

$$\zeta_i(x, y, t) = A(x) \cos(S_i(x, y, t)). \quad (2.51)$$

This result of ray theory is based on the "geometric wave approximation" (2.40), and is supposed to hold for all angles of incidence. However, it only

Jin E. Zhang et al.

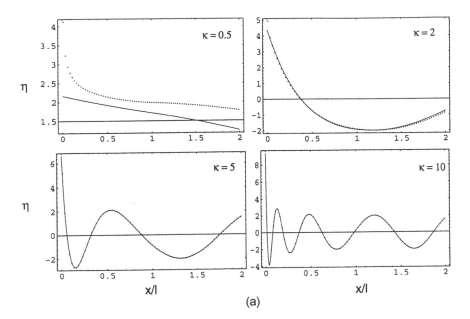

FIG. 7. Variations of wave elevation η over a range of x/l for $A = 1$; (a) $\beta = 45°$, $\kappa = 0.5$, 2, 5, 10.

covers the incident wave since the ray theory by itself is incapable of predicting reflected waves. The reason is because the ray theory does not possess the detailed structures in the solution required for satisfying the boundary condition at the seabed effecting the refraction *and* reflection of the waves. In contrast, the present linear theory in its comprehensive form (without further expansion) predicts the resultant motion, which is finite up to the waterline but is not explicit in separating the incoming and reflected waves. Only when the asymptotic ray expansion is acquired from the linear solution can we separate the incoming and reflected waves. But we should note that the expansion is not uniformly valid since the wave amplitude diverges like $h^{-1/4}(x)$ as $x \to 0$. It is in this asymptotic expansion of the linear solution that the ray theory is found to be in complete agreement with the *incident wave component* of the expansion, as indicated by (2.51) and (2.46)–(2.47). We further note that the ray theory cannot predict the run-up since it is singular at the waterline.

Figure 7 shows η as a function of x for $A_0 = 1$ on a sloping plane beach, in which (a) is for $\beta = 45°$ and $\kappa = 0.5$, 2, 5, 10; and (b) is for $\kappa = 10$ and

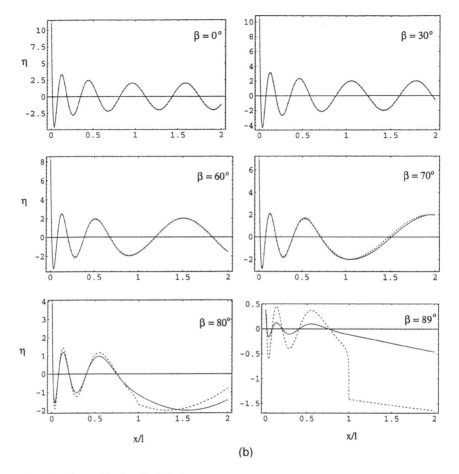

FIG. 7. (b) $\kappa = 10$, $\beta = 0°$, $30°$, $60°$, $70°$, $80°$, $89°$. Solid line, present theory; dashed line, asymptotic formula (2.46b,c).

$\beta = 0°$, $30°$, $60°$, $70°$, $80°$, $89°$. The results of this comparative study show that under condition (2.40), the asymptotic expansion is in excellent agreement with the linear theory for $\kappa > 2$ and $\beta < 60°$ except in a small neighborhood of waterline with $0 < x < 0.02$, and becomes poorer for smaller κ and larger β. The top left panel of Figure 7a demonstrates the margin of departure of the asymptotic expansion from the linear theory for κ as small as $\kappa = 0.5$. The discrepancy between the linear theory and its ray expansion becomes greater, the closer the incidence angle β approaches the grazing limit of 90°, this being apparently due to the singular behavior of the trapped modes of waves at resonance. These comparative results therefore imply that *the criteria of validity for the ray theory will be the same as that of the asymptotic expansion because of the complete agreement between the two on the incident component of the wave system.* These are the new conspicuous features that now qualify the ray theory.

Figure 8 shows the phase lines of incident and reflected waves on a sloping plane beach for $S_i = -2\pi, 0, 2\pi, 4\pi, 6\pi, 8\pi, S_r = 4\pi, 6\pi, 8\pi, 10\pi, 12\pi, 14\pi$, and the incident and reflected rays for $\kappa = 10, \beta = 60°$. It is of interest to note that at the waterline, the reflected wave actually has a phase lead given by (2.47) as

$$S_r(0, y, t) - S_i(0, y, t) = -\pi/2, \qquad (2.52)$$

as shown in the inset of Figure 8.

G. Bidirectional Model for Normal Run-Up

An alternative method for determining beach waves is to consider the bidirectional system of incoming and reflected waves. For normal incidence and run-up on beach, $\beta = 0$, we start with the LNDL equations

$$\zeta_t + (h(x)\,u)_x = 0, \qquad (2.53)$$

$$u_t + \zeta_x = 0. \qquad (2.54)$$

Taking the wave speed $c(x) = \sqrt{h(x)}$ as specified, assuming

$$\zeta(x, t) = \zeta_+(x, t) + \zeta_-(x, t), \qquad (2.55)$$

$$u(x, t) = \frac{1}{c(x)}(\zeta_+(x, t) - \zeta_-(x, t)), \qquad (2.56)$$

and substituting these relations into (2.53) and (2.54) yields two equations:

$$\partial_t\zeta_+ + \partial_t\zeta_- + [c(\zeta_+ - \zeta_-)]_x = 0, \qquad (2.57)$$

$$\frac{1}{c}\partial_t\zeta_+ - \frac{1}{c}\partial_t\zeta_- + \partial_x\zeta_+ + \partial_x\zeta_- = 0. \qquad (2.58)$$

From them we obtain, after some straightforward algebra, the bidirectional

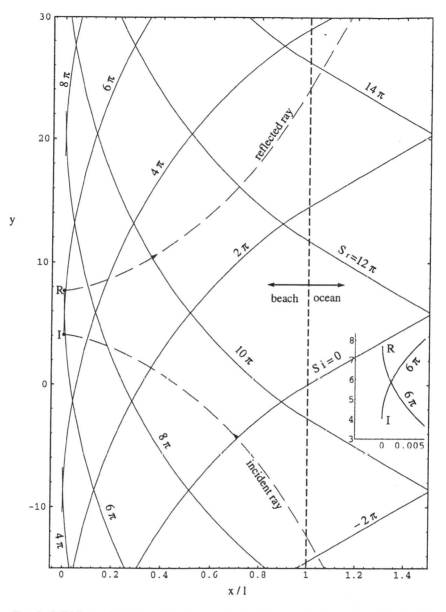

FIG. 8. Solid lines, phase lines of incident, and reflected waves for $S_i = -2\pi$, 0, 2π, 4π, 6π, 8π, $S_r = 4\pi$, 6π, 8π, 10π, 12π, 14π; dashed lines, incident and reflected ray tracks for $\kappa = 10$, $\beta = 60°$. The phase lag between the incident wave (point I) and reflected wave (point R) at $x = 0$ is $\pi/2$ [see eq. (2.52)], i.e., $y_R - y_I = \pi/(2k_2)$.

model equations as

$$\partial_t \zeta_+ + c \partial_x \zeta_+ + \frac{1}{2}(\zeta_+ - \zeta_-) c_x = 0, \tag{2.59}$$

$$\partial_t \zeta_- - c \partial_x \zeta_- + \frac{1}{2}(\zeta_+ - \zeta_-) c_x = 0. \tag{2.60}$$

The bidirectional model is exact with respect to the original equations, (2.53) and (2.54).

III. Lagrangian–Eulerian Hybrid Numerical Method for Run-Up Computation

A new numerical method is introduced for computation of two-dimensional run-up of ocean waves on inclined beach by combining the Lagrangian description of the moving waterline with the Eulerian description of the interior flow field based on the shallow-water equations, both in their exact form in this hybrid method. The numerical computation is further simplified by applying a geometric transformation of the time-dependent flow domain into a fixed region of computation. The results of this exact method are found to be in excellent agreement with known analytical solutions available, as shown in the case studies given subsequently.

A. THE NONLINEAR SHALLOW-WATER WAVE MODEL

The linear theory presented in Section II is based on the following assumptions:

$$(h + \zeta) \simeq h \quad \text{and} \quad u \ll c_0 = \sqrt{gh_0} = 1, \tag{3.1}$$

which require that the wave amplitude be everywhere small compared with water depth. But this theory evidently breaks down around the waterline since $(h + \zeta)$ itself vanishes there as a whole. Actually, it is beyond linear theory to describe how waves go across the rest waterline point [$x = 0$, and $h(0) = 0$]. Partly this is owing to the interpretation that the wave speed on linear theory is $\sqrt{h}$, which increasingly loses its significance as a wave approaches the rest waterline $h = 0$. Further, waves may be reflected before reaching, not physically beyond the fixed point $h = 0$ on linear theory. It is therefore clear that linear theory fails in delineating the wave run-up and run-down processes on beach. The remedy must come with an appropriate account of the nonlinear effects, with which the wave speed will be more like $\sqrt{h + \zeta}$ (relative to local fluid bulk) than $\sqrt{h}$. The speed $\sqrt{h + \zeta}$ vanishes only at the real waterline point, $h(x) + \zeta(x, t) = 0$, which marks the time-varying boundary of the flow domain enclosing the wave run-up and run-down stretch on beach, instead of just stopping at $x = 0$, where $h(0) = 0$.

We should therefore note the distinction between linear and nonlinear theories concerning run-up of waves. On physical ground, we define the run-up of waves, for inviscid nonlinear theory, by the wave elevation of the moving waterline itself [i.e., at $h(x) + \zeta(x, t) = 0$],

$$R = \zeta|_{h + \zeta = 0}. \tag{3.2}$$

So for run-up comparisons between linear and nonlinear theories, we actually compare the wave elevation at $h = 0$ for linear theory and that at $h + \zeta = 0$ for nonlinear theory.

To proceed, we first adopt the shallow-water equations (also known as Airy's model), which is the nonlinear nondispersive long wave model, for evaluating two-dimensional run-ups of water waves on sloping beaches,

$$\zeta_t + [(h + \zeta)u]_x = 0, \tag{3.3}$$

$$u_t + uu_x + \zeta_x = 0. \tag{3.4}$$

Equations (3.3) and (3.4) have the mathematical characteristics given by

$$\frac{dx}{dt} = u \pm \sqrt{h + \zeta}, \tag{3.5}$$

$$\frac{d\zeta}{dt} \pm \sqrt{h + \zeta}\frac{du}{dt} = -uh_x. \tag{3.6}$$

At a waterline, where $h + \zeta = 0$, the characteristic equations (3.5) and (3.6) reduce to

$$\frac{dx}{dt} = u, \tag{3.7}$$

$$\frac{d\zeta}{dt} = -uh_x. \tag{3.8}$$

B. THE LAGRANGIAN WATERLINE EQUATIONS

A moving two-dimensional waterline can be specified in the Lagrangian variable $x = X(t)$, which defines the time-varying position of the fluid particles at the waterline with $h(x) + \zeta(x, t) = 0$. To achieve a new method of simple and exact calculation of two-dimensional moving waterlines, we introduce their Lagrangian material description in totality as

$$h(X(t)) + \zeta(X(t), t) = 0, \tag{3.9}$$

$$\frac{dX}{dt} = u(X(t), t) \equiv U(t), \tag{3.10}$$

$$\frac{dU}{dt} = -\zeta_x. \tag{3.11}$$

Equation (3.9) signifies that the total water depth vanishes at the moving waterline. Equation (3.10) gives it the Lagrangian velocity $U(t)$, which coincides with the degenerated mathematical characteristic line given by (3.5). And eq. (3.11) is simply the Lagrangian material form of the momentum eq. (3.4). Thus, these three relations are geometrical in (3.9), kinematical in (3.10), and dynamical in (3.11); they are necessary and sufficient as the *exact* Lagrangian equations characterizing two-dimensional waterlines.

For given $X(t)$ and $h(x)$, (3.9)–(3.11) can provide $U(t)$, $\zeta(x, t)$, and ζ_x at the waterline for making analytic continuations of solution into the interior flow domain. For the interior flow, we retain the original field form of Euler's, namely, by (3.3) and (3.4), because they are simpler than the full Lagrangian counterpart which gives rise to more nonlinear terms. Indeed, the present Eulerian–Lagrangian hybrid scheme is designed to exploit the advantages of both descriptions, namely, in the simplicity of Euler's field form for the interior flow and in the precision of capturing Lagrange's material identity of the moving waterline. These joint advantages will be demonstrated below by numerical applications.

C. NUMERICAL SCHEME FOR FIXED-REGION COMPUTATION

To further simplify the algorithm of numerical computation, we introduce a new approach of converting the time-varying flow domain into a fixed region of computation by adopting the transformation

$$x = (1 + X/L)x' + X \quad \text{and} \quad t = t', \tag{3.12}$$

where $x = X(t)$ locates the moving waterline and $x = -L$ is the other open end of the computation region. By (3.12), the time-varying region $-L \leqslant x \leqslant X(t)$ is transformed into a fixed region $-L \leqslant x' \leqslant 0$ (L being an arbitrary constant). In the (x', t') coordinates, we can conveniently take all the grid points to be fixed with t'.

Under transformation (3.12), the shallow-water equations (3.3) and (3.4) become

$$\zeta_t - c_1 U \zeta_x + c_2[(h + \zeta)u]_x = 0, \tag{3.13}$$

$$u_t - c_1 U u_x + c_2(uu_x + \zeta_x) - 0, \tag{3.14}$$

$$c_1 = c_1(x, t) = \frac{1 + x/L}{1 + X/L}, \qquad c_2 = c_2(t) = \frac{1}{1 + X/L}, \tag{3.15}$$

where c_1 and c_2 result from the transformation, the primes being omitted for brevity.

We use the Richtmyer two-step Lax–Wendroff scheme, of which the first step reads

$$\zeta_{i+1/2}^{n+1/2} = \frac{1}{2}(\zeta_i^n + \zeta_{i+1}^n) + \frac{\Delta t}{2\Delta x} c_1 U(\zeta_{i+1}^n - \zeta_i^n)$$

$$- \frac{\Delta t}{2\Delta x} c_2[(h_{i+1} + \zeta_{i+1}^n)u_{i+1}^n - (h_i + \zeta_i^n)u_i^n], \tag{3.16}$$

$$u_{i+1/2}^{n+1/2} = \frac{1}{2}(u_i^n + u_{i+1}^n) + \frac{\Delta t}{2\Delta x} c_1 U(u_{i+1}^n - u_i^n)$$

$$- \frac{\Delta t}{2\Delta x} c_2 \left[\frac{1}{2}(u_{i+1}^n)^2 + \zeta_{i+1}^n - \frac{1}{2}(u_i^n)^2 - \zeta_i^n \right]; \tag{3.17}$$

in which $c_1 U$ and c_2 are evaluated at grid points $(i + 1/2, n)$. And the second step reads

$$\zeta_i^{n+1} = \zeta_i^n + \frac{\Delta t}{\Delta x} c_1 U(\zeta_{i+1/2}^{n+1/2} - \zeta_{i-1/2}^{n+1/2})$$

$$- \frac{\Delta t}{\Delta x} c_2[(h_{i+1/2} + \zeta_{i+1/2}^{n+1/2})u_{i+1/2}^{n+1/2} - (h_{i-1/2} + \zeta_{i-1/2}^{n+1/2})u_{i-1/2}^{n+1/2}], \tag{3.18}$$

$$u_i^{n+1} = u_i^n + \frac{\Delta t}{\Delta x} c_1 U(u_{i+1/2}^{n+1/2} - u_{i-1/2}^{n+1/2})$$

$$- \frac{\Delta t}{\Delta x} c_2 \left[\frac{1}{2}(u_{i+1/2}^{n+1/2})^2 + \zeta_{i+1/2}^{n+1/2} - \frac{1}{2}(u_{i-1/2}^{n+1/2})^2 - \zeta_{i-1/2}^{n+1/2} \right], \tag{3.19}$$

in which $c_1 U$ and c_2 are evaluated at the grid points (i, n).

At the waterline point $(I, n + 1)$, which is located at the Ith (the last) grid point along the x axis and at the $(n + 1)$th time step, we use the leap-frog scheme for integrating (3.11) to give for the velocity U the formula:

$$U_I^{n+1} = U_I^{n-1} - \frac{\Delta t}{\Delta x}(3\zeta_I^n - 4\zeta_{I-1}^n + \zeta_{I-2}^n). \tag{3.20}$$

For integrating (3.10) for the waterline location X, we use

$$X_I^{n+1} = X_I^{n-1} + 2\Delta t\, U_I^n. \tag{3.21}$$

These schemes are second order in space and time.

D. CASE STUDY 1: PERIODIC WAVE RUN-UP ON SLOPING PLANE BEACH

Here we consider the two-dimensional run-up of periodic waves on a beach of uniform slope α, $h = -\alpha x$. The beach slope can always be scaled to $\alpha = 1$ since by $x' = \alpha x$, $t' = \alpha t$, (3.3) and (3.4) remain invariant but the beach then becomes $h = -x'$. For convenience, we take $\alpha = 1$ and omit the primes.

For this case ($h = -x$), Carrier and Greenspan's classical solution (1958) to (3.3) and (3.4) for periodic wave of elevation $\zeta(x, t)$ and horizontal velocity $u(x, t)$ can be expressed in parametric form as

$$\phi = AJ_0(\sigma)\cos\lambda, \tag{3.22}$$

$$u = \frac{1}{\sigma}\phi_\sigma = -\frac{A}{\sigma}J_1(\sigma)\cos\lambda, \tag{3.23}$$

$$\zeta = \frac{1}{4}\phi_\lambda - \frac{1}{2}u^2 = -\frac{A}{4}J_0(\sigma)\sin\lambda - \frac{A^2}{2\sigma^2}J_1^2(\sigma)\cos^2\lambda, \tag{3.24}$$

$$x = \zeta - \sigma^2/16 \quad \text{and} \quad t = \lambda/2 - u, \tag{3.25}$$

where ϕ satisfies, with $\sigma = 4(\zeta - x)^{1/2}$ and $\lambda = 2(t + u)$, the linear equation

$$(\sigma\phi_\sigma)_\sigma - \sigma\phi_{\lambda\lambda} = 0. \tag{3.26}$$

Thus we have the solution (3.22) for ϕ, (3.23) and (3.24) for u and ζ, and (3.25) for x and t. This beach-wave solution is time periodic for $0 < A < 1$, breaks (with infinite slope) at the waterline at the run-down extreme for the critical case of $A = 1$, and becomes multivalued for $A > 1$.

Figures 9 and 10 show Carrier–Greenspan's exact solution for the case of periodic waves and the corresponding numerical results of the present

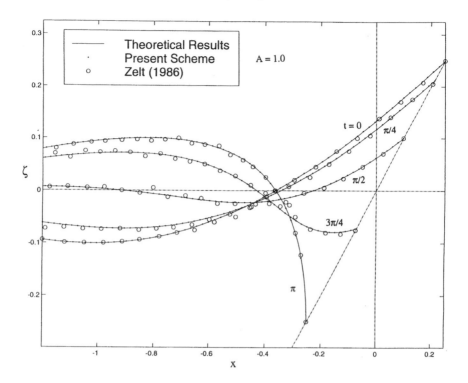

FIG. 9. Run-up and run-down of a periodic wave with amplitude $A = 1.0$ and period $T = 2\pi$, with the wave profile shown versus x at five time steps: solid line, exact solution; solid dots, present numerical scheme; open circles, numerical results of Zelt (1986). Notice the wave profile reaching vertical slope at the run-down limit.

Lagrangian–Eulerian hybrid numerical method, both based on (3.3) and (3.4). Also shown therein for comparison are the corresponding results obtained by Zelt (1986) and Hibberd and Peregrine (1979). For the numerical work, two initial conditions are required. From Carrier–Greenspan's solution (3.22)–(3.24) at $t_0 = \lambda/2 = 3\pi/4$, we obtain $u(x, t_0) = 0$ and $\zeta(x, t_0)$ implicitly by (3.24) and (3.25) as the two initial conditions. As for the boundary conditions on the ocean side, say, at $x = x_0$ of the computation region, we have two options, one being to use $\zeta(x_0, t)$ and $u(x_0, t)$ numerically computed, for each time step, from (3.22)–(3.25) for the boundary conditions. The other way is to compute ζ and u over an extended domain, with its open boundary bounded by a vertical solid wall, placed so far away that no disturbances reflected back from this wall will reach the

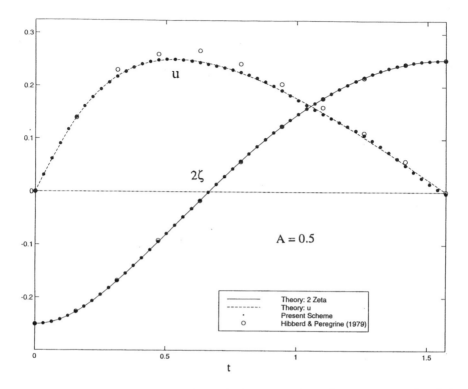

FIG. 10. Run-up and run-down of a periodic wave with amplitude $A = 0.5$ and period $T = 2\pi$, with the wave elevation ζ and velocity u at waterline shown as a function of t: solid line, exact solution for ζ; dashed line, exact solution for u; solid circles, present numerical scheme; open circles, numerical results of Hibberd and Peregrine (1979). For continuation, regard ζ as even and u odd in $(t - \pi/2)$.

inner original domain with magnitudes greater than a specified error limit during the entire numerical experiment. We chose the latter approach for its simplicity.

By comparison, the present numerical results are found in excellent agreement with the exact analytical solution. Taking $\Delta x = 1/100$ and $\Delta t/\Delta x = \pi/10$ for grid size, we find the relative error uniformly less than 1.3×10^{-3} over the entire flow domain of computation, including the waterline during one-half period from the highest run-up to the lowest run-down. For $\Delta x = 1/200$, and with $\Delta t/\Delta x$ unchanged, the relative error becomes 4.0×10^{-4}, about $1/4$ of the previous one, thus qualifying the scheme as being of second order. This excellent accuracy surpasses the other numerical methods known for solving this problem. Further, it only consumes a CPU time of 3.5 seconds on a Sun Ultra-Enterprise-10000 computer for each run with $\Delta x = 1/200$ for the results shown in Figure 9.

After examining the error distributions of the numerical results, we find that the error near the moving waterline is less than 10^{-3}, whereas the error around the ocean-side open boundary of the domain of computation is much smaller, being only less than 10^{-6}, a result suggesting that most of the error comes from the numerical evaluation of large values of ζ_x near the waterline. In general, however, such a uniformly high accuracy is thought to suffice, especially because the error can be further reduced, for instance, by decreasing the grid size.

E. CASE STUDIES 2 AND 3: WATER RUN-DOWN ON SLOPING PLANE BEACH

The same problem of wave run-up on sloping plane beach as described above has been solved by Tuck and Hwang (1972) using the transformation

$$\zeta^* = \zeta + u^2/2, \quad u^* = u, \quad x^* = x - \zeta, \quad t^* = t + u, \qquad (3.27)$$

by which the nonlinear equations (3.3) and (3.4) are converted into a set of linear equations:

$$\zeta_{t^*}^* + [hu^*]_{x^*} = 0, \qquad h = -x^*, \qquad (3.28)$$

$$u_{t^*}^* + \zeta_{x^*}^* = 0, \qquad (3.29)$$

for beach slope $\alpha = 1$. Tuck and Hwang (1972) applied their theory to obtain an exact solution of (3.3) and (3.4) for a particular initial mass release problem, with zero initial velocity everywhere and with the initial wave elevation $\zeta(x, 0)$ implicitly given by

$$\text{Tuck and Hwang:} \qquad \alpha x = \alpha b \log(\zeta/a) + \zeta, \qquad (3.30)$$

where α is the original beach slope. For $\alpha = 1$, their exact solution to (3.3) and (3.4) reads:

$$u = -\frac{2ab}{\sqrt{-x^*}} \int_0^\infty \kappa e^{-b\kappa^2} J_1(2\kappa\sqrt{-x^*}) \sin(\kappa t^*) \, d\kappa, \qquad (3.31)$$

$$\zeta = 2ab \int_0^\infty \kappa e^{-b\kappa^2} J_0(2\kappa\sqrt{-x^*}) \cos(\kappa t^*) \, d\kappa - \frac{1}{2} u^2, \qquad (3.32)$$

$$x = x^* + \zeta \qquad \text{and} \qquad t = t^* - u. \qquad (3.33)$$

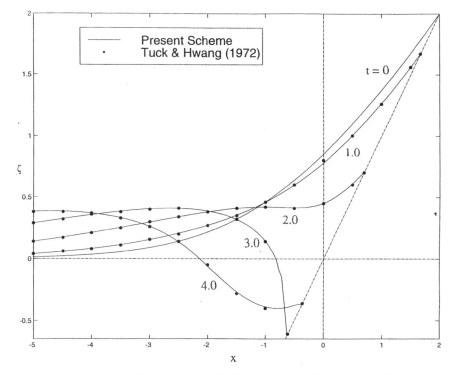

Fig. 11. Comparison with the results of Tuck and Hwang (1972), for $a = 2$, $b = 1$, $\alpha = 1$: variation of wave elevation versus x for five time steps indicated. Solid line, present numerical result; solid circles, results of Tuck and Hwang.

Figure 11 shows that our numerical results are also in excellent agreement with Tuck and Hwang's theoretical ones shown in their Figure 1.

For our third case study, we consider Spielvogel's (1975) solution to the same problem of initial mass release as stated above, now obtained by superposition of Carrier–Greenspan's periodic solutions, with zero initial velocity and with $\zeta(x, 0)$ given by

$$\text{Spielvogel:} \qquad \alpha x = \frac{H}{\log(R/H)} \log(\zeta/R) + \zeta, \qquad (3.34)$$

where the parameters involved in (3.30) and (3.34) are related by $a = R$ and $\alpha b = H/\log(R/H)$. Compared with Spielvogel (1975), our numerical results are also found in good agreement. For instance, our results in Figure 12 (for $\alpha = 1/20$, $R = 0.1$, and $H = 0.05$) differ from the corresponding ones in Spielvogel's Figure 4 by a margin of less than 2%.

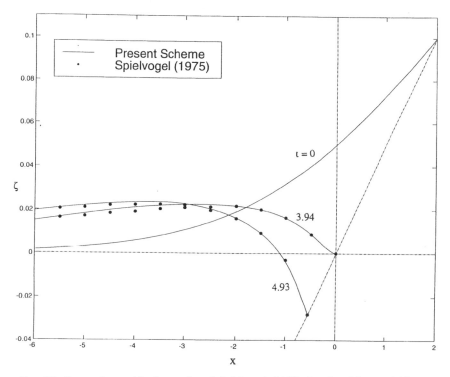

Fɪɢ. 12. Comparison with the results of Spielvogel (1975), for $R = 0.1$, $H = 0.05$: wave elevation as a function of x for three time steps indicated. Solid line, present numerical result; solid circles, results of Spielvogel.

IV. Nonlinear Effects on Run-Up of Solitary Waves on Parabolic Beaches

We now apply the Lagrangian–Eulerian hybrid method developed in the last section to compute two-dimensional run-ups of solitary waves on beaches, here based on the nonlinear nondispersive shallow-water equations (3.3) and (3.4). In our first numerical experiment, we take an incoming solitary wave, of initial amplitude $a = 0.1$ in the open ocean, impinging on a parabolic beach with the bathymetric depth function defined by (2.1a) and the beach width taken to be $l = 4$. By varying parameter m in the beach-wave function (2.33), we can change the curvature of the parabolic beach, e.g., the beach is concave upward for $1 < m < 2$, convex for $0 < m < 1$, and becoming a plane beach for $m = 1$. We impose the quiescent initial condition at $t = 0$ and for the boundary condition at the junction point between the ocean and the beach at $x = l$ by the following relation:

$$\zeta = a \operatorname{sech}^2 \sqrt{3a/4}(l + ct - 32) \qquad (c = 1 + a/2,\ l = 4), \qquad (4.1)$$

for the wave elevation, which is an exact solution of the KdV equation, and

$$u = -\frac{c\zeta}{1 + \zeta} \qquad (4.2)$$

for the velocity at $x = l$, which is the first integral of the mass conservation law for a solitary wave with speed c. The wave is propagating from the right to the left, and the initial domain of interest is $x \in [0, l]$. The left boundary of the domain will become negative when the wave starts to run up.

For the corresponding results on linear theory, we adopt the linear theory developed in Section II based on (2.7a,b) and utilize Fourier synthesis to construct the required solution. For simplicity, we confine our analysis to the case of normal incidence of a solitary wave on parabolic beaches.

Suppose we have in the open ocean an incoming solitary wave of the KdV form

$$\zeta_0(x, s) = a \operatorname{sech}^2 \sqrt{3a/4}[-t - (x - l)] \qquad (x > l), \qquad (4.3)$$

which also satisfies (2.7b) in the open ocean where it can assume the Fourier integral form (Zhang and Wu, 1999) as

$$\zeta_0(x, s) = \operatorname{Re} \int_0^\infty A_0(k) \exp\{ik[-t - (x - l)]\}\, dk \qquad (x > l), \quad (4.4a)$$

$$A_0(k) = \frac{4}{3} k \operatorname{csch}\left(\frac{\pi k}{\sqrt{3a}}\right), \qquad (4.4b)$$

Therefore, by Fourier synthesis of the simple harmonic wave solutions (2.9) and (2.10), we obtain for the resultant wave field the solution

$$\zeta(x, t) = \operatorname{Re} \int_0^\infty A_0(k) R(\kappa) e^{-ikt + i\Delta} F(x/l; \kappa, m)\, dk \qquad (0 < x < l), \qquad (4.5a)$$

$$\zeta(x, t) = \operatorname{Re} \int_0^\infty A_0(k) \left[e^{-ik(x - l + t)} - e^{i[k(x - l - t) + 2\Delta]}\right] dk \qquad (x > l), \qquad (4.5b)$$

where $A_0(k)$ is given by (4.4b), $F(x/l; \kappa, m)$ by (2.34a), and $R(\kappa)$ and $\Delta(\kappa)$ by (2.21).

For this case of normal incidence on plane beaches, we present the following numerical results with $\beta = 0$, $a_0 = 0.1$, and $\alpha^{-1} = 1, 2, 3, 4, 5, 6$

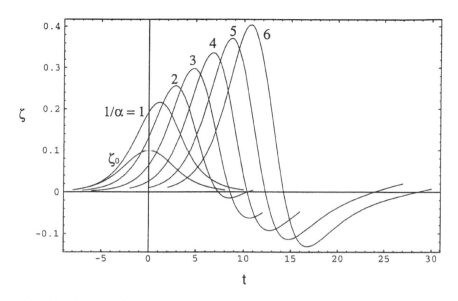

FIG. 13. Time records of run-up, at $x = 0$, of a solitary wave with $a_0 = 0.1$ at normal incidence ($\beta = 0°$) on a sloping plane beach for $1/\alpha = 1, 2, 3, 4, 5, 6$.

for increasingly smaller slopes of the beach. The wave profiles are numerically integrated from (4.5a,b) and shown in Figure 13 for $\zeta(0, t)$ at $x = 0$ and in Figure 14 for the reflected wave $\zeta_r(l, t)$ at $x = l$, both versus time t. It is of interest to note that at the waterline, the run-up increases monotonically with decreasing $\alpha(< 1)$ and there appears, after reaching the maximum run-up, a negative run-down on the beach whose magnitude also increases with decreasing α. The increasing run-down is evidently playing an active role in resulting in an increasing departure from the fore-and-aft symmetry for the reflected wave, as depicted in Figure 14, while giving rise to a dipped tail with its magnitude increasing with decreasing α.

Table 1 shows the maximum run-up of a solitary wave on a family of parabolic beaches with the depth function given by (2.33) and parabolic shape extended to the negative region of the x axis (see inset in Figure 15), and the results are shown in Figure 15.

Table 2 shows the maximum run-up of a solitary wave on a family of parabolic beaches with the depth function given by (2.33) and a tangent line extended to the negative region of the x axis (see inset in Figure 16), and the results are shown in Figure 16.

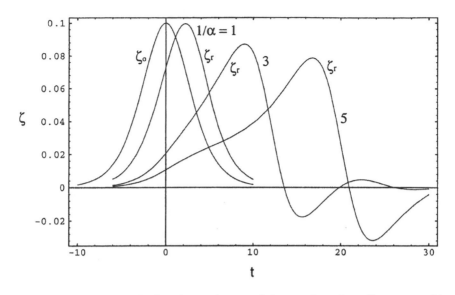

FIG. 14. Time records of reflected wave ζ_r at $x = l$, from an incoming solitary wave with $a_0 = 0.1$ at normal incidence ($\beta = 0°$) on a sloping plane beach for $1/\alpha = 1, 3, 5$.

TABLE 1
RUN-UP OF SOLITARY WAVE OF AMPLITUDE $a/h = 0.1$ ON PARABOLIC
BEACH (2.33) OF A ONE-PARAMETER FAMILY IN m AND SEAWARD
LENGTH $x = l = 4$, AS PREDICTED BY THE LINEAR THEORY AND THE
NONLINEAR MODEL (BASED ON AIRY'S SHALLOW-WATER EQUATIONS),
WITH ZERO INITIAL CONDITIONS AND WITH BOUNDARY CONDITIONS
(4.1, 4.2) IMPOSED AT THE OCEAN BORDER AT $x = l$

Parameter m	l	Maximum run-up R_l, by linear theory	R_n, by nonlinear model
0.7	4.0	0.3828	0.3987
0.8	4.0	0.3649	0.3713
0.9	4.0	0.3499	0.3540
1.0	4.0	0.3372	0.3408
1.1	4.0	0.3261	0.3294
1.2	4.0	0.3164	0.3203
1.3	4.0	0.3078	0.3115
1.4	4.0	0.3000	0.3046
1.5	4.0	0.2931	0.2976
1.6	4.0	0.2867	0.2923

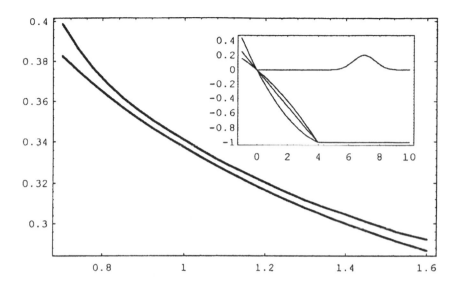

FIG. 15. Maximum run-up of a solitary wave of amplitude $a = 0.1$ as a function of m, on a family of parabolic beach $h(x) = mx/l + (1 - m)(x/l)^2$ (see the inset), with $l = 4$, over $0.7 < m < 1.6$. Upper line, nonlinear shallow-water wave model; lower line, linear theory.

TABLE 2

RUN-UP OF SOLITARY WAVE OF AMPLITUDE $a/h = 0.1$ ON PARABOLIC BEACH (2.33) OF A FIXED SLOPE $s_1 = m/l = 0.25$ AT INITIAL WATERLINE LOCATION ($x = 0$), AS PREDICTED BY THE LINEAR THEORY AND THE NONLINEAR MODEL (BASED ON AIRY'S SHALLOW-WATER EQUATIONS), WITH ZERO INITIAL CONDITIONS AND WITH BOUNDARY CONDITIONS (4.1, 4.2) IMPOSED AT THE OCEAN BORDER AT $x = l$

Parameter m	l	Maximum run-up R_l, by linear theory	R_n, by nonlinear model
0.7	2.8	0.3231	0.3361
0.8	3.2	0.3293	0.3311
0.9	3.6	0.3338	0.3371
1.0	4.0	0.3372	0.3408
1.1	4.4	0.3396	0.3438
1.2	4.8	0.3415	0.3466
1.3	5.2	0.3429	0.3491
1.4	5.6	0.3439	0.3513
1.5	6.0	0.3446	0.3533
1.6	6.4	0.3451	0.3549

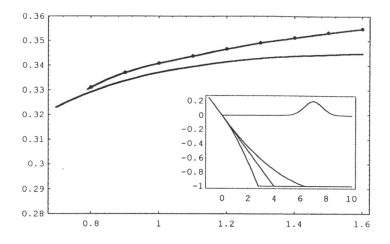

Fig. 16. Maximum run-up of a solitary wave of amplitude $a = 0.1$ as a function of m, on a family of parabolic beach $h(x) = mx/l + (1 - m)(x/l)^2$ (see the inset), with $m/l = 0.25$, over $0.7 < m < 1.6$. Upper line, nonlinear shallow-water wave model; lower line, linear theory.

From the numerical results in Tables 1 and 2, we observe that the maximum run-up of a solitary wave is very sensitive to the local slope at the initial waterline, and less sensitive to the shape of the beach bathymetry beyond the waterline. This agrees with the results from our linear theory in Section II.

It is important to note that the shallow-water equations do not support solutions with waves of permanent form traveling in water of uniform depth. Actually, according to that model, a wave always becomes increasingly more steepened forward with increasing time due to lack of the dispersive effects to counterbalance the nonlinear wave-steepening effects. It follows that if we impose the boundary condition

$$\zeta = a \operatorname{sech}^2 \sqrt{\frac{3a}{4}}(x_0 + ct - 32), \qquad u = -\frac{c\zeta}{1 + \zeta}, \quad (c = 1 + a/2), \quad (4.6)$$

at different places for x_0, we should expect to obtain somewhat different run-up results because the wave-steepening effects would bring the same initial waves placed at different initial locations to waves evolved to different heights at the waterline.

TABLE 3
RUN-UP OF A SOLITARY WAVE OF AMPLITUDE $a/h = 0.1$, PREDICTED BY THE SHALLOW-WATER EQUATIONS AND LINEAR THEORY IN II, WITH THE BOUNDARY CONDITION (4.6) IMPOSED AT x_0, RUNNING UP FROM AN OCEAN OF DEPTH $h = 1$ TO A PLANE BEACH OF SLOPE $\alpha = 1/4$

	x_0	Maximum run-up R	Time to reach the maximum run-up	Time from beach foot to the maximum run-up
	−16	0.3816	33.44	
	−14	0.3743	33.56	
	−12	0.3666	33.68	
Shallow-water	−10	0.3594	33.80	
equations	−8	0.3526	33.96	
	−6	0.3464	34.12	
	−4	0.3408	34.24	7.57
Linear theory	−4	0.3366	33.43	6.76

In our second numerical experiment, we take the same $a = 0.1$ for the initial wave amplitude and $\alpha = 1/4$ for the beach slope as before, but impose the boundary condition (4.6) at different places with $x_0 = -16, -12, -10, -8, -6, -4$, for each of which we compute the maximum run-up. The results are shown in Table 3. Figure 17 shows the maximum run-up as a function of x_0. The maximum run-up increases with increasing x_0, as should be expected. Actually, if we stand at the point $x = -4$, we see different waves passing through this point in the different cases due to the nonlinear steepening effect inherent to the shallow-water equations. The larger the x_0 taken for imposing the same boundary condition, the more steepened the wave becomes at the beach foot and therefore the greater the maximum run-up.

We take $\Delta x = 1/20$, $\Delta t = 1/100$ in both numerical experiments. It takes 1 to 2 min of CPU time for a Sun station to compute one run-up case.

We can now draw the following conclusion: The maxmum run-up of a solitary wave predicted by the shallow-water equations depends on the position for prescribing the initial solitary wave. The farther the same initial solitary wave of the KdV form is imposed from the beach, the larger the maximum run-up will reach. An intent of this particular study is to find the pertinent causes for producing differences between linear and nonlinear values of run-up, i.e.,

$$R_n \neq R_l. \tag{4.7}$$

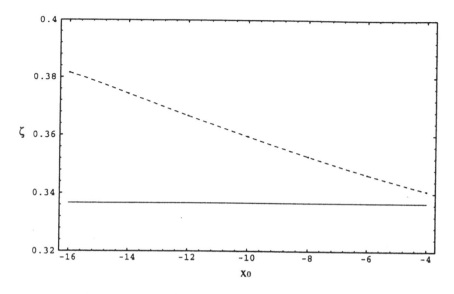

FIG. 17. Maximum run-up as a function of x_0, the station for setting off an initial solitary wave.

Having attained some clarification on the nonlinear effects in ocean wave run-up on beaches, we proceed to examine the dispersive effects so far neglected in the governing equations. The solitary wave changes its shape when it propagates from the ocean of uniform depth to the beach if the physical effects of dispersion are neglected. It is therefore of interest to determine the correction factor by taking the dispersive effects into account, e.g., by using the Boussinesq equations as our model. This issue will be examined in the next section.

V. Dispersive Effects on Run-Up of Waves on Beach

In Section IV, we have found that the run-up of solitary waves predicted by the shallow-water equations depends on the initial position at which the solitary wave is prescribed. This leads us to examine the role played by the dispersive effects during the entire run-up process.

In Section V.A, we first study the dispersive effects on run-up based on the linear dispersive long-wave model. Next we will evaluate in Section V.B the importance of the dispersive effects on Carrier–Greenspan's exact solution for periodic waves on sloping beach. In Section V.C we will perform numerical experiments on two-dimensional run-up of a solitary wave using the generalized Boussinesq model and compare the results with that predicted by the shallow-water equations. In Section V.D we will study the problem of two-dimensional run-ups of a solitary wave on beach with variable downward slope.

A. LINEAR DISPERSIVE MODEL

On physical ground, dispersive effects have both long-time and local instantaneous aspects, with the former coming from accumulation during wave propagation, especially in water of variable depth, and the latter arising from finite to large values of wave slope and its higher derivatives. According to geometric wave theory, the dispersion relation on linear theory, $\omega^2 = \mathrm{g}k \tanh kh$ for waves of circular frequency ω and wavenumber k progressing in water of depth h becomes $\omega^2 = ghk^2(1 - k^2h^2/3)$ for long waves ($kh \ll 1$), which implies that k varies like $1/\sqrt{h}$ in virtue of ω being a constant of motion for free waves.

Now we study again the problem given in Section II by adopting the linear dispersive model

$$\zeta_t + \nabla \cdot (h\boldsymbol{u}) = 0, \tag{5.1}$$

$$\boldsymbol{u}_t + \nabla\zeta = \frac{h}{2}\nabla[\nabla \cdot (h\boldsymbol{u}_t)] - \frac{h^2}{6}\nabla^2\boldsymbol{u}_t, \tag{5.2a}$$

which comes from the generalized Boussinesq model (2.2) and (2.3) by neglecting the nonlinear effects. The bathymetry of the beach and the outside open ocean is again given by (2.1). For slowly varying $h(x)$, (5.2a) can be written as

$$\boldsymbol{u}_t + \nabla\zeta = \frac{h}{3}\nabla[\nabla \cdot (h\boldsymbol{u}_t)] + \frac{1}{6}\nabla h\nabla \cdot (h\boldsymbol{u}_t) + \frac{1}{6}\nabla h \cdot \nabla(h\boldsymbol{u}_t), \tag{5.2b}$$

and by eliminating $\boldsymbol{u}$ from (5.1) and (5.2b) we obtain for ζ the equation:

$$\zeta_{tt} - (h\zeta_x)_x - h\zeta_{yy} - \frac{1}{3}h^2(\zeta_{xxtt} + \zeta_{yytt}) - hh_x\zeta_{xtt} = 0 \qquad (0 < x < l). \tag{5.3a}$$

$$\zeta_{tt} - \zeta_{xx} - \zeta_{yy} - \frac{1}{3}(\zeta_{xxtt} + \zeta_{yytt}) = 0 \qquad (x > l, \ h = 1), \tag{5.3b}$$

As in Section II, we also use the new scaling:

$$x' = x/l, \quad y' = y/l, \quad t' = t/l, \tag{5.3c}$$

which converts (5.3a,b) into the primed variables, but then with a new factor α^2 (where $\alpha = h_0/l$) for all the dispersion terms (i.e., with the third and fourth derivatives). We do, however, omit the prime for brevity but replace k by $\kappa = kl$ and ω by $\tilde{\omega} = \omega l$. In the open ocean ($x > 1$), we again have an incident and reflected sinusoidal wave train,

$$\zeta = A(k) \exp[i(\kappa_2 y + \kappa_1(x - 1) - \tilde{\omega}t)]$$
$$+ B(k) \exp[i(\kappa_2 y - \kappa_1(x - 1) - \tilde{\omega}t)], \tag{5.4a}$$

where (κ_1, κ_2) is the wavenumber vector, $\tilde{\omega}$ is the circular frequency,

$$\kappa_1 = -\kappa \cos\beta, \qquad \kappa_2 = \kappa \sin\beta, \tag{5.4b}$$

$$\tilde{\omega} = \kappa/\sqrt{1 + \alpha^2\kappa^2/3}, \qquad \alpha = (h_0/l) \quad (h_0 = 1 \text{ here}). \tag{5.4c}$$

This long-wave dispersion relation is correct and in a form desirable for computational purposes.

Within the inclined beach ($0 < x < 1$), the solution can assume the form

$$\zeta(x, y, t) = C_2(k) F(x; \kappa, \beta, \alpha) \exp(i\kappa s) \quad (s = y \sin\beta - \tilde{\omega}t/\kappa, 0 < x < 1). \tag{5.5}$$

Substituting (5.5) into (5.3a) gives for the *beach-wave function* $F(x; \kappa, \beta, \alpha)$ the equation

$$F'' + \frac{1}{x} p(x) F' + \frac{1}{x^2} q(x) F = 0, \tag{5.6a}$$

where

$$p(x) = xh_x(1 - \alpha^2\tilde{\omega}^2 h)/[h(1 - \alpha^2\tilde{\omega}^2 h/3)], \tag{5.6b}$$

$$q(x) = x^2(\tilde{\omega}^2 - \kappa^2(h - \alpha^2\tilde{\omega}^2 h^2/3) \sin^2\beta)/[h(1 - \alpha^2\tilde{\omega}^2 h/3)]. \tag{5.6c}$$

For the present case, $h \leqslant 1$ over the whole field, so we have

$$1 - \alpha^2\tilde{\omega}^2 h/3 \geqslant 1 - \alpha^2\tilde{\omega}^2/3 = (1 + \alpha^2\kappa^2/3)^{-1} > 0, \tag{5.7}$$

for all κ. We further assume that $p(x)$ and $q(x)$ are both analytic and regular in $0 \leqslant x \leqslant 1$, and hence can again be represented by series expansion (2.13), for which $x = 0$ is again a regular singular point. By Frobenius's theory,

(5.6a) has a solution

$$F(x) = \sum_{n=0}^{\infty} b_n x^{n+\nu},$$ (5.8a)

where ν satisfies the indicial equation

$$\nu^2 + (p_0 - 1)\nu + q_0 = 0.$$ (5.8b)

Because $p_0 = 1$ and $q_0 = 0$, as already argued in Section II, we again have $\nu_1 = \nu_2 = 0$.

Thus we obtain a beach-wave function $F = F(x; \kappa, \beta, \alpha)$ for the root $\nu_1 = 0$ and discard the logarithmically singular solution corresponding to the double root $\nu_2 = 0$ to give

$$F = F(x; \kappa, \beta, \alpha) = \sum_{n=0}^{\infty} b_n x^n,$$ (5.9)

where $b_0 = 1$ by normalization [see (2.11b)], and b_n are determined by the recurrence formula,

$$n^2 b_n = - \sum_{r=0}^{n-1} (r p_{n-r} + q_{n-r}) b_r \qquad (n = 1, 2, \ldots),$$ (5.10)

with $p_0 = 1$ and $q_0 = 0$ being understood. We note that in addition to β for wave incidence and κ for wavenumber, the new dispersion relation (5.4c) for $\tilde{\omega} = \omega l$ is a known function of κ and α. So in the present case for dispersive waves, the parameter α appears both in (5.4c) and in the argument of the depth function $h(x)$ [see (2.11c)], in contrast with the nondispersive case treated in Section II in which α has only the latter entry. Finally, application of matching conditions (2.8a,b) determines the coefficients $B(k)$ in (5.4a) and $C_2(k)$ in (5.5) as

$$B(k) = A(k) \exp[2i\Delta(\kappa, \beta, \alpha)],$$ (5.11a)

$$C_2(k) = A(k) R(\kappa, \beta, \alpha) \exp[i\Delta(\kappa, \beta, \alpha)],$$ (5.11b)

$$R(\kappa, \beta, \alpha) = 2(M^2 + N^2)^{-1/2}, \qquad \Delta(\kappa, \beta, \alpha) = \arg(M + iN),$$ (5.11c)

$$M(\kappa, \beta, \alpha) = F(1; \kappa, \beta, \alpha) = \sum_{n=0}^{\infty} b_n, \quad N(\kappa, \beta, \alpha) = \frac{1}{\kappa_1} F_x(1; \kappa, \beta, \alpha) = \frac{1}{\kappa_1} \sum_{n=0}^{\infty} n b_n,$$

(5.11d)

in which $\kappa_1 = -\kappa \cos \beta$, $\kappa = kl$ as before. The run-up function $R(\kappa, \beta, \alpha)$ gives the relative wave run-up on the beach and the phase function $\Delta(\kappa, \beta, \alpha)$

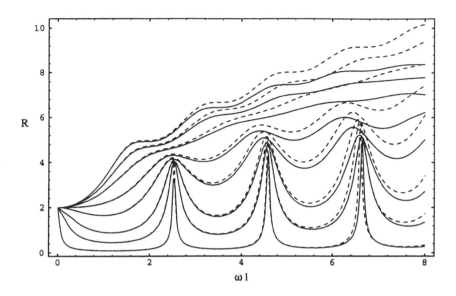

FIG. 18. Variation of the run-up function, $R(\tilde{\omega}, \beta, \alpha)$, versus $\tilde{\omega}$ for waves incident on a plane beach $h(x) = x/l$, $l = 10$ for a set of different incidence angles, $\beta = 0°$, $30°$, $45°$, $60°$, $70°$, $80°$, $85°$, $89°$, from the top to the bottom. Solid line, linear dispersive model; dashed line, linear nondispersive model.

gives the phase lag of the waves in the beach region and the phase lag 2Δ for the waves reflected.

Figure 18 shows the variation of the run-up function, $R(\kappa, \beta, \alpha = 0.1)$, versus $\tilde{\omega}$ for waves incident on a plane beach $h(x) = x/l$, $l = 10$, for a set of incidence angles β.

We have a sinusoidal wave train with a given frequency propagating from the open ocean of uniform depth $h = 1$ to the sloping plane beach. The maximum run-up of this wave train predicted by the linear dispersive model is smaller than the prediction of the linear nondispersive model. Hence, qualitatively, *dispersion reduces run-up*.

B. DISPERSIVE EFFECTS ON CARRIER–GREENSPAN'S SOLUTION FOR PERIODIC WAVES

In Section III, Carrier–Greenspan's exact solution is given for a periodic train of standing waves running up and down an inclined plane beach of infinite extent; the solution is based on the shallow-water equations, i.e., without the dispersive effects. We now proceed to assess the dispersive effects on Carrier–Greenspan's periodic wave solution based on the generalized Boussinesq model (or the gB model; Wu, 1979), which is a weakly nonlinear and weakly dispersive model) by comparing the dispersive term therein with the linear term.

For the two-dimensional case, the gB model reads

$$\zeta_t + [(h + \zeta)u]_x = 0, \tag{5.12}$$

$$u_t + uu_x + \zeta_x = \frac{h}{2}(hu_t)_{xx} - \frac{h^2}{6}u_{xxt} \equiv D, \tag{5.13a}$$

with an error of $O(\tilde{a}\varepsilon^{5/2}, \varepsilon^{7/2})$. Its validity and error estimate is based on assumptions (2.4).

To facilitate using this model up to the waterline, the dispersive term is modified as follows:

$$
\begin{aligned}
D &= \frac{1}{3}h^2 u_{xxt} + hh_x u_{xt} + \frac{1}{2}hh_{xx}u_t \\[4pt]
&= \frac{1}{3}(h + \zeta)^2 u_{xxt} + (h + \zeta)h_x u_{xt} + \frac{1}{2}(h + \zeta)h_{xx}u_t \\[4pt]
&= -\frac{1}{3}(h + \zeta)^2 \zeta_{xxx} - (h + \zeta)h_x \zeta_{xx} - \frac{1}{2}(h + \zeta)h_{xx}\zeta_x, \tag{5.14a}
\end{aligned}
$$

in which the modifications are made by applying the lower order relations to D without changing the order of error estimate. We further assume the variation of $h(x)$ to be small,

$$h_x = O(\varepsilon), \qquad h_{xx} = O(\varepsilon^{3/2}). \tag{5.15}$$

Accordingly, for assessment purposes, the dispersive term can be reduced to

$$D = -\frac{1}{3}(h + \zeta)^2 \zeta_{xxx}. \tag{5.14b}$$

We point out that this improved form of the dispersion term reached by reasoning argument has since been confirmed to be completely valid by the fully nonlinear fully dispersive (FNFD) model introduced by Wu in Chapter 1 of this volume.

Referring to (3.22)–(3.25), we have at the maximum run-down moment, $t = \pi/4$, $\lambda = \pi/2$, and u vanishing everywhere. Consider the critical case of

$A = 1$ (with $\zeta_x = \infty$ at the waterline) so that at $t = \pi/4$,

$$\zeta = -\frac{1}{4}J_0(\sigma) = -\frac{1}{4} + \frac{1}{16}\sigma^2 - \frac{1}{256}\sigma^4 + o(\sigma^6), \qquad (5.16a)$$

$$x = -\frac{1}{16}\sigma^2 - \frac{1}{4}J_0(\sigma) = -\frac{1}{4} - \frac{1}{256}\sigma^4 + o(\sigma^6). \qquad (5.16b)$$

Therefore,

$$\zeta = -\frac{1}{4} + \left(-x - \frac{1}{4}\right)^{1/2} + O\left(-x - \frac{1}{4}\right), \qquad (5.16c)$$

$$\zeta_x = \frac{1}{2}\left(-x - \frac{1}{4}\right)^{-1/2} + \cdots, \qquad (5.16d)$$

$$\zeta_{xx} = -\frac{1}{4}\left(-x - \frac{1}{4}\right)^{-3/2} + \cdots, \qquad (5.16e)$$

$$\zeta_{xxx} = -\frac{3}{8}\left(-x - \frac{1}{4}\right)^{-5/2} + \cdots. \qquad (5.16f)$$

The leading order dispersion term on the right-hand side of eq. (5.13b) is

$$D = -\frac{1}{3}(h + \zeta)^2\zeta_{xxx} = -\frac{1}{3}(-x + \zeta)^2\zeta_{xxx} = \frac{1}{8}\left(-x - \frac{1}{4}\right)^{-3/2} \gg |\zeta_x|.$$

$$(5.14c)$$

This shows that the dispersion term dominates the linear term ζ_x near the singular point $x = -1/4$, and is therefore important near the waterline region.

Figures 19 and 20 show their values as functions of x at times $t = 3\pi/4$, $\pi/4$, which correspond to the end of run-up and run-down time instants, respectively. The slope of the beach is taken as 1/4. Compared with the linear term $-\zeta_x$, the dispersion term D is very small at $t = 3\pi/4$. This result seems natural since the wave is very flat during the peak of run-up, with $\partial(\cdot)/\partial x \ll 1$ and $|\zeta_{xxx}| \ll |\zeta_x|$. But at $t = \pi/4$, i.e., the extreme of run-down, the wave profile appears so much more curved that $\zeta_x \to \infty$ at $x = -1/4$, maximum run-down point, and $\zeta_{xxx} \to \infty$ faster than ζ_x. The dispersive effect thus becomes very important around the waterline at this time. It can become larger than the linear term ζ_x retained therein. This result shows that the dispersive effect is important in run-up calculations, especially when the wave profile becomes relatively steep.

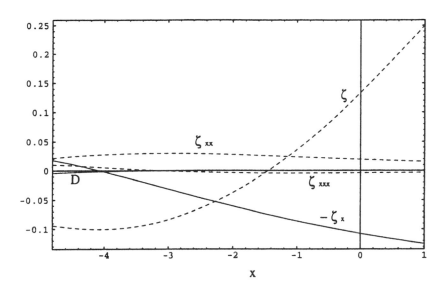

FIG. 19. Wave elevation ζ, its derivatives $-\zeta_x$, ζ_{xx}, ζ_{xxx}, and the value of dispersive term $D = -\frac{1}{3}(h + \zeta)^2\,\zeta_{xxx}$ of Carrier–Greenspan exact solution at the maximum run-up time $t = 3\pi/4$ for $A = 1.0$, $l = 4$.

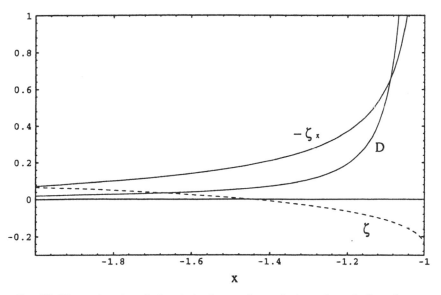

FIG. 20. Wave elevation ζ, its derivatives $-\zeta_x$, and the value of dispersive term $D = -\frac{1}{3}(h + \zeta)^2\,\zeta_{xxx}$ of Carrier–Greenspan exact solution at the maximum run-down time $t = \pi/4$ for $A = 1.0$, $l = 4$.

C. Numerical Solution of Wave Run-Up by the gB Model

Having made an asymptotic estimate of the dispersive effects on nonlinear run-up given by the shallow-water (nondispersive) model and their circumstantial dominating importance, we now present further numerical results of two-dimensional wave run-up using the gB model

$$\zeta_t + [(h + \zeta)u]_x = 0, \tag{5.12}$$

$$u_t + uu_x + \zeta_x = D = \frac{1}{3}(h + \zeta)^2 u_{xxt}, \tag{5.13b}$$

which has the nonlinear and dispersive effects both covered to $O(\varepsilon^{5/2})$ according to (2.4), with an error of $O(\varepsilon^{7/2})$. Since

$$D|_{h+\zeta=0} = 0, \tag{5.14d}$$

the dispersive term vanishes at the waterline; we therefore have the same equations governing the waterline as (3.9)–(3.11). Hence we can also use the same scheme for numerical computation of waterline as (3.20) and (3.21). After applying the waterline transformation (3.12), the governing equations for ζ and u in (x', t') become (after omitting the primes)

$$\zeta_t - c_1 U \zeta_x + c_2[(h + \zeta)u]_x = 0, \tag{5.17}$$

$$u_t - c_1 U u_x + c_2(uu_x + \zeta_x) = \frac{1}{3}(h + \zeta)^2 u_{xxt}, \tag{5.18}$$

$$c_1 = c_1(x, t) = \frac{1 + x/L}{1 + X/L}, \qquad c_2 = c_2(t) = \frac{1}{1 + X/L}, \tag{5.19}$$

where $c_1(x, t)$ and $c_2(t)$ are the coefficients of transformation. This system has an error of $O(\varepsilon^{7/2})$.

The leapfrog scheme is used to perform the numerical computation, i.e.,

$$\zeta_i^{n+1} = \zeta_i^{n-1} + \frac{\Delta t}{\Delta x} c_1 U(\zeta_{i+1}^n - \zeta_{i-1}^n) - \frac{\Delta t}{\Delta x} c_2\{[(h + \zeta)u]_{i+1}^n - [(h + \zeta)u]_{i-1}^n\}, \tag{5.20}$$

$$-\frac{1}{3(\Delta x)^2} c_2^2(h_i + \zeta_i^n)^2 u_{i-1}^{n+1} + \left[1 + \frac{2}{3(\Delta x)^2} c_2^2(h_i + \zeta_i^n)^2\right] u_i^{n+1}$$

$$-\frac{1}{3(\Delta x)^2} c_2^2(h_i + \zeta_i^n)^2 u_{i+1}^{n+1} = u_i^{n-1} + \frac{\Delta t}{\Delta x} c_1 U(u_{i+1}^n - u_{i-1}^n)$$

$$-\frac{\Delta t}{3\Delta x}c_2(u_{i+1}^n + u_i^n + u_{i-1}^n)(u_{i+1}^n - u_{i-1}^n) - \frac{\Delta t}{\Delta x}c_2(\zeta_{i+1}^n - \zeta_{i-1}^n)$$

$$+\frac{1}{3(\Delta x)^2}c_2^2(h_i + \zeta_i^n)^2(u_{i+1}^{n-1} - 2u_i^{n-1} + u_{i-1}^{n-1}). \tag{5.21}$$

The presence scheme is one-step, explicit for ζ, and implicit for u. They are second order in space and time.

The first numerical experiment is to evaluate the propagation of a right-going solitary wave on shallow water of uniform depth (Figure 21), here based on the gB model (5.12)–(5.13b) and our hybrid numerical method. The amplitude of the initial solitary wave is taken to be $a = 0.1$, scaled by the water depth, with the initial conditions

$$\zeta(x, t = 0) = a\,\text{sech}^2\sqrt{3a/4}(x + 24), \tag{5.22}$$

$$u = c\zeta/(1 + \zeta) \qquad (c = 1 + a/2) \tag{5.23}$$

for the wave elevation and velocity, while the initial wave phase velocity is that of the KdV equation. These conditions have been explained (for left-going) soliton in (4.1) and (4.2). For the numerical computation, we take $\Delta t = 0.04$, $\Delta x = 0.05$, and use the leapfrog scheme. Figure 21 shows that the

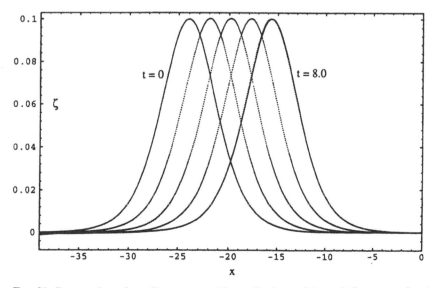

FIG. 21. Propagation of a solitary wave with amplitude $a = 0.1$ on shallow water $h = 1$. Solid line, exact solution of the generalized Boussinesq (gB) equation; dotted line, numerical solution by present scheme, $\Delta x = 0.05$, $\Delta t = 0.04$.

scheme works very well for solitary waves of permanent form. The amplitude of the solitary wave at time $t = 8.0$ is $a = 0.1001$, having increased only 0.1%; the speed of the solitary wave is $c = (24 - 15.687)/8.0 = 1.039$, which is slightly smaller than the wave speed of the KdV solution, namely, 1.05. The solitary wave speed for the gB model is known to be smaller than that for the KdV equation, as pointed out by Teng and Wu (1992). Computation here reconfirms their result.

The second numerical experiment is to compare the wave run-up predicted by the (nonlinear) shallow-water equations and the gB model, with an objective to investigate the role played by the dispersive effect during wave run-ups. Figure 22 shows the run-up process of a solitary wave on a sloping beach. The initial conditions are the same as those adopted in the former experiment, except they are now initially centered at $x = -16$, i.e.,

$$\zeta(x, t = 0) = a \operatorname{sech}^2 \sqrt{3a/4}(x + 16), \tag{5.24}$$

approaching the beach of length l ($l = 4$) is taken in this case). To deal with the dispersive term, which is a third-order derivative, the conjunction point

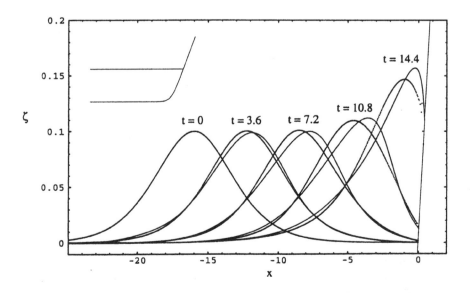

FIG. 22. Comparison between run-up of a solitary wave predicted by shallow-water equations and that predicted by gB model. Solid line, shallow-water equation; dotted line, gB model, $\Delta x = 0.05$, $\Delta t = 0.04$.

between the ocean and beach is smoothened by using

$$h(x) = 1 - (0.1) \log[1 + \exp 10(x/l + 1)] \tag{5.25}$$

as the local depth function, which has a rounded-off corner about $x = -l$ with a small radius of curvature and a rapid transition asymptotically to:

$$h(x \to \infty) = -x/l \quad \text{and} \quad h(x \to -\infty) = 1. \tag{5.26}$$

We recall that based on the shallow-water equations, a solitary wave will steepen with time and, hence, the higher the run-up, the longer the run to the beach. But for the gB model, a solitary wave will maintain its shape in water of uniform depth and result in a unique run-up on a sloping beach, which is also less in height than that predicted by the shallow-water equations. The dispersive effect plays an important role in the evolution of solitary waves. Notice that the wave seems to start fluctuating at $t = 14.4$ and tends to blow up at $t = 15.6$. This behavior can be attributed to the dispersive effects making the wave become shorter around the waterline, and in turn making the dispersive term become stronger; thus the cycle magnifies.

According to the assumptions for the gB model, the dispersive term is always of the same order as the nonlinear term, and smaller than a single linear term. To make the numerical computation convergent, we multiply the dispersive term D by the following factor:

$$f(x) = \tanh^4(10\, x/l) \tag{5.27}$$

where x is the new coordinate after the waterline transformation. This factor is almost 1 when $x < -l$, i.e., in the ocean. The factor is 0.95 at $x = l/4$ (Figure 23). It suppresses the dispersive effect only in the region very close to waterline.

Figure 24 shows the run-up of the solitary wave predicted by the gB model without or with the suppression factor. We see that at $t = 15.6$ the wave does not blow up if we include the suppressing factor. In the following computation, we will always include the factor as understood.

Figure 25a shows the wave elevation during run-up at time $t = 13.04$, 14.64, 16.24, 17.84, 19.44. The solitary wave reaches the maximum run-up $R = 0.3400$ at time $t = 19.44$. Figure 25b shows the velocity during run-up at the same time instants as listed in Figure 25a. The maximum speed during this run-up is as high as 0.2697, which occurs at the waterline at $t = 15.88$.

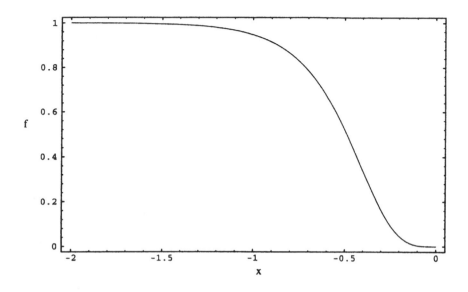

FIG. 23. Dispersion suppressing factor $f(x)$, for $l = 4$.

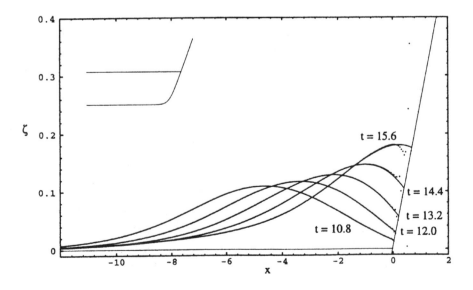

FIG. 24. Run-up predicted by gB model. Solid line, with dispersion suppressing factor; dotted line, without the factor.

Figure 25c shows the wave elevation during the backwash at time $t = 19.44$, 21.04, 22.64, 24.24, and Figure 25d shows the velocity during the backwash at the same time instants as listed in Figure 25c. The maximum speed reached during backwash is as high as 0.6972, which occurs at the waterline at $t = 24.40$.

For a solitary wave of amplitude $a = 0.1$, initially centered at $x = -16$, the maximum run-ups on a plane beach with slope 0.25, predicted by the linear theory, the shallow-water equation and the gB model, are given in Table 4. The results show that both nonlinear and dispersive effects are important during the wave run-up. Figure 26 shows the comparison of wave run-up predicted by the three models.

We have carried out a few more numerical experiments on the cases, which have appeared in publications, and made comparison between these results in Table 5. For more comparisons, see Appendix A.

D. Run-Up of Solitary Wave on Arbitrary Beach

Here, we present the results of a few numerical experiments based on the gB model using the hybrid Lagrangian–Eulerian method. The first experiment is to calculate the run-up of a solitary wave of amplitude $a = 0.1$, initially centered at $x = -16$, on a beach with the following geometry:

$$h(x) = 1 - (0.1) \log\{1 + \exp[10(1 + x/l) + 3 \sin^2(\pi x/l)]\}. \quad (5.28)$$

The corresponding run-up and run-down results are shown in Figures 27a and 27b. The maximum run-up $R - 0.3408$ occurs at time $t = 17.84$.

The second experiment is on the run-up of a solitary wave of amplitude $a = 0.1$, initially centered at $x = -16$, on a beach with following depth variation:

$$h(x) = 1 - (0.1) \log\{1 + \exp[10(1 + x/l) - 3 \sin^2(\pi x/l)]\}. \quad (5.29)$$

The run-up results are shown in Figure 28. The maximum run-up is $R = 0.3735$, which occurs at time $t = 19.84$.

The third experiment is on the run-up of a solitary wave of amplitude $a = 0.1$, initially centered at $x = -16$, on a step beach given by

$$h(x) = \begin{cases} 1 - (0.1) \log\{1 + \exp[10(1 + x/l)]\} & x < 0, \\ -(0.1) \tanh(10x/l) & x > 0. \end{cases} \quad (5.30)$$

The resultant wave elevations during the wave run-up are shown in Figure 29.

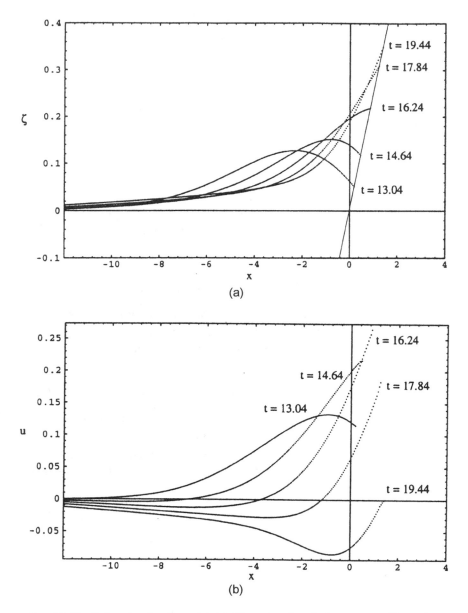

FIG. 25. Predictions by the gB model: (a) Wave elevation during run-up, with maximum run-up of 0.3400 appearing at $t = 19.44$. (b) Velocity during run-up.

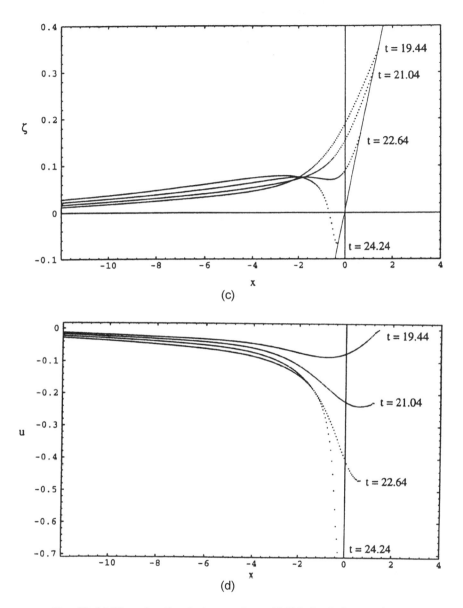

FIG. 25. (c) Wave elevation during run-down. (d) Velocity during run-down.

TABLE 4

RUN-UP OF A SOLITARY WAVE OF AMPLITUDE $a/h = 0.1$, PREDICTED BY LINEAR THEORY, THE
SHALLOW-WATER EQUATIONS AND THE gB MODEL, WITH THE SOLITARY WAVE CENTERED
AT $x = -16$ INITIALLY (SEE EQS. 5.23, 5.24), RUNNING UP FROM AN OCEAN OF DEPTH $h = 1$ TO A
PLANE BEACH OF SLOPE $\alpha = 1/4$

Models	Maximum run-up R	Time to reach maximum run-up
Linear theory	0.3366	18.76
Shallow-water equation	0.3816	18.20
gB model	0.3400	19.44

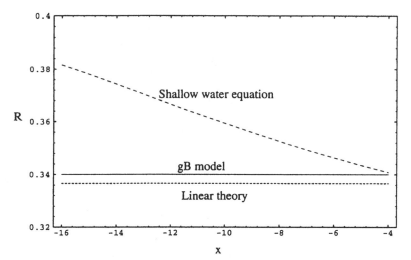

FIG. 26. Maximum run-up of a solitary wave with $a = 0.1$, centered at different places x, on a plane beach of slope 0.25.

TABLE 5

COMPARISON BETWEEN NUMERICAL RESULTS OF THE PRESENT SCHEME AND OTHERS' RESULTS

Source[a]	Beach length l	a	Others' results	Results by present scheme	Laboratory experiments[b]
HH	10	0.05	0.180	0.2428	na
KLL	3.732	0.05	0.135	0.1783	0.173
KLL	3.732	0.1	0.308	0.3408	0.281
KLL	3.732	0.2	0.766	0.7502	0.599
HH	3.333	0.1	0.310	0.3245	na
PG	2.747	0.098	0.275	0.2866	0.252
PG	2.747	0.193	0.599	0.6441	0.552
PG	2.747	0.294	0.958	1.0309	0.898
KLL	1.0	0.2	0.504	0.4991	0.454
KLL	1.0	0.480	1.610	1.2909	1.270

[a]HH: Heitner and Housner (1970); KKL: Kim et al. (1983); PG: Pedersen and Gjevik (1983).
[b]The laboratory experiments column shows results derived from the interpolation of Hall and Watts (1953) data set, when possible.

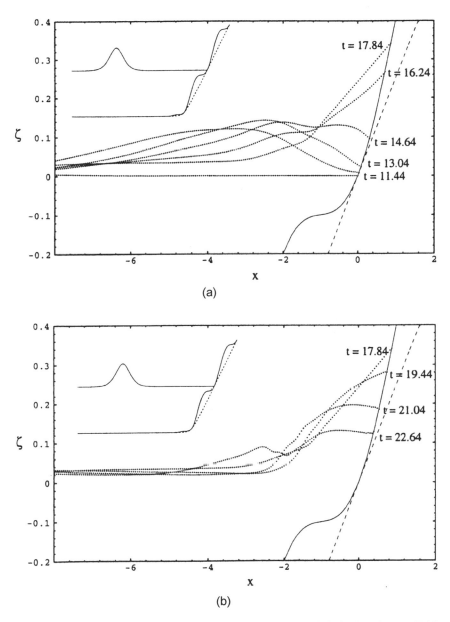

FIG. 27. (a) Run-up of a solitary wave with $a = 0.1$ on beach $h(x)$ given by eq. (5.28), predicted by the gB model with the present hybrid scheme, maximum run-up $R = 0.3408$ appearing at $t = 17.84$. (b) Continued run-down of the same solitary wave.

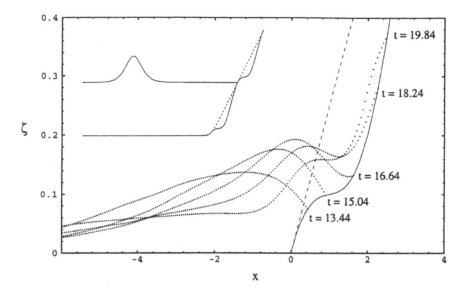

FIG. 28. Run-up of a solitary wave with $a = 0.1$ on beach $h(x)$ given by eq. (5.29), predicted by the gB model with the hybrid scheme, maximum run-up $R = 0.3735$ at $t = 19.84$.

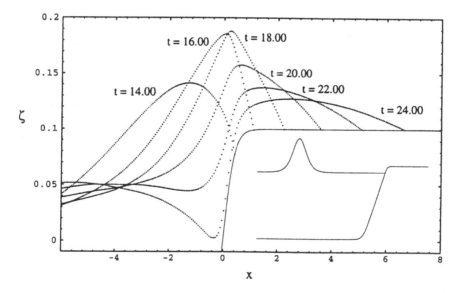

FIG. 29. Run-up of a solitary wave with $a = 0.1$ on a step beach $h(x)$ given by eq. (5.30), predicted by the gB model with the hybrid scheme.

VI. Wave-Induced Longshore Current

According to the present linear long-wave approximation, the horizontal projection of the pathline of a fluid particle is given by the integral

$$\mathbf{x}(\xi, t) = \xi + \int_0^t \mathbf{u}_L(\xi, \tau)\, d\tau, \tag{6.1a}$$

where $\xi = \mathbf{x}(\xi, t = 0)$ is the initial position of the fluid particle at $\mathbf{x} = (x, y, t)$ at time t, and $\mathbf{u}_L(\xi, t) = (u_L, v_L, 0)$ is the Lagrangian particle velocity, projected onto a horizontal plane, of the particle designated by $\xi = (\xi_1, \xi_2, 0)$. The Lagrangian drift velocity is related to the Eulerian field velocity, $\mathbf{u}(\mathbf{x}, t)$, for small $\Delta\xi = \mathbf{x} - \xi$, by

$$\mathbf{u}_L(\xi, t) = \mathbf{u}(\mathbf{x}, t) = \mathbf{u}(\xi, t) + \left(\int_0^t \mathbf{u}(\xi, \tau)\, d\tau \right) \cdot \nabla \mathbf{u}(\xi, t) + O(\Delta\xi)^2. \tag{6.1b}$$

For the case of a uniform beach, with ζ expressed as in (2.10) and (2.11a), integration of (2.6) yields

$$\mathbf{u} = (u, v, 0) = (\zeta_x/ik, \zeta \sin\beta, 0) \qquad (0 < x < \infty). \tag{6.2}$$

To obtain the time average of $\mathbf{u}_L$, we use the theorem that if $f = ae^{i\omega t}$ and $g = be^{i\omega t}$, a and b being complex constant, then

$$\overline{(\mathrm{Re}f)(\mathrm{Re}g)} = \frac{1}{2}\,\mathrm{Re}(fg^*). \tag{6.3}$$

where the overhead bar denotes the time averaged and $*$ indicates the complex conjugate. The mean current velocity is thus found from (6.1)–(6.3) to have the x component as

$$\overline{u_L}(x) = \frac{1}{2}\,\mathrm{Re}\left\{ u_x\left(\frac{1}{ik} u^*\right) + u_y\left(\frac{1}{ik} v^*\right) \right\} = \frac{1}{2k}\,\mathrm{Im}\left\{ \frac{1}{k^2} \eta_{xx}\eta_x^* + \eta_x\eta^* \sin^2\beta \right\}$$

$$= 0, \tag{6.4}$$

which vanishes since each quantity in the brackets is purely real in view of using (2.9) for $x > l$ and (2.10) for $0 < x < l$. The longshore (y component) current has the mean

$$\overline{v_L}(x) = \frac{1}{2}\,\mathrm{Re}\left\{ v_x\left(\frac{1}{ik} u^*\right) + v_y\left(\frac{1}{ik} v^*\right) \right\} = \frac{1}{2}\left(\frac{1}{k^2} |\eta_x|^2 + |\eta|^2 \sin^2\beta \right) \sin\beta,$$

$$\tag{6.5}$$

which is positive definite for $0 < \beta < \pi/2$ unless $|\eta|^2 = 0$, and is symmetric about $\beta = 0$. This formula holds for the general case of uniform beach of arbitrary downward slope.

At waterline $x = 0$ of a sloping plane beach, use of (2.10), (2.26), and (6.5) yields the current velocity

$$\overline{v_L}(x = 0; \kappa, \beta) = \frac{1}{2} A_0^2 R^2 (\kappa^2 + \sin^2 \beta) \sin \beta, \qquad (6.6a)$$

where $R = R(\kappa, \beta)$ is the relative run-up as before. For large κ, substituting expansion (2.45d) for R in (6.6a) yields

$$\overline{v_L}(0; \kappa, \beta) = A_0^2 \pi \kappa^3 \sin 2\beta + O(\kappa^{3/2}). \qquad (6.6b)$$

This asymptotic limit shows that $\overline{v_L}(0)$ increases like κ^3 for κ large and reaches, for κ fixed, its maximum at $\beta = 45°$. For the open ocean, $x > l$, use of (2.9) in (6.5) yields the result:

$$\overline{v_L} = A^2 \sin \beta \{1 - \cos 2\beta \cos 2[k(x - l) \cos \beta + \Delta]\}, \qquad (6.6c)$$

to which the incident and reflected waves each contribute one-half of the total. For sloping plane beach, the mean drift velocity can be calculated from (6.5).

As a typical example, Figure 30a shows the variations of η, η_x, mean longshore current $\overline{v_L}$, and the longshore discharge flux $q = h\overline{v_L}$, with increasing shore distance x/l, for the case with $\beta = 30°$, $A_0 = 1$, and $\kappa = 2$. The longshore current increases shoreward quite rapidly to 23.86 at the shoreline, whereas the longshore mass flux has a maximum of 1.867 at $x/l = 0.205$. Figure 30b shows the variations of mean longshore current $\overline{v_L}$ with increasing shore distance x/l, for $A_0 = 1$, $\kappa = 2$, $\beta = 10°$, $20°$, $45°$. Figure 31 shows the variation of mean longshore current function $\overline{v_L}$ $(x = 0; \kappa, \beta)$ at waterline with increasing κ for different β. From these results we see that the longshore current velocity on the beach, $\overline{v_L}$, increases with β, for fixed κ, up to about $\beta = 45°$, and within this range of β, increases with κ like κ^3. In the range $45° < \beta < 90°$, the outstanding feature is that $\overline{v_L}$ is dominated by the eigenmodes of the trapped waves.

For a fluid particle, initially at $(x_0, -\Delta/k_2)$, its trajectory $(x(t),\ y(t))$ satisfies the following equations:

$$\frac{dx}{dt} = \frac{1}{k} \eta_x \sin[k(y \sin \beta - t) + \Delta], \qquad (6.7a)$$

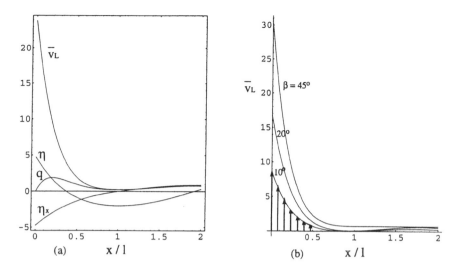

FIG. 30. (a) Variations of η, η_x, mean longshore current $\overline{v_L}$, and longshore discharge flux $q = h\overline{v_L}$, with shore distance x/l, for $\beta = 30°$, $A_0 = 1$, and $\kappa = 2$. (b) Distributions of mean longshore current $\overline{v_L}$ over shore distance x/l, for $A_0 = 1$, $\kappa = 2$, and $\beta = 10°$, $20°$, $45°$.

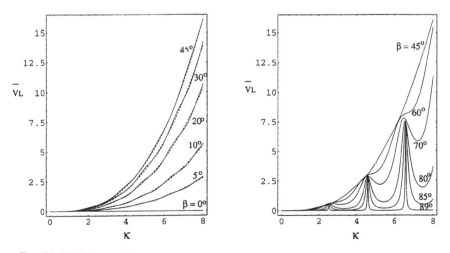

FIG. 31. Variations of the mean longshore current $\overline{v_L}$ at waterline with increasing κ for different β. Dashed line, asymptotic formula [eq. (6.6b)].

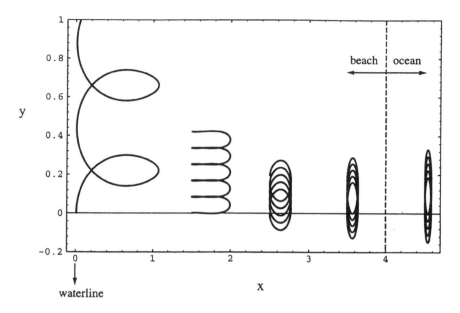

FIG. 32. Horizontal projections of pathlines of water particles at different places on the beach and in the ocean, for $\alpha = 1/4$, $\beta = 60°$, $k = 0.5$, and $A = 0.05$. These tracks are the pathlines traversed by the water particles in five time periods except for the one at the waterline, which covers only about two periods in this inviscid irrotational flow.

$$\frac{dy}{dt} = \eta \sin \beta \cos[k(y \sin \beta - t) + \Delta], \qquad (6.7b)$$

$$x(0) = x_0, \qquad y(0) = -\Delta/k_2. \qquad (6.7c)$$

Numerical results of pathlines are obtained by integrating (6.7) for $k = 0.5$, $\beta = 60°$, $l = 4$, and $A = 0.05$, on the beach

$$h(x) = x \quad \eta(x) = 0.1835 \exp(-\sqrt{3x}/4) M(-0.6547, 1, \sqrt{3}x/2), \quad 0 < x < l,$$
$$(6.8a)$$

and in the open sea, with

$$h(x) = 1, \quad \eta(x) = 0.1 \cos(0.5x + 0.583), \quad x > l. \qquad (6.8b)$$

Figure 32 shows the horizontal projection of several trajectories of water particles at different places on the beach and in the sea, for $\alpha = 1/4$, $\beta = 60°$, $k = 0.5$, and $A = 0.05$. This figure shows clearly how the coiling of pathlines traversed by fluid particles becomes increasingly stretched out in both longshore and seaward directions as the longshore current is amplified toward the waterline. This varying pattern of pathlines is thought to play a basic role in the processes of sand suspension and sandbar formation. The eventual processes of coastal mass transport will require taking the viscous and dissipative effects into account.

VII. Conclusion

The primary objective of this study is to develop a comprehensive approach toward modeling ocean waves on the shelves and beaches, taking into account such important physical factors as three-dimensional topography of bathymetry and the nonlinear and dispersive effects playing their leading roles. Through this study we have explored the physical processes of evolution of a train of sinusoidal ocean waves obliquely incident, from an open ocean of uniform depth, on a uniform beach of variable offshore slope as evaluated on the linear and nonlinear shallow-water wave models and, in addition, on the generalized Boussinesq equations. It begins with the linear theory, based on which the fundamental solution is determined in Frobenius's series by Fourier synthesis for a general family of beach configurations. The Fourier spectra so assembled are useful for constructing solutions to general initial-boundary problems. This is illustrated by the solution for oblique run-up of solitary waves on parabolic beaches, which brings forth salient features of the wave field regarding the effects of wave incidence angle and variations in beach slope and curvature. For incidences close to the grazing angle, run-ups of the periodic waves become very small except in a spectrum of eigenmodes that characterizes the system at resonance. This is a variation of the phenomenon found by Eckart (1951) for edge waves. From these salient features of the wave field we have gained from the linear theory clear insights into the basic physical mechanisms underlying the complicated wave phenomenon under study. We have also identified the special topic on trapped eigenmode waves at grazing incidence as worthy of further research to determine its impact on coastal environmental quality involving coastal ocean circulation and long-term interactions between ocean and land as well as possible connections with river wave hydrodynamics.

The hybrid Lagrangian–Eulerian method, developed here for computation of solutions based on the nonlinear shallow-water wave equations (or

Airy's model), is exact and very simple to implement. From the three case studies, this method is found to have the virtue of being very stable and fast in convergence on iterations, using a CPU time that is only a small fraction of that required for other known numerical methods, yet achieving considerably higher accuracy. The results so obtained for this fully nonlinear but nondispersive Airy's model show that the run-up of ocean waves on a given sloping plane beach is generally increased due to the nonlinear effects. For the family of inclined parabolic beaches with mild curvature, the run-up depends primarily on the beach slope at the waterline.

When the dispersive effects are incorporated into the linear shallow-water wave equations to form a linear dispersive long-wave model, the original beach-wave function is extended to render this model readily useful. Applications of this solution to specific cases show that the run-up on a sloping plane beach is decreased due to the added dispersive effect by a margin somewhat less than the increment ascribed to the nonlinear effect alone. For example, for a solitary wave with an amplitude of one-tenth of the water depth propagating from a distance of 16 times water depth toward a beach with a breadth of 4 times water depth, the run-up on sloping plane beach is decreased by 10% due to the added dispersive effect. It is found that the dispersive effects can become important under certain criteria (Zhang, 1996). For instance, for predicting a solitary wave arriving at a station on the ocean-beach border junction after travel in an open ocean of uniform depth, it is found on the nonlinear shallow-water wave theory to depend on the length of travel in the ocean because the wave would keep growing according to the theory, thereby rendering the terminal run-up prediction nonunique. To overcome such imbalances, the combined influence of the nonlinear and dispersive effects based on the generalized Boussinesq model is found to yield all-around results that provide a comprehensive representation of all the primary effects well in balance.

Two further developments have been made of the fundamental solution of the linear theory. In one, a *beach-wave ray theory* is derived by the WKB asymptotic expansion of the fundamental solution to give formulas, in closed form, for the general case of arbitrary beach slope distribution. This ray theory provides a solution describing the refraction of ray lines and amplification of wave amplitude not only for the incoming waves but also the reflected ones. Coverage of the latter is owing to the existence of the general solution for the wave field on beach that can afford the key decomposition [see (2.47)] into the incoming and reflected waves; this is a capability that the classical ray-tracing technique lacks. By comparison, this ray theory has established a criterion, including the incidence range $0 < \beta < 60°$, under which the classical and the new theories are found in excellent agreement, and beyond which the classical theory becomes poor.

The linear theory solution has also been used to derive the material drift of water particles and the resulting longshore current, to leading order, under the ideal condition of no wave breaking along the uniform beach. The remarkable magnitude of the longshore current predicted here for incidence angles near the optimum obliquity and the rather irregular behavior of the longshore current about the eigenmode waves in the grazing angle range are results with conspicuous features. It should be of great interest to carry out experimental investigations to further explore this interesting phenomenon.

The present study is an extension of Zhang (1996) and Zhang and Wu (1999), studies that were devoted to linear theoretical considerations. With the important effects of nonlinearity and dispersion included, the present results have established the linear theory with precise qualifications of its accuracy in various circumstances. These new results have provided a sound foundation for further extension and improvement by including the remaining physical factors regarding the fluid viscous and dissipative losses, especially in the presence of wave breaking, droplets, and foaming bubble formation and in more general and complicated cases for broad applications.

Appendix A. More Comparisons between the Present Results and Other Theories

In Section II, we obtained a solution for solitary waves running from the ocean of uniform depth to a sloping plane beach by using linear nondispersive theory. On the beach, the wave elevation obtained from (2.28ab) and (4.5a) is given by

$$\zeta(x, t) = \text{Re} \int_0^\infty A_0(k) \, R_0(\kappa) \, e^{-ikt + i\Delta_0} J_0(2k \sqrt{lx}) \, dk \qquad (0 < x < l), \qquad \text{(A.1)}$$

where $A_0(k)$ is given by (4.4b), and $R_0(\kappa)$ and $\Delta_0(\kappa)$ by (2.29ab). Then the run-up, i.e., the wave elevation at $x = 0$, is

$$R(t) = \text{Re} \int_0^\infty A_0(k) \, R_0(\kappa) \, e^{-ikt + i\Delta_0} \, dk. \qquad \text{(A.2)}$$

By using an asymptotic expansion method, Synolakis (1987) found an approximate formula for the maximum run-up

$$R = \max R(t) = 2.831\sqrt{l}\,a^{5/4}, \tag{A.3}$$

where $1/l$ is the beach slope and a is the amplitude of the solitary wave. The formula is given a range of validity for

$$a \gg (0.288)/l)^2. \tag{A.4}$$

We have numerically evaluated the maximum R of eq. (A.2) and compare it with formula (A.3) in Table 6 and Figure 33. For $a = 0.1$ and $l = 1$, Synolakis's approximate formula has an error of about 25%.

We have carried out numerical computation on the run-up of solitary waves by using the generalized Boussinesq model (Wu, 1979) with the scheme developed in Section V.C. We compare our numerical results attained from formula (A.3) with experimental results available. The results are shown in Tables 7 and 8 and Figures 34 and 35.

For the first case, $l = 19.85$, formula (A.3) is supposedly correct when $a \gg 0.0002$ according to (A.4). The comparison is shown for amplitude a in between 0.0052 and 0.0280. For this mild beach, our numerical results agree with formula (A.3) quite well, the relative difference is less than 5%. Both are larger than the experimental results of Synolakis (1987).

For the second case, $l = 1$, formula (A.3) is supposedly correct when $a \gg 0.08$ as claimed by (A.4). We compare for amplitude a in between 0.2 and 0.5. For this steep beach, the relative difference between our numerical results and formula (A.3) is around 20%. Formula (A.3) is valid for $l \gg 1$.

TABLE 6

RUN-UP PREDICTED BY LINEAR NONDISPERSIVE THEORY, FOR A SOLITARY WAVE OF AMPLITUDE $a/h = 0.1$, RUNNING UP FROM AN OCEAN OF DEPTH $h = 1$ TO A PLANE BEACH OF SLOPE $\alpha = 1/l$

Beach length l	Approximate formula (A.3) $R = 2.831\sqrt{l}\,a^{5/4}$	Numerical evaluation of the maximum run-up R by (A.2)
1.0	0.1592	0.2163
2.0	0.2251	0.2561
3.0	0.2757	0.2981
4.0	0.3184	0.3366
5.0	0.3560	0.3717
6.0	0.3900	0.4039

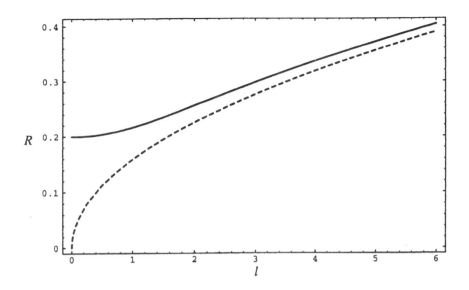

FIG. 33. Maximum run-up of solitary wave of amplitude $a - 0.1$ as a function of l evaluated by linear nondispersive theory. Solid line, numerical results; dashed line, formula (A.3).

TABLE 7

RUN-UP OF NONBREAKING SOLITARY WAVES PREDICTED BY THE gB MODEL WITH PRESENT
SCHEME, BY AN APPROXIMATE FORMULA AND EXPERIMENTAL RESULTS (SYNOLAKIS, 1987)

Beach length l	Amplitude a	Formula (A.3) $R = 2.831\sqrt{l}\,a^{5/4}$	Results by gB with present scheme	Laboratory experiments
19.85	0.0052	0.018	0.0187	0.019
19.85	0.0065	0.023	0.0245	0.022
19.85	0.0071	0.026	0.0272	0.026
19.85	0.0080	0.030	0.0315	0.029
19.85	0.0092	0.036	0.0373	0.036
19.85	0.0095	0.037	0.0388	0.041
19.85	0.0097	0.038	0.0398	0.038
19.85	0.0129	0.055	0.0564	0.048
19.85	0.0141	0.061	0.0629	0.052
19.85	0.0144	0.063	0.0646	0.049
19.85	0.0170	0.077	0.0792	0.063
19.85	0.0180	0.083	0.0849	0.074
19.85	0.0190	0.089	0.0908	0.077
19.85	0.0210	0.101	0.1029	0.075
19.85	0.0220	0.107	0.1091	0.098
19.85	0.0230	0.113	0.1155	0.087
19.85	0.0250	0.125	0.1289	0.100
19.85	0.0270	0.138	0.1435	0.108
19.85	0.0280	0.144	0.1514	0.123

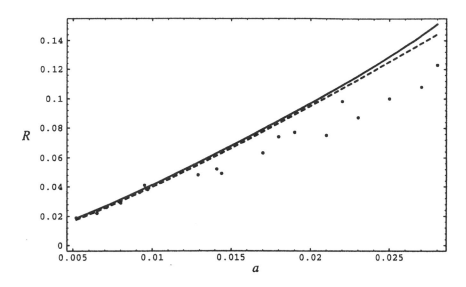

Fig. 34. Maximum run-up of solitary waves running from an ocean with uniform depth up to a plane beach of slope $\alpha = 1/19.85$, as a function of wave amplitude a. Solid line, our numerical results; dashed line, formula (A.3); dots, experimental data (Synolakis, 1987).

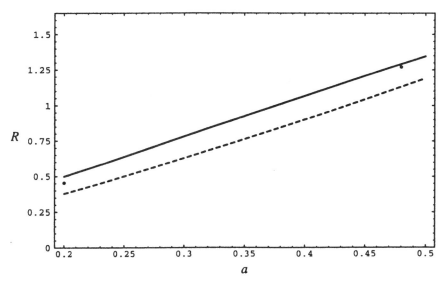

Fig. 35. Maximum run-up of solitary waves running from an ocean with uniform depth up to a plane beach of slope $\alpha = 1$, as a function of wave amplitude a. Solid line, our numerical results; dashed line, formula (A.3); dots, experimental data (Hall and Watts, 1953).

TABLE 8
RUN-UP OF NONBREAKING SOLITARY WAVES PREDICTED BY THE gB MODEL WITH
PRESENT SCHEME IN COMPARISON WITH THE APPROXIMATE FORMULA OF SYNOLAKIS (1987)

Beach length l	Amplitude a	Formula (A.3) $R = 2.831\sqrt{l}a^{5/4}$	Results by gB with present scheme
1.00	0.200	0.3786	0.4991
1.00	0.210	0.4025	0.5263
1.00	0.220	0.4265	0.5538
1.00	0.230	0.4509	0.5816
1.00	0.240	0.4756	0.6097
1.00	0.250	0.5005	0.6379
1.00	0.260	0.5256	0.6663
1.00	0.270	0.5510	0.6949
1.00	0.280	0.5766	0.7236
1.00	0.290	0.6025	0.7523
1.00	0.300	0.6285	0.7811
1.00	0.310	0.6549	0.8098
1.00	0.320	0.6814	0.8385
1.00	0.330	0.7081	0.8671
1.00	0.340	0.7350	0.8958
1.00	0.350	0.7621	0.9244
1.00	0.360	0.7894	0.9530
1.00	0.370	0.8169	0.9815
1.00	0.380	0.8446	1.0099
1.00	0.390	0.8725	1.0384
1.00	0.400	0.9006	1.0667
1.00	0.410	0.9288	1.0950
1.00	0.420	0.9572	1.1232
1.00	0.430	0.9858	1.1513
1.00	0.440	1.0145	1.1794
1.00	0.450	1.0434	1.2074
1.00	0.460	1.0725	1.2353
1.00	0.470	1.1017	1.2631
1.00	0.480	1.1311	1.2909
1.00	0.490	1.1606	1.3185
1.00	0.500	1.1903	1.3461

Acknowledgments

The authors would like to thank Wendong Qu for providing excellent assistance for this study. The work of TYW has been supported by National Science Foundation grants CMS-9503620 and CMS-9615897. The work of JEZ has been partially supported by City University of Hong Kong. JEZ kindly acknowledges the support from a Powell Fellowship granted by the California Institute of Technology.

References

Abramowitz, M., and Stegun, L. A. (eds.) (1964). *Handbook of Mathematical Functions.* National Bureau of Standards, Washington, DC.

Brocchini, M., and Peregrine, D. H. (1996). Integral flow properties of swash zone and averaging. *J. Fluid Mech.* **317**, 241–273.

Carrier, G. F., and Greenspan, H. P. (1958). Water waves of finite amplitude on a sloping beach. *J. Fluid Mech.* **4**, 97–109.

Carrier, G. F., and Noiseux, C. F. (1983). The reflection of obliquely incident tsunamis. *J. Fluid Mech.* **133**, 147–160.

Copson, E. T. (1948). *Theory of Functions of a Complex Variable.* Oxford Univ. Press, London.

Eckart, C. (1951). Surface waves in water of variable depth, Wave Report 100-99. Marine Physical Lab. of Scripps Institute of Oceanography.

Grilli, S. (1997). Fully nonlinear potential flow models used for long wave runup prediction. In *Long-Wave Runup Models* (H. Yeh, P. Liu, and C. Synolakis, eds.) pp. 116–180. World Scientific, Singapore.

Grilli, S., Subramanya, R., Svendsen, I.A. and Veeramony, J. (1994). Shoaling of solitary waves on plane beaches. *J. Wtrwy. Port Coast. Ocean Engrg.* **120**(6), 609–628.

Grilli, S. T., Svendsen, I. A., and Subramanya, R. (1997). Breaking criterion and characteristics for solitary waves on slope. *J. Wtrwy. Port Coast. Ocean Engrg.* **123**(3), 102–112.

Hall, J. V., and Watts, J. W. (1953). Laboratory investigation of the vertical rise of solitary waves on impermeable slopes, Tech. Memo. 33. Beach Erosion Board, US Army Corps of Engineers.

Heitner, K. L., and Housner, G. W. (1970). Numerical model for tsunami run-up. *Proc. ASCE* **WW3**, 701–719.

Hibberd, S., and Peregrine, D. H. (1979). Surf and run-up on a beach: A uniform bore. *J. Fluid Mech.* **95**, 323–345.

Kanoglu, U., and Synolakis, C. E. (1998). Long wave runup on piecewise linear topographies. *J. Fluid Mech.* **374**, 1–28.

Keller, J. B. (1961). Tsunami — water waves produced by earthquake. In *Proc. Tsunami Meetings 10th Pacific Science Congress* (D. C. Cox, ed.), IUGG Monograph, vol. 24, pp. 154–166.

Kim, S. K., Liu, P. L.-F., and Ligget, J. A. (1983). Boundary integral equation solutions for solitary wave generation, propagation and run-up. *Coastal Engng.* **7**, 299–317.

Lamb, S. H. (1932). *Hydrodynamics*, 6th ed. Cambridge Univ. Press, Cambridge.

Li, Y., and Raichlen, F. (1999). Solitary wave run-up on plane slopes. Research report, W. M. Keck Laboratory of Hydraulics, and Water Resources, California Institute of Technology, Pasadena.

Liu, P. L.-F., Cho, Y.-S., Briggs, M. J., Kanoglu, U., and Synolakis, C. E. (1995). Runup of solitary waves on a circular island. *J. Fluid Mech.* **302**, 259–285.

Pedersen, G., and Gjevik, B. (1983). Run-up of solitary waves. *J. Fluid Mech.* **135**, 283–299.

Spielvogel, L. Q. (1975). Single wave run-up on sloping beaches. *J. Fluid Mech.* **74**, 685–694.

Synolakis, C. E. (1987). The runup of solitary waves. *J. Fluid Mech.* **185**, 523–545.

Teng, M. H., and Wu, T. Y. (1992). Nonlinear water waves in channels of arbitrary shape. *J. Fluid Mech.* **242**, 211–233.

Tuba Ozkan-Haller, H., and Kirby, J. T. (1997). A Fourier–Chebyshev collocation method for the shallow water equations including shoreline runup. *Appl. Ocean Res.* **19**, 21–34.

Tuck E. O., and Hwang, L.-S. (1972). Long wave generation on a sloping beach. *J. Fluid Mech.* **51**, 449–461.

Whitham, G. B. (1974). *Linear and Nonlinear Waves.* Wiley-Interscience, New York.

Wu, T. Y. (1979). Tsunamis. In *Proc. National Science Foundation Workshop,* May 7–9, 1979, pp. 110–149. Tetra Tech Inc., Pasadena.

Wu, T. Y. (1981). Long waves in ocean and coastal waters. *J. Engng. Mech. Div. ASCE* **107**, 501–522.

Yeh, H., Ghazali, A., and Marton, I. (1989). Experimental study of bore run-up. *J. Fluid Mech.*, **206**, 563–578.

Yeh, H., Liu, P. L.-F., and Synolakis, C. E. (eds.). (1997). *Long Wave Runup Models.* World Scientific, Singapore.

Zelt, J. A. (1986). Tsunamis: The response of harbours with sloping boundaries to long wave excitation. Ph.D. thesis, California Institute of Technology, Pasadena.

Zelt, J. A. (1991). The run-up of nonbreaking and breaking solitary waves. *Coastal Engrg.* **15**(3), 205–246.

Zelt, J. A., and Raichlen, F. (1990). A Lagrangian model for wave-induced harbor oscillations. *J. Fluid Mech.* **213**, 203–225.

Zhang, J. E. (1996). I. Run-up of ocean waves on beaches, II. Nonlinear waves in a fluid-filled elastic tube. Ph.D. thesis, California Institute of Technology, Pasadena.

Zhang, J. E., and Wu, T. Y. (1999). Oblique long waves on beach and induced longshore current. *J. Eng. Mech. ASCE* **125**(7), 812–826.

ADVANCES IN APPLIED MECHANICS, VOLUME 37

Onset of Oscillatory Interfacial Instability and Wave Motions in Bénard Layers*

MANUEL G. VELARDE

Instituto Pluridisciplinar
Universidad Complutense
Madrid, Spain

ALEXANDER A. NEPOMNYASHCHY

Instituto Pluridisciplinar
Universidad Complutense
Madrid, Spain
and
Department of Mathematics
Technion-Israel Institute of Technology
Haifa, Israel

and

MARCEL HENNENBERG

Instituto Pluridisciplinar
Universidad Complutense
Madrid, Spain
and
Microgravity Research Center
Faculte des Sciences Appliquées
Université Libre de Bruxelles
Brussels, Belgium

*Dedicated to Professor Milton Van Dyke and to the memory of Professor David G. Crighton.

ADVANCES IN APPLIED MECHANICS, VOL. 37
ISBN 0-12-002037-8
ISSN 0065-2165/01 $35.00

I. Introduction

In Bénard's experiments (1900, 1901) with shallow liquid layers (0.5- to 1-mm depth for a horizontal extension of 20 cm) uniformly heated from below, steady convective hexagonal cells appeared with a spacing of 3.27 or more times the liquid depth (Figure 1). Inspired by Bénard's findings, Lord Rayleigh's theoretical analysis (1916) of buoyancy-driven natural convection predicted that if hexagonal cells were to form, the ratio of their spacing to the layer depth was about 3.28. The agreement was impressive yet we now know it was fortuitous. Lord Rayleigh's masterly theory — although providing a genuine and seminal discovery — had little to do with the surface tension gradient-driven convection (Marangoni effect; see, e.g., Scriven and Sternling, 1960) which was the essence of Bénard's convection. Bénard noted the role of surface tension but did not properly elaborate on it, perhaps overwhelmed by the simplicity and clearcut predictive power of Rayleigh's theory.

Around the same time, Dauzère (1908, 1912) showed that a monolayer of stearic acid was capable of inhibiting the onset of convection when poured over thin layers of melted wax. Low (1929) demonstrated that solid silicone monolayers prevented the occurrence of Bénard cells in drying paint films. On the other hand, the Langmuirs (1927) had noted that monolayers of several partially soluble surfactants, i.e., surface-active substances like oleic acid and cetyl alcohol, inhibited the convection that normally accompanies the evaporation of aqueous ether solutions.

The role of surface traction was dramatically shown in the 1950s with Block's (1956) qualitative experiments on heating or cooling a shallow liquid layer from below. In a very sketchy note he described having observed Bénard cells in exceedingly shallow layers and found no indication of a critical depth (he went down to 50 μm = 5×10^{-2} mm). He also confirmed what Bénard and also Hershey (1939) had noted: When heating a layer from below, the colder point at the surface were the higher. This was theoretically explained later by Scriven and Sternling (1964) whose rheological model of

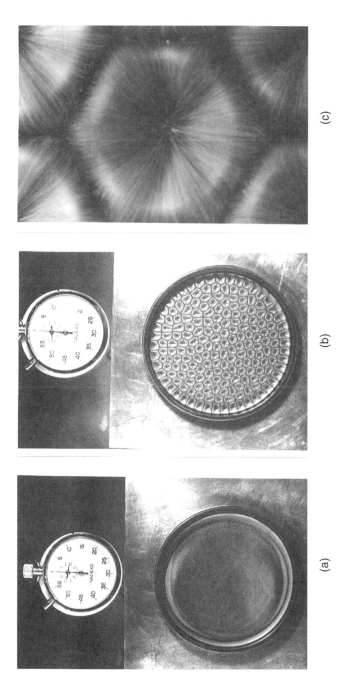

(a)

(b)

(c)

FIG. 1. Top view of Bénard patterns in a silicone oil (viscosity 1 Stokes) layer 1 mm deep: (a) Reaching the threshold. (b) The steady hexagonal structure with plenty of defects. (c) Enlarged view taken with exposure time of 10 s, whereas liquid moves across the cell from the center to the edge in 2 s (flow velocity, 7.5×10^{-2} cm/s; temperature difference, 7°C).

169

the interface leads "to upflow beneath depression and to downflow beneath elevation of the free surface." As this characterizes the flow in the liquid, flow along the surface is from the trough toward the crest so that the trough is at a lower surface tension and thus at a higher temperature than the crest. When cooling the layer from below, an experiment that Block did for a liquid depth of 0.8 mm, he observed convective motions but gave no specification of steady or time-dependent motions, as he only said "cellular patterns were observed, the more regular patterns having the appearance of Bénard cells." He and other authors also confirmed the inhibitory action of relatively insoluble surfactants (Tanford, 1989). In the 1960s, Koschmieder (1966, 1967; for a comprehensive account of his work, see Koschmieder, 1993) took up Bénard convection and definitively established the different role played by surface tension tractions and buoyancy in triggering instability and sustaining thermoconvection. Thus, with an appropriate choice of boundaries and geometries, he experimentally studied the two variants of Bénard convection, e.g., the surface tension gradient- and buoyancy-driven called, respectively, in the modern jargon, *Bénard–Marangoni* and *Rayleigh–Bénard instabilities* (Busse, 1978; Chandrasekhar, 1961; Drazin and Reid, 1981; Koschmieder, 1993; Normand *et al.*, 1977).

Pearson (1958) was the first author to address in a proper theoretical way the question of surface tension forces in Bénard convection and to provide a sound yet simple theory. Many authors have attempted generalizations of his analysis. Unfortunately, some of the possible generalizations that lead to seemingly new results and hence possibly new physical effects are actually ill founded and are eventually *misleading* extensions of the earlier theory. A problem or a theory or the solution of a problem is not automatically more relevant or more general because more parameters are incorporated in the model or when the space of solutions or numerical results is enlarged. There is the question of whether or not the model is physically acceptable or relevant and well founded. Furthermore, results should not explicitly (or implicitly) contradict earlier stated assumptions of the model. With this perspective in mind, we comment in the present review on results found in some *generalizations* from Pearson's and Rayleigh's seminal papers with particular emphasis on oscillatory interfacial instability leading to waves in Bénard convection. We also provide a succinct account of recent findings about patterned convection, Bénard cells (see also Davis, 1987; Oron *et al.*, 1997; Ostrach, 1982).

This article is organized as follows. In Section II we recall the well-accepted results obtained by scientists working until the end of the 1960s.

Then Section III is devoted to a succinct account of the basic equations and approximations needed to study Bénard convection with heat or mass transfer and Marangoni stresses. In Section IV we discuss the possible generalizations of the earlier theories, and hence the various attempts to go beyond the Boussinesq approximation to the hydrodynamic equations. We comment on their successes and failures. In Section V a succint review is given of recent results obtained on steady patterned Bénard convection. In Section VI we provide heuristic arguments about oscillatory convection and wave phenomena expected in Bénard layers. Then in Section VII very recent theoretical findings about (nonlinear) waves together with a succinct account of experimental results, old and recent, are presented. Finally, in Section VIII we comment on the potential of the field of surface tension gradient-driven convection.

II. Summary of Results and Limitations of the Classical Theories

The pioneering work of Pearson (1958) dealt with the linear stability analysis of a liquid layer open to air and uniformly heated from the bottom, so that in the absence of convection one obtains a constant vertical temperature gradient, β, in the layer. Pearson's theory is valid also in the case of a constant gradient of surfactant concentration. In the simplest case of a system satisfying the Boussinesq–Oberbeck approximation (we shall come back to this question further below), the time evolution of small disturbances of velocity, $\mathbf{v}$, pressure, p, and temperature, T, is governed by the following dimensionless (linear) evolution and boundary value problem. The Navier–Stokes equations:

$$(1/\mathrm{Pr})\partial \mathbf{v}/\partial t = -\nabla p + \Delta \mathbf{v}, \tag{2.1}$$

the Fourier equation (or its counterpart, the Fick diffusion equation for mass transfer problems):

$$\partial T/\partial t - v_z = \Delta T, \tag{2.2}$$

the continuity equation for an incompressible liquid:

$$\nabla \cdot \mathbf{v} = 0, \tag{2.3}$$

the mechanical and thermal conditions at the solid support located at $z = -1$:

$$\mathbf{v} = 0 \quad \text{and} \quad T = 0, \tag{2.4}$$

and at the open surface of the liquid (or the corresponding conditions at the interface between two fluids) located at $z = 0$:

$$\partial v_x/\partial z + \partial v_z/\partial x = \text{Ma } \partial T/\partial x, \qquad (2.5)$$

$$\partial v_y/\partial z + \partial v_z/\partial y = \text{Ma } \partial T/\partial y, \qquad (2.6)$$

$$v_z = 0, \qquad (2.7)$$

$$\partial T/\partial z = -\text{Bi } T, \qquad (2.8)$$

where ∇ denotes the gradient operator, $\Delta = \nabla^2$, $\text{Pr} = \nu/\kappa$ is the Prandtl number, with ν being the kinematic viscosity and κ the thermal diffusivity (for mass transfer, the corresponding group is the Schmidt number, $\text{Sc} = \nu/D$, with D being the mass diffusion coefficient), and Bi is the Biot number, which characterizes the heat transfer at the air–liquid interface. An air–liquid interface is well approximated by a vanishing Biot number. The Marangoni number, Ma, is the ratio of surface tension forces to the two dissipative forces due to viscosity and heat. Each of these agents has a characteristic time and their combination is reflected in the mathematical definition, which is $\text{Ma} = -(\partial \sigma/\partial T)\beta d^2/\eta\kappa$, where σ is the surface tension and $\partial \sigma/\partial T$ is the temperature variation of the surface tension; β is taken, in what follows, to be positive when the heating is from the liquid, i.e., from the support side; d is the liquid layer thickness; and η is the shear or dynamic viscosity ($\eta = \rho\nu$ with ρ the density). For most liquids, the surface tension σ decreases when increasing the temperature and it has the same behavior when increasing the concentration of surfactant. There are noticeable exceptions such as water–(high) alcohol solutions, dodecylammonium (DAC), some binary mixtures, some liquid crystals, and some other liquid mixtures that have a minimum in their variation for surface tension with temperature. Hence, Ma is a suitable dimensionless measure of the temperature gradient across the liquid layer, and it is the parameter defining the driving force for interfacial thermocapillary convection. It appears as a Peclet number in the heat equations when defining the Reynolds number using the Marangoni-driven surface velocity, $v^* = (\partial \sigma/\partial T)\beta d/\eta$; hence, $\text{Re} = v^*d/\nu$ and $\text{Pe} = \text{Re}\,\text{Pr} = \text{Ma}$.

According to Pearson's theory, a liquid layer open to *passive* air ($\eta_{\text{air}} \ll \eta_{\text{liquid}}$) is unstable to a well-defined short-wave planform of steady cellular convection for a *critical* value of the Marangoni number when the heating is from the liquid side.

Since eqs. (2.1)–(2.3) and the boundary conditions are invariant to translations in time and horizontal spatial coordinates, x and y, the solutions of the linear boundary value problem, (2.1)–(2.8), can be written as a superposition of (Fourier) *normal modes* with exponential dependence on t, x, and y:

$$(\mathbf{v}, p, T) = (v(z),\ p(z),\ T(z))\ \exp(i\mathbf{k}_1 \cdot \mathbf{x}_1 + \lambda t), \qquad (2.9)$$

$$\mathbf{x}_1 = (x, y). \qquad (2.10)$$

The quantity λ is, generally, complex and its real part, λ_r, gives the (dimensionless) growth rate of eventually unstable modes of corresponding wave vector, $\mathbf{k}_1$. Neutral modes ($\lambda_r = 0$) may exist as monotonic (Im $\lambda = 0$, steady) or oscillatory (Im $\lambda \neq 0$, wave) disturbances. In the latter case the imaginary part of λ provides the corresponding frequency, and hence the dispersion relation for a given k (the subscript and boldface characters are only used when strictly needed to avoid confusion). In a naive approach, overstability or oscillatory convection may be pictured physically as one in which flow proceeds for a time, comes to a halt, reverses itself, and so on indefinitely. If insoluble surfactant molecules were present at the surface of the liquid, they would be alternatively swept together and swept apart by the underlying liquid, thereby generating forces counter to the flow. We shall see that surfactants eventually play a more subtle role and that overstability can develop in seemingly complex albeit periodic flows as nonlinear waves.

Pearson developed his model problem with these given assumptions: (1) No buoyancy was taken into account, (2) the open surface was deformable, and (3) the analysis was restricted to steady instability, i.e., no oscillatory or otherwise time-dependent disturbances were studied ($\lambda = 0$). The main result of Pearson's theory is the following analytical expression for the neutral curve Ma $=$ Ma$_0(k)$ corresponding to the condition $\lambda(\text{Ma}_0, k) = 0$:

$$\text{Ma}_0(k) = [8k(k \cosh k + \text{Bi} \sinh k)(\sinh k \cosh k - k)]/(\sinh^3 k - k^3 \cosh k).$$

$$(2.11)$$

Because there is no privileged direction in the model other than the one imposed by the gradient, Pearson's result demands only that the heating, normal to the free surface, be done from the liquid side. Pearson obtained one single finite critical Marangoni number at a finite wavelength. The immediate extension of Pearson's model is to consider that both the free undeformable surface and the heating rigid wall are poor heat conductors, i.e., by introducing another Biot number in eq. (2.4) and having it vanish. Then the theoretical prediction is that the critical wavenumber drops to zero. Koschmieder (1967) and Koschmieder and Biggerstaff (1986) have neatly confirmed these predictions.

Nield (1964) extended Pearson's theory by adding buoyancy to eq. (2.1) and hence providing the first linear stability analysis of a liquid layer submitted to the coupled effects of surface tension stresses and buoyancy, the Bénard–Rayleigh–Marangoni problem, but still with an undeformable open surface. He also restricted his analysis to steady cellular convective instability and showed how instability was enhanced due to the combined action of buoyancy and surface tension tractions. The Rayleigh number, $Ra = \alpha\beta gd^4/\nu\kappa$, with α being the volume thermal expansion coefficient and g the acceleration of gravity, is the ratio of the three characteristic timescales involved in the problem as it is the ratio of buoyancy to the dissipative viscous and heat "forces." It is also a (dimensionless) measure of the thermal gradient across the layer. Note that water has vanishing α at $4°C$ and 1 atm, hence a negative value between freezing and 4 degrees. Because the Rayleigh and Marangoni numbers have in their definition the layer depth to different powers, an experimental check could be and was obtained concerning the relevance of buoyancy to surface tension stresses in a problem (Berg *et al.*, 1966a,b; Koschmieder, 1988). Imaishi and Fujinawa (1974a,b) extended the study of Nield to a two-layer system, with either sign of the Marangoni and Rayleigh numbers, assuming undeformable surfaces and ignoring oscillatory disturbances. Their results emphasized the already known stabilizing role of surface contamination. Reference to related works by other authors is provided further below.

For the mathematically equivalent problem of solute transport, e.g., mass diffusion of a surfactant, in the case of two semi-infinite phases and no buoyancy, Sternling and Scriven (1959) have shown that oscillatory instability is possible for gradients of the opposite sense if due account is taken of the dynamics of both the upper and lower phases as for the case of an interface between two liquids with transport from either side. This and other predictions made by Sternling and Scriven have been verified by Linde *et al.* (1964, 1979) for heat and mass transfer in liquid layers open to air and in two-liquid-layer systems with an interface. In particular for an air–liquid interface, oscillatory instability with practically planar waves appeared only when heating from the air side. This complies with Sternling and Scriven's (1959) *necessary* conditions for instability, $D_{from} > D_{to}$ and $\nu_{from} > \nu_{to}$, in the short-wavelength limit with semi-infinite layers. However, their mathematical model neglected the possible difference between the incoming and outgoing fluxes along the interface, which is a major feature for partially soluble surfactants. The theory of Sternling and Scriven did not incorporate adsorption/desorption processes or surface accumulation, a problem which together with surface deformation was taken up later on by Sanfeld and collaborators (1979) and several other authors to whom we refer below.

Subsequently, for the simpler case of a liquid layer heated from the solid support at a finite distance from the free surface that was open to passive air, Scriven and Sternling (1964) allowed the surface to deform. Their model included capillarity (Laplace overpressure) but not gravity, hence no hydrostatic contribution at the deformable interface. They introduced the capillary number, $Ca = \eta\kappa/\sigma d$ (also called the crispation number), as a suitable dimensionless measure of the surface deformation: large values of Ca (or low values of surface tension) allow a large surface deformation (usually, $Ca \sim 10^{-5}-10^{-7}$). The absence of gravity was shown to yield no threshold for long wavelength instability, i.e., at vanishing wavenumber.

Smith (1966) extended Pearson's (1958) and Scriven and Sternling's (1964) theories by incorporating gravity in the boundary conditions describing a deformable interface, between two adjacent phases with a given finite depth and bounded by perfectly conducting surfaces. Taking into account the hydrostatic pressure, he introduced the *static* Bond number, $Bo = (\Delta\rho)gd^2/\sigma$, earlier called the Eötvös number. This Bond number is the ratio of the displaced weight to the surface tension forces, hence it is of relevance when the level of the gravity acceleration g varies as in experiments in parabolic flights or aboard space ships. The term $\Delta\rho$ accounts for a density difference as in the case of a liquid–liquid interface, while for the case of a liquid open to air, it can be approximated by ρ, the liquid density. However, Smith did not consider the influence of buoyancy in the volume. Nor did he consider the possibility of oscillatory disturbances. He reobtained, in the appropriate limiting cases, the results of the previous authors and, furthermore, proved that monotonic instability was possible for gradients of opposite signs, providing estimates of thresholds for various significant cases.

Smith (1966) predicted the possibility of *high* and *low* wavenumber modes of steady convection. He showed that for positive Marangoni numbers two minima appear, i.e., two unstable modes exist depending on the value of the capillary number (Figure 2). However, for negative Marangoni numbers he obtained only one instability mode. From Smith's neutral stability curves, it is clear that the *long*-wave instability corresponds to a critical Marangoni

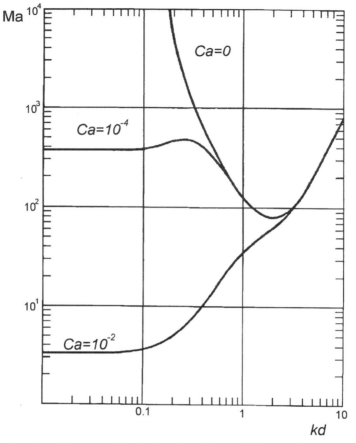

FIG. 2. Neutral stability curves (Ma, kd) in a liquid layer with deformable open surface and hence varying capillary number, Ca = 0, 10^{-4}, and 10^{-2}. (Redrawn after Smith, 1966, with permission of Cambridge University Press.)

number proportional to the Galileo number, $Ga = Bo/Ca = gd^3/\nu\kappa$, when *surface deformation* is important (at variance to Pearson's result for an *undeformable surface* between perfectly insulating boundaries). For instance, in the limit $Bi \to 0$,

$$Ma_c = (2/3)Ga. \tag{2.12}$$

For the *short*-wave instability, the critical Marangoni number is almost independent of the Galileo number. Thus Smith was the first author to predict the possible coexistence or competition between two monotonically unstable modes at threshold. If the Smith (1966) model is suitably generalized to incorporate a new element like the finite horizontal extent of the layer capable of stabilizing the liquid layer, i.e., raising the Marangoni number at zero wavenumber, a second minimum in the neutral curve would appear at a nonvanishing, different and lower wavenumber than Pearson's value, 2. Long-wave instability has recently been observed (Van Hook *et al.*, 1995, 1997) in experiments with very low values of the Galileo number. They are very difficult to reach on earth, since they demand that the *dynamic* Bond number, Ga/Ma, be lower than 1.5.

Berg and Acrivos (1965; for a comprehensive study see Berg *et al.*, 1966a,b) extended Pearson's analysis to account for the role of insoluble surfactants, and in agreement with earlier results showed that even trace amounts of such materials would exert a highly stabilizing influence on the monotonic instability of the liquid layer. The stabilization depended primarily on the slowness with which concentration variations of the insoluble monolayer can be diminished by surface diffusion. The sustained nonuniformity in surface monolayer concentration maintains surface tension gradients opposing the destabilizing disturbing forces.

Brian (1971) was, to our knowledge, the first to extend the Bénard–Marangoni problem to the isothermal transfer of a *soluble* surfactant from the rigid wall, taking into account the steady surface excess concentration, Γ, although not the deformation of the open surface. He numerically showed that the Marangoni number was much higher than when Γ vanishes, hence illustrating the stabilizing action of a very small concentration of contaminant. He was able to find the critical Γ beyond which the monotonic Marangoni-driven instability disappears. Also, since the dimensionless surface elasticity has to be positive, monotonic instability cannot exist for surfactant transfer from the gas phase. Brian's results were analytically obtained by Chu and Chen (1987) and by Kozhoukharova and Slavchev (1992). From a practical engineering purpose, the models of both Pearson and Brian were too simple to be used to estimate the enhancement of mass transfer due to convection. Imaishi *et al.* (1983) assumed a linear concentration gradient very near to the interface and, moreover, also assumed that beneath that region, whose width was linked to the unperturbed Fick equation, the liquid layer has a constant concentration. They found then quite good agreement between experiment and theory, even if they showed the linear theory to be insufficient to explain all their findings.

Palmer and Berg (1972) extended the study made by Berg and Acrivos (1965). They heated from below a binary mixture that contained a dilute, nonvolatile surface active solute. Their liquid density changed with tempera-

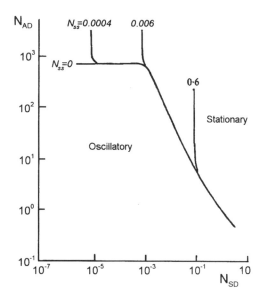

FIG. 3. The boundaries between the regions of oscillatory and monotonic (stationary) instabilities for a surfactant solution heated from below. Parameters: Adsorption number, $N_{AD} = D_b/k_1d$; surface diffusion number, $N_{SD} = k_1D_S/k_{-1}D_bd$; surface Schmidt number, $N_{SS} = k_1v/k_{-1}D_bd$; D_b is the bulk diffusion coefficient, D_S is the surface diffusion coefficient, k_1 is the adsorption rate, and k_{-1} is the desorption rate. Pr = 10, Sc = 10^3, and Bi = 0. (Redrawn after Palmer and Berg, 1972, with permission of Cambridge University Press.)

ture only, but surface tension depended on both temperature and surfactant concentration. Their linear stability analysis showed that the earlier accepted stabilizing effect of a surfactant, even for an undeformable surface, drastically depended on the interfacial properties, surfactant concentration in the volume, and the process and rate of surfactant transfer between the bulk and the surface, the value of the surfactant surface concentration being the strongest stabilizing element. A new type of oscillatory instability was discovered caused by the competition of tangential forces originated by the temperature and surfactant fields across the interface (Figure 3).

In view of the results found in these classical studies, one expects that when surface deformation and buoyancy are simultaneously incorporated in the model a more complete theory will emerge. However, between the *possible* and the *actual* generalization we shall see that life is not easy. In the next section we show that the influence of buoyancy is usually negligible, at least to a quantitative level, when the surface deformation is important for Bénard layers.

III. The Boussinesq Approximation and Surface Deformation

Most of the stability analyses of Bénard convection, which have been done taking into account simultaneously buoyancy (Rayleigh–Bénard convection) and surface tension gradients (Bénard–Marangoni convection), use the Boussinesq approximation (the name of Oberbeck should also be added to define this approximation but in view of the standard use in the literature, and for simplicity, here we shall refer to Boussinesq only; see, e.g., Mihaljan, 1962; Pérez-Cordón and Velarde, 1975; Velarde and Pérez-Cordón, 1976; de Boer, 1984, 1986; Zeytounian, 1990), either for a Newtonian liquid layer open to an inert gaseous phase or for two Newtonian liquid layers separated by an interface. The Boussinesq approximation corresponds to the vanishingly small value of $\varepsilon = \alpha\beta d \ll 1$, which is a measure of α and of the temperature difference, βd, and hence both must be small enough, ruling out compressibility and property variation effects with temperature. The layer thickness, d, must also be small enough to avoid the influence of the liquid compressibility through the hydrostatic pressure. Then a recipe for a quick check of the validity of the Boussinesq approximation is Ra ~ 1, Ga $=$ Ra/$\varepsilon \gg 1$, and thus Ra $\ll$ Ga.

If the open surface deformation of an otherwise incompressible liquid is measured by a dimensionless quantity, ζ, then disregarding surface tension we have

$$\zeta \approx (\varepsilon/\text{Ra})(\text{Pr}^{-1}p - 2\partial w/\partial z), \tag{3.1}$$

where p denotes the pressure jump across the interface, and z and w refer to vertical coordinate and the vertical component of the velocity, respectively. Thus, in the Boussinesq approximation we have a vanishing ζ with vanishing small ε, while the value of the constraint Ra remains finite ($\varepsilon \to 0$, Ra $=$ const.). Accordingly, if interfacial deformation is added to the model, we have to be careful about neglecting other non-Boussinesq corrections as long ago emphasized by Davis and Segel (1968) and by Scanlon and Segel (1967). For instance, continuity equation (2.3) becomes

$$\text{div}\, v = \rho^{-1}(\partial\rho/\partial t + v\,\text{grad}\,\rho) \approx -\alpha(\partial T/\partial t + v\,\text{grad}\,T) \sim \varepsilon. \tag{3.2}$$

Besides compressibility and surface deformation, non-Boussinesq effects are also a temperature-dependent viscosity, a nonlinear temperature profile, or, equivalently, a temperature-dependent heat diffusivity, the anomalous thermal expansion behavior of water around 4°C, etc. (Normand *et al.*, 1977). These effects, even when they are quantitatively small, can be qualitatively important. For instance, it was established by Hoard *et al.* (1970) that a temperature-dependent kinematic viscosity is responsible for the appearance of hexagonal cells rather than rolls in (Rayleigh–Bénard) buoyancy-driven convection, thus confirming a prediction earlier made by Palm (1960) and thoroughly studied by Busse (1967, 1978). In a particular experiment some of these corrections may quantitatively be negligible or less important than others.

To make the presentation complete, let us discuss a typical difficulty with the Boussinesq approximation that appeared in the analysis done by Izakson (1969; see also Izakson and Yudovich, 1968). He considered a Bénard layer uniformly heated from below and open to air with a deformable interface. Although he did not consider oscillatory instability and the Marangoni effect was not taken into account, Izakson's paper is worth considering in order to neatly clarify the contradiction between assumptions made at the beginning of a paper and the conclusions reached at the end, leading to the prediction of a spurious monotonic long-wave instability. Indeed, assuming the Boussinesq approximation, a convective instability was found for a critical Rayleigh number, $\mathrm{Ra}_c = (40/11)\,\mathrm{Ga}$, corresponding to a vanishing wavenumber. Thus at threshold, $\varepsilon = 40/11$, and, consequently, there is an obvious violation of the earlier *explicitly* assumed Boussinesq approximation ($\varepsilon \ll 1$). This intrinsic contradiction was noticed by Rasenat *et al.* (1989). The deformability of the surface was found to have a destabilizing effect that was maximum when there is no surface tension (diverging static Bond number), thus illustrating the crucial role played by the Galileo number. Further details can be found in Gershuni and Zhukhovitsky (1976, pp. 46–51).

One could expect that on considering Izakson's problem with no restriction to the Boussinesq approximation, the predicted long-wavelength instability would still exist with just a different and, possibly, more accurate value for the critical Rayleigh number due to the other neglected non-Boussinesq effects. Nepomnyashchy (1983) investigated this issue using eq. (3.2) as the continuity equation, putting a temperature-dependent viscosity in the Navier–Stokes equations (2.1) and a temperature-dependent heat diffusivity in Fourier's equation (2.2). He found an explicit expression for the growth rate, λ_r, of the surface deformation in the limit of vanishing wavenumber, k:

$$\lambda_r = -k^2 \left[\int_{-d}^{0} \rho(z)a(z)\ dz \Big/ \int_{-d}^{0} \rho(z)\ dz \right], \qquad (3.3)$$

where $\rho(z)$ is the actual equilibrium density distribution in the liquid layer, $-d < z < 0$; and $a(z)$ is a positive function accounting for the equilibrium stratification and shear viscosity. Then eq. (3.3) shows that no long-wave monotonic instability is possible because the growth rate of the predicted instability is always negative if the density is positive everywhere in the liquid layer, as it must be. If a linear density variation is assumed in order to satisfy the Boussinesq approximation, $\rho(z) = \rho_0(1 + \alpha\beta z)$, then Izakson's instability is recovered with the density *negative* in the lower part of the liquid layer, which contradicts the assumptions made. We have here a case where a new qualitative feature, a new instability, is a mere artifact of the improper use and ill-founded generalization of the Boussinesq approximation.

Another case, also with contradictory results, is the study by Gouesbet *et al.* (1990). It must first be said that these authors have provided quite a useful computer package to easily get instability thresholds. With the particular model problem they solved in the Boussinesq approximation with buoyancy and surface deformation and allowing for overstability, they aimed at generalizing the studies by Smith (1966) and by Takashima (1981a,b) whose work we discuss below. Unfortunately, Gouesbet *et al.* fell short of this aim because a careful analysis of their results shows that a predicted influence of the Rayleigh number on the critical Marangoni number is valid when Ga is of the order of the Rayleigh number or less, hence violating the initially assumed Boussinesq approximation. The works of Perez-Garcia and Carneiro (1991) and Wilson and Thess (1997) suffer from the same contradiction. Their results systematically violate the earlier assumed Boussinesq approximation. For instance, the prediction by Perez-Garcia and Carneiro of the possible coexistence or competition of two steady modes with different wavenumbers and with a common critical Rayleigh number occurs for Ra greater or equal to Ga. Also their prediction that when the surface deformation is relevant the instability threshold is at zero wavenumber is indeed the long-wavelength instability expected at a critical Marangoni number that practically does not depend on the Rayleigh number. A further numerical prediction about the possible coexistence or competition between two different oscillatory modes also violates the Boussinesq approximation since it occurs only for $Ra > 10^2 > Ga$. Consequently, their apparently fashionable and appealing analogies with other problems are therefore meaningless. Although they did not elaborate on this, it turns out that some of their results are relevant to microgravity conditions when buoyancy is negligible (see García-Ybarra and Velarde, 1987).

The same devilish trap appears in the paper of Char and Chiang (1996) who considered double diffusive convective instability. In the absence of buoyancy, they found the long wavenumber mode and the finite wave-number mode already derived by Smith (1966). When buoyancy was added, they got results that numerically sound very promising. But the physical interpretation is doubtful, in region of finite surface deformation, since the Rayleigh number is clearly higher than the Galileo number. The same problem arises with most of the results found by Chang and Chiang (1998). The Boussinesq approximation is not a "holy grail" and, in fact, is far from it. But if a model problem falls in its realm one has to be consistent, carefully cross-checking the domain of parameter values where the results obtained are expected to be acceptable.

A rather subtle difficulty of the Boussinesq approximation appears in a series of studies by Depassier and collaborators (Depassier, 1984; Benguria and Depassier, 1987, 1989). They considered (Rayleigh–Bénard) buoyancy-driven convection with a stress-free support and open deformable surface. Their results predicted a new instability, not a limiting case of earlier discoveries. Besides the fact that they considered different boundary temperatures, the main point of their 1987 paper is their prediction, within the range of validity of the Boussinesq approximation, of a zero wavenumber oscillatory instability at a critical Rayleigh number $Ra_c = 30$. This value is much lower than the earlier known critical values for steady convection, hence its interest. No experimenter has seen such an instability although many have done quite disparate experiments with a variety of liquids and boundary condition. Thus, the question arises about the validity of their results. From their own data (e.g., Benguria and Depassier, 1987, Fig. 5, p. 1681), the growth rate for the instability is

$$\lambda_r = a(Ra - Ra_c)k^2 - b\,Ga\,k^4, \qquad (3.4)$$

where a and b are some constants, albeit Prandtl number dependent. Formally, the condition $Ra \ll Ga$ is not violated for large Ga. However, the neglected contribution of the non-Boussinesq term in eq. (3.4) is of the same order of magnitude as the terms that are taken into account, hence the prediction of instability for $Ra_c = 30$ is a spurious result. The further model extension made by Kraenkel *et al.* (1992a,b), however interesting from the mathematical viewpoint, lacks physical relevance. Indeed, Porubov and Samsonov (1995) considered this very problem and took into account the non-Boussinesq corrections, thus making use of the methodology earlier used by Nepomnyashchy (1983). They found that oscillatory instability never appears when heating from below in a buoyancy-driven liquid layer with open deformable surface. As mentioned earlier and further discussed below, a long-wavelength oscillatory instability does indeed exist in the realm of the Boussinesq approximation but for surface tension (Bénard–Marangoni) gradient-driven convection (Levchenko and Chernyakov, 1981; Takashima, 1981b; García-Ybarra and Velarde, 1987). The formulas described by Depassier (1984; Benguria and Depassier, 1987, 1989), by Kraenkel *et al.* (1992a,b), and by Garazo and Velarde (1991) for the case of Rayleigh–Marangoni–Bénard convection are only correct in the absence of buoyancy, hence for Ra = 0. As already pointed out, neglecting non-Boussinesq corrections may on occasion not be so dramatic because one specific non-Boussinesq effect may dominate the others in quantitative value.

IV. Appropriate Generalizations of the Classical Theories

Many appropriate attempts to generalize the models studied by Sternling and Scriven (1959), no buoyancy, no surface deformation; by Scriven and Sternling (1964), no buoyancy and even no gravity at the deformable interface; by Nield (1964), finite liquid layer with buoyancy open to a semi-infinite air layer with the further neglect of surface deformation; and by Smith (1966), no buoyancy but gravity at the deformable interface, have been made over the years. We can distinguish two main lines of thought according to whether the emphasis is on buoyancy or on surface tension effects. The work of Zeren and Reynolds (1972) is a good example of the latter. They considered, for both ways of heat or mass transfer across the interface (and no surface adsorption) a finite two-layer problem incorporating gravity in both buoyancy (inside the fluid) and surface deformation. They restricted consideration to steady disturbances only and, consequently, they did not find instability to (transverse or longitudinal) interfacial wave motions of which we shall say more below. To satisfy Sternling and Scriven's (1959) necessary conditions for instability, with suitable ratios of kinematic viscosities and diffusivities, they made the choice of a two-layer system with benzene over water, which are immiscible liquids with nearby densities $(1 - \rho_u/\rho_l \approx 0.1)$.

As noted earlier, with interface deformation and buoyancy simultaneously incorporated, care must be taken in order to list only meaningful results, i.e., results consistent with the imposed Boussinesq approximation. Indeed, for steady convection the results of Zeren and Reynolds (1972) are undoubtedly correct in the short-wave region where the interfacial deformation is of minor importance. Their numerical results based on a realistic system did give in the long-wave region the expected instability. They found the relationship

$$\text{Ma} = C_1 \, \text{Ga}(1 - \rho_u/\rho_l) + C_2 \, \text{Ra}, \tag{4.1}$$

where the C_i ($i = 1, 2$) depend on the ratios of physical parameters and the thickness of each layer, and ρ_u and ρ_l are the densities of the upper and lower layer, respectively. The first term of the right-hand side (RHS) of eq. (4.1) corresponds to the deformation of the boundary and the second term accounts for buoyancy. For the cases considered by Zeren and Reynolds, with an experimental Galileo number about 5×10^5, the Boussinesq approximation, Ra $\ll$ Ga, is never violated. One could expect that the influence of buoyancy is negligible. Actually, in some cases, the influence of the first term of the RHS of eq. (4.1) can be reduced due to the small value of C_1 $(1 - \rho_u/\rho_l)$. Hence from a physical point of view, for small but finite ε, including the second term, C_2 Ra, is acceptable, though in the limit $\varepsilon \to 0$ is asymptotically smaller than the first one. The work of Zeren and Reynolds is an example of a successful investigation of the coupling between buoyancy and deformation without violation of the basic assumption of the model. Some typical results of this work are shown in Figures 4 and 5.

Although, for both ways of transport, instability is predicted by theory, Zeren and Reynolds (1972) confirmed experimentally the existence of monotonic instability when heating from below but they did not find any instability when heating from above. Contamination of the interface was invoked as a possible inhibitory mechanism that strongly stabilized the interface. Relative to Smith's (1966) findings, these authors established that adding buoyancy, a stable density gradient plays a stabilizing effect on moderate and short-wavelength disturbances, with apparently very little buoyancy influence on the instability threshold for long-wavelength convection.

There is one more interesting quantitative aspect, however, which should not pass unnoticed. Zeren and Reynolds (1972) considered immiscible liquids, where not only the densities are close but also the viscosities, the heat conductivities, and the heat diffusivities. When heating from below and for a particular range of layer depth ratios, the heavier water layer was rather passive, leading to no buoyancy-driven convection. At the common interface, the slightly lighter benzene layer is submitted to the Rayleigh–Bénard instability that pushes surface fluids toward the hot spot. However, since the free surface is underlying the benzene, the surface tension as well

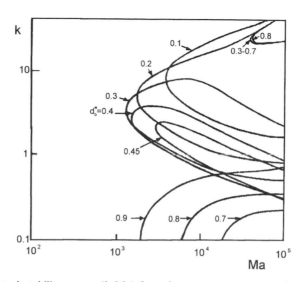

FIG. 4. Neutral stability curves (k, Ma) for a benzene–water system when heating from above. Total depth, 2 mm. The Marangoni number, Ma, is defined using the parameters of the lower liquid and the total depth. The dimensionless wavenumber, k, is defined using the total depth; d_b^* is the ratio of the lower layer depth and the total depth. (Redrawn after Zeren and Reynolds, 1972, with permission of Cambridge University Press.)

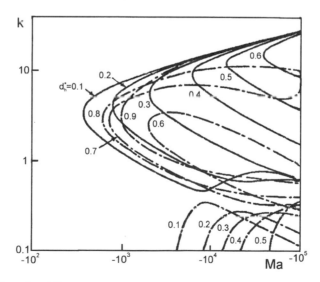

FIG. 5. Neutral stability curves (k, Ma) for the benzene–water system when heating from below. Total depth, 2 mm. Parameters and definitions as in Fig. 4. (Redrawn after Zeren and Reynolds, 1972, with permission of Cambridge University Press.)

as the buoyancy of the lower water phase tend to pull away liquid from those hot spots, thus leading, in the steady state, to two counter-rotating cells in the benzene while there is a single cell in the water (mechanical coupling). As the middle cell is entirely due to the exchange of shear stress at the interface, it is somehow crashed between the two external bulk cells, which are due to buoyancy and which keep co-rotating in the same direction (thermal coupling).

Years later several authors predicted oscillatory instabilities for Rayleigh–Bénard convection with no Marangoni effect but with some interfacial phenomena incorporated. Gershuni and Zhukhovitsky (1982) proved that in buoyancy-driven convection the presence of an interface within the layer and thus a transverse gradient imposed on a two-immiscible liquid layer brings the possibility of overstable motions even if the interface is not deformable and there is no Marangoni traction. It suffices to heat from below appropriate liquids. Because they considered as lower liquid the denser of the two, a Rayleigh–Taylor instability was excluded. A sound physical explanation of the newly discovered instability was given. One of the two layers will become unstable and induce the steady motion of the two liquids. This means that we have to consider for each set of the physical parameters, two loci of Ra as a function of the wavenumber. A possible intersection of these two curves is meaningless, as a steady instability, so that the two-layer system responds by becoming oscillatory.

Rasenat *et al.* (1989) also studied the onset of instability of finite two-liquid layers of almost equal densities. Their analysis extended those of Zeren and Reynolds (1972) and of Gershuni and Zhukhovitsky (1982). They started neglecting the Marangoni effect that appeared in the tangential shear stress balance at the interface. They did it precisely because the surface distortion was multiplied by $Ga(1 - \rho_u/\rho_l)$, which is a very large quantity in the Boussinesq approximation. When the Rayleigh numbers of both layers are very different from one another, the onset of instability in one of the layers drives the other, hence the appearance of two counter-rotating cells (mechanical coupling). When both Rayleigh numbers approach a common value, for about the same critical wavenumbers, the situation is more complex. They numerically obtained that the two-layer system could develop two kinds of steady cells. Either they had upward flow in the upper cell above uprising in the lower layer, or uprising in the upper cell above downflow in the lower layer for other values of the Rayleigh number. They experimentally verified their theoretical predictions and noticed that, past the instability threshold, the thermal coupling was, generally, getting dominant. They also showed numerically but did not experimentally observe that coupled oscillatory instabilities are also possible for appropriate values of the parameters, extending previous results of Gershuni and Zhukhovitsky (1982). They then included in their study the surface deformation still neglecting the surface tension. Thus, for lower values of $Ga(1 - \rho_u/\rho_l)$, which satisfied the Boussinesq approximation when $\varepsilon > (1 - \rho_u/\rho_l)$, Rasenat *et al.* predicted the possible existence of an oscillatory interfacial instability due to the density difference of the extreme part of each layer far away from the distorted interface. Indeed, in those extreme regions, there could be cases where the upper density would be higher than the lower one. A Rayleigh–Taylor mechanism may appear there although balanced by the stable density jump at the interface together with all other dissipative effects.

An extension of the problem treated by Rasenat *et al.* (1989) was studied by Renardy and Joseph (1985) and, subsequently, in a further generalization by Renardy and Renardy (1985) and Renardy (1986). Renardy and Joseph considered a finite two-liquid layer system with emphasis on the case of two liquids with almost equal properties. They had deformable interface but no Marangoni effect. Their approach is valid for density ratios $|1 - \rho_u/\rho_l| \sim \varepsilon$, and in the numerical case they treated $\varepsilon = 10^{-3}$ with $\rho_u/\rho_l = 1$. They proved that, for such a two-layer problem, oscillatory instabilities are possible due to the competition between the unstable thermal gradient and the stabilizing role played by the interface. Renardy and Renardy (1988) also studied nonlinear convective motions in a two-layer system.

As mentioned earlier, the experimental results of Rasenat *et al.* (1989) showed that the viscous coupling may yield to thermal coupling. When convection was strong enough, these results were also obtained numerically for very different viscosities in disagreement with the experimental findings. Trying to solve this contradiction, Cardin *et al.* (1991) studied, experimentally and numerically, the two-layer system taking into account the deformability of the surface and introducing the interfacial viscosities used by Sternling and Scriven (1959). They used silicone oil over glycerol, whose surface tension increases with increasing temperature, an anomalous case, hence *negative* Marangoni number when heating from the bottom. Their *ad hoc* assumption of surface viscosities, whose precise origin was not clarified by the authors, gave the desired results. In the absence of surface viscosities, their linear analysis was only given viscous coupling at the threshold while beyond a critical value, the surface viscosities were always giving rise to the thermal coupling, for similar bulk viscosities.

Wahal and Bose (1988) attempted an extension of the work by Zeren and Reynolds (1972) and by Renardy and Joseph (1985) by considering a finite two-liquid layer with the upper top boundary, open and deformable. They also incorporated the deformability of the interface between the two liquids and the action of Marangoni stresses hence acting at two opposite surfaces. Unfortunately, the consideration of the top boundary as a deformable open surface is incompatible with the Boussinesq condition, Ra/Ga ≪ 1. However, while not discussing the kind of coupling they observed, their conclusion about oscillatory instability is correct as the Galileo number of the bottom liquid is always very high (about 10^6) and the density ratio of the two liquids is about unity. The role of the lower Rayleigh number can be essential. In retrospect, the extension of the model to consideration of an upper top open deformable surface really adds nothing. Wahal and Bose described the known stabilizing role played by insoluble surfactants and recalled the results of Berg and Acrivos (1965). They also found that the action of the surfactant yields instability, with the most unstable disturbance changing from monotonic to oscillatory, as predicted by Palmer and Berg (1972).

Reichenbach and Linde (1981) built on the results produced by Sternling and Scriven (1959) for both ways of heat and/or mass transfer, though disregarding surface excess concentration and buoyancy in the bulk but not gravity at the interface. They considered both monotonic and oscillatory disturbances, for a two-layer system whose external boundaries were perfect heat conductors separated by a finite distance. They showed that, when lowering the surface tension, the threshold for instability fell down. On the other hand, for the case of a two-layer system with undeformable interface they reobtained that overstability was indeed possible, and hence instability as longitudinal or dilational waves, a result of the Sternling and Scriven (1959) theory that was experimentally confirmed by Linde *et al.* (1979). However, due to their neglect of interfacial deformation in the overstable case, they missed the instability threshold for transverse capillary-gravity waves found by Levchenko and Chernyakov (1981) and Takashima (1981b) and later on by García-Ybarra and Velarde (1987). To be noted are the much higher values of the critical Marangoni number for the onset of overstability and hence waves (10^5 and above) relative to the onset of patterned convection (Benard cells; 10^2 or less), which in terms of thermal gradients amount to a hundred degrees per cm in the former case and a few degrees per cm in the latter.

The problem of Reichenbach and Linde (1981) was also studied by Nepomnyashchy and Simanovsky (1983, 1986) with an undeformable interface, but including Marangoni stresses, for the cases of zero and nonzero Rayleigh numbers, respectively. More recently, the same problem has been taken up by Birikh and Rudakov (1996) who have considered the deformability of the two outer boundaries of a floating liquid layer in zero gravity. In the first case Nepomnyashchy and Simanovsky provided neutral curves with application to the system transformer oil on formic acid for which they estimated the influence of depth ratio in the two-layer system. For the case of a nonzero Rayleigh number they considered the system of water over silicone oil and recovered Reichenbach's and Linde's instability predictions. They also proved that with finite-depth liquid layers, an oscillatory instability is possible where the criteria of Sternling and Scriven, obtained in the small-wavelength limit, predict stability or at most steady convection. Again the problem of Bénard–Marangoni and Rayleigh–Bénard instabilities with an undeformable interface was taken up by Gilev *et al.* (1987). For a layer heated from below they provided various neutral curves for both monotonic and oscillatory instabilities (Figure 6). For a layer of water open to air, they recovered Nield's result about steady cellular instability due to the combined action of buoyancy and surface tension gradients. For the case of water–silicone oil, they showed that competition between buoyancy and Marangoni stresses led to convective instability mainly in the upper liquid. For the system transformer oil–formic acid, they predicted various possible cases of the influence of the Marangoni number on the onset of overstability. They also showed that when the two-liquid-layer system is heated from below, longitudinal oscillations are expected when there is competition between Marangoni and Rayleigh forces, provided the Rayleigh number of the upper liquid is higher than the Rayleigh number of the lower liquid. The role of surface deformation is yet to be assessed.

Progress in our understanding of the physics of oscillatory interfacial instabilities was made by Levchenko and Chernyakov (1981) who considered the stability of a liquid layer heated from the air side. In the limiting case of a deep liquid layer subjected to a prescribed constant heat flux from the air side, they obtained a general dispersion relation for surface waves. Thus they showed that for *high-frequency* motions, two kinds of excitable surface waves are possible: transverse, gravity-capillary waves and viscous, longtitudinal or dilational waves (Lamb, 1945; Lucassen, 1968). Threshold values were provided and the possible crossover from one mode of instability to the other was also predicted. Levchenko and Chernyakov observed that the instability growth rate is drastically enhanced near the points of a resonant mixing of both modes (Figure 7). We discuss this phenomenon in more detail in Section VI. The prediction about the onset of dilational waves

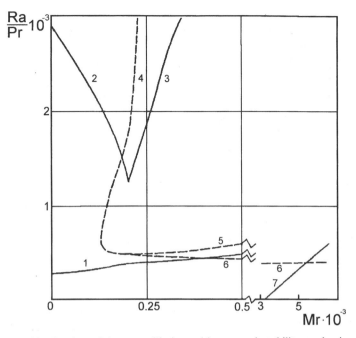

FIG. 6. Combined action of thermocapillarity and buoyancy instability mechanisms for the water–silicone oil (DC 200) system, $d_1/d_2 = 1.6$. The critical Grashof number, $Gr_1 = Ra_1/Pr_1$, is shown as a function of the parameter $Mr = (Ma_1/Pr_1)(\mu_1/\mu_2)$ for the lower monotonic neutral curve (lines 1 and 7), the upper monotonic neutral curve (lines 2 and 3), and the oscillatory neutral curve (line 6) [subscript 1 (2) corresponds to the upper (lower) liquid]. Lines 4 and 5 correspond to the long-wave and short-wave (codimension-2) points where the frequency of oscillations tends to zero. (Redrawn after Simanovsky and Nepomnyashchy, 1993.) Reprinted with permission of Gordon and Breach Publishers.

when heating the layer from the air side confirmed earlier findings made by Sternling and Scriven (1959) and at about the same time by Reichenbach and Linde (1981).

Simultaneous to the work of Levchenko and Chernyakov is the work of Takashima (1981a), who aimed at extending Smith's theory (1966). As a straightforward generalization of Pearson's analysis (1958), Takashima provided an extensive numerical exploration of Bénard convection with a free deformable open surface. He disregarded buoyancy because he restricted consideration to thin layers where buoyancy is not expected to be relevant. However, he did not disregard gravity because he, like some earlier authors, incorporated hydrostatic and capillarity contributions in the boundary conditions. Takashima showed that surface deformation is rel-

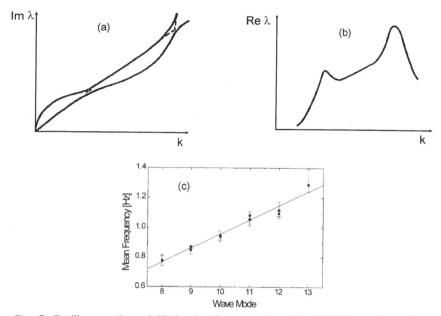

FIG. 7. Capillary-gravity and dilational surface waves in a Bénard liquid layer heated from the air side: (a) frequency, Im λ, and (b) instability growth rate, Re λ (redrawn after Levchenko and Chernyakov, 1981). (c) Experimental data on dilational waves (redrawn after Wierschem, Velarde, and Linde, unpublished).

evant only when Ga < 120, a result earlier found by Smith (1966). From this value it follows that surface deformation is relevant for a layer when its depth is less than $d^* = (120\nu\kappa/g)^{1/3}$, e.g., 0.012 cm for air–water, 0.018 cm for air–mercury, and 0.11 cm for air–glycerine, which are all depths of a millimeter or less. As mentioned earlier, the later prediction was verified experimentally only very recently (Van Hook *et al.*, 1995, 1997).

Takashima showed the marked destabilizing role played by the deformation of the interface and by the Galileo number, hence for a given capillary number by the *static* Bond number. He also predicted the possibility of stabilizing a hanging liquid layer with the open deformable surface in the underside, which corresponds to negative static Bond numbers, due to the combined action of gravity on the underlying, open deformable interface, and Marangoni stresses. In an experiment, because the liquid is overlying air which is lighter, the layer may first be unstable to Rayleigh–Taylor instability. Thus Takashima predicted the possible stabilization of a mechanically unstable layer by heating it from the solid support to which the liquid film adhered.

Takashima also found that on decreasing the surface tension, hence appreciably increasing the role of surface deformation, for high enough capillary numbers the convective planform should abruptly pass from a critical value $k_c = 2$ to a vanishing wavenumber, thus reobtaining a long-wave monotonic instability threshold, another of Smith's findings. Figures 1 and 3 of Takashima's (1981a) paper point toward the possible coexistence or competition of two steady modes of convection at very low wavenumber and wavenumber 2, respectively, and hence the coexistence of Pearson's instability with another, long-wavelength monotonic instability. In retrospect, we see that, generally, all results found by Takashima (1981a) are contained in Smith's paper (1966).

Takashima's real extension of Smith's work came with his later (1981b) paper where he searched for overstability in Bénard convection. He found that provided the heating is from the air side, oscillatory modes of instability were predicted for both positive and negative static Bond numbers, i.e., for a liquid resting on a solid support or hanging from it, provided the wall was cooled enough relative to the ambient air (Figure 8). Thus Takashima predicted the possibility of exciting and sustaining capillary-gravity waves at *negative* Marangoni numbers as, independently, and almost simultaneously, found by Levchenko and Chernyakov (1981). As his single-layer theory predicted no overstability with no liquid surface deformation, in agreement with an earlier result by Vidal and Acrivos (1966), Takashima missed the finding of a threshold for the viscous, longitudinal or dilational waves found in two-layer theories by Sternling and Scriven (1959) and by Reichenbach and Linde (1981).

Takashima's emphasis was on results for varying the static Bond number at given capillary number and vice versa. For a buoyancy-free layer he made the interesting discovery that on varying the static Bond and the Prandtl numbers, coexistence of two critical values of the Marangoni number, corresponding to respectively different wavenumbers for oscillatory interfacial instability, is possible. This finding extended to oscillatory modes the earlier result, found by Smith and others, for monotonic disturbances. Because Takashima did not consider buoyancy, he missed the overstability with the Brunt–Väisälä frequency expected in a stably stratified layer when heating from the air side hence from above (Rednikov et al., 1998c). We discuss this internal wave motion further below and its relation to surface tension gradient-driven instability.

Takashima also rightly remarked that while for a liquid resting on a solid support or hanging from it, Pearson's result on steady instability is recovered as a unique limiting value, at vanishing small capillary number, this is not true for overstable motions where positive or negative static Bond

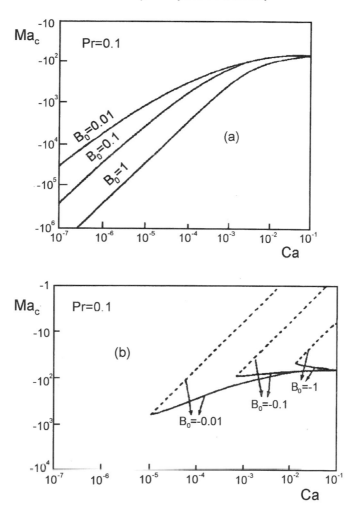

FIG. 8. Critical Marangoni number, Ma_c, for the onset of oscillatory instability as a function of capillary number, Ca, for (a) liquid resting on a solid support (Bo > 0) and for (b) a liquid hanging from a solid support (Bo < 0). (Redrawn after Takashima, 1981b, with permission of The Physical Society of Japan.)

numbers in the limit of zero deformation lead to different limit values, as a consequence of Rayleigh–Taylor instability. As has been noted by Limat (1993), in the isothermal case the Rayleigh–Taylor stability problem of a film of finite thickness depends on the ratio of two characteristic lengths, the capillary length $l_c = (\sigma/\rho g)^{1/2}$ and the Galileo length $l_g = (\nu\kappa/g)^{1/3}$. The interest of the experiment reported by Van Hook *et al.* (1995, 1997) is that they went down to a viscous thin layer, for which $l_c \gg l_g$.

Although Takashima discussed the role of the layer depth, he left room for a discussion, which starting with a remark made by Block (1956), had not been duly analyzed in the literature. Indeed, contrary to the usual approach where the geometry in an experiment is held fixed and the temperature difference is allowed to change, Goussis and Kelly (1990) focused attention on the explicit role played by the layer depth while the temperature difference is held fixed. Generally, one can use neutral stability curves with a fixed Galileo number. As already mentioned, the form of these curves was established by Smith (1966) for two-liquid layer systems and as a limiting case for a standard Bénard layer, thus recovering the earlier result for vanishing Ga, found by Scriven and Sternling (1959). If the temperature difference is held constant and the layer depth is allowed to vary moderately, an alternative albeit equivalent reading can be made of these earlier theoretical predictions. With the definitions of the Marangoni, Galileo (called after Archimedes by Goussis and Kelly), and Prandl numbers, from the results already available in the literature it follows that $Ma = 2^{1/3} M_{GK} Pr Ga^{1/3}$, with $M_{GK} = -(d\sigma/dT)(\Delta T/\eta)(2/vg)^{1/3}$. The quantity M_{GK} is the Marangoni number used by Goussis and Kelly. We can also write $Ma = CaGa^{1/3}$ with $Ca = M_{GK} Pr2^{-1/3}$. If Ca is smaller than some Ca* considering the crossing of the curve $Ma = CaGa^{1/3}$ with the corresponding long-wave and short-wave instability boundaries, there is instability for, respectively, small enough and large enough liquid depths, d. When Ca > Ca*, there is instability for all depths. As already noted, the values of GaCa* and Ca* are results already known in the literature. It suffices to compare Pearson's results for short-wave instability and the general formula provided by Takashima for steady long-wave instability. Strictly speaking, this is the summary of the report by Goussis and Kelly. However, it seems interesting indeed to explicitly alert the experimenters to the existence of two instability regions obtained by changing the liquid layer depth and keeping the temperature difference constant (see, however, Velarde and Perez-Cordon, 1976; de Boer, 1984).

Aware of Takashima's (1981b) results but not of the work of Levchenko and Cherniakov (1981), García-Ybarra and Velarde (1987) considered the case of a semi-infinite Bénard layer open to air with a deformable surface, for a single- and a two-component liquid mixture including the Soret-cross-transport effect. This model problem was in the spirit of the analyses developed by Sternling and Scriven (1959) and by Scanlon and Segel (1967). Air was taken passive. Further, García-Ybarra and Velarde considered only high-frequency interfacial disturbances, which amounts to "zooming" the interface, the neglect of buoyancy and other volume effects. However, in line with earlier works they incorporated gravity in the boundary conditions and used the capillary length as the natural length scale of the problem. They analytically showed and provided physical insight about the possibility of exciting and sustaining capillary-gravity waves in a Bénard layer with a deformable open surface when the Marangoni number is *negative*, as found by Sternling and Scriven (1959), Levchenko and Cherniakov (1981), and Takashima (1981b). Moreover, García-Ybarra and Velarde proved the relevance of this surface tension gradient-driven oscillatory interfacial instability to experiments in variable gravity and microgravity conditions. For instance, the striking discovery was made that a drastic lowering of the instability threshold occurred when the effective gravity goes down drastically, four or five orders of magnitude. They missed the prediction of overstable longitudinal or dilational waves because in their analysis surface deformation was as essential as in Takashima's (1981b) work. Finally, as a by-product of their results they recovered under appropriate limiting conditions the results obtained for monotonic instability by Pearson (1958) for layers with undeformable surface and by Scriven and Sternling (1964) for zero gravity but deformable surfaces. Velarde and García-Ybarra (1987) and later on Chu and Velarde (1988) linked the instability leading to capillary-gravity waves to the role played by the Marangoni effect in getting a π-phase lag between surface deformation and the temperature, and a $\pi/2$-phase lag between surface velocity and surface deformation. The instability does not bring the hottest or coldest points on the extrema, minima and maxima, respectively, of the surface deformation.

The essence of the above-mentioned instability can be explained in the following way. At the air–liquid interface, for a 2D geometry, eq. (2.1) reduces to

$$\partial w/\partial t = -(1/\rho)(\partial p/\partial z) + \nu(\partial^2 w/\partial z^2), \qquad (4.2)$$

where as defined earlier t, x, and z denote time, horizontal, and vertical coordinates, respectively, and ρ and p are density and pressure, respectively. To simplify notation again w denotes the vertical velocity. Due to the surface deformation, $\xi(x, t)$, eq. (2.7) becomes

$$w = \partial \xi/\partial t, \qquad (4.3)$$

which is the *kinematic* condition that provides a direct relationship between a point at the geometrical interface and a (material) liquid point at all times. Moreover, at the interface there is a dynamic balance for the normal and tangential components of the stress tensor, according to the appropriate extension of eqs. (2.5)–(2.7). We have

$$p - \rho g \xi + \sigma \partial^2 \xi / \partial x^2 = 2\rho v(\partial w/\partial z), \qquad (4.4)$$

$$\rho v(\partial w/\partial x) = (\partial \sigma/\partial T)[(\partial T/\partial x) - \beta(\partial \xi/\partial x)]. \qquad (4.5)$$

Focusing on the interface, in the high-frequency limit ($\mathrm{Im}\,\lambda \equiv \Omega \gg 1$), using (4.3)–(4.5), eq. (4.2) becomes

$$d^2\xi/dt^2 + k^2[4v - k\beta(\partial\sigma/\partial T)(2\kappa)^{1/2}/\rho\Omega^{3/2}](\partial\xi/\partial t)$$

$$+ [(gk + \sigma k^3/\rho) + (\partial\sigma/\partial T)\beta k^3\kappa/2\Omega^{1/2}\rho]\xi = 0, \qquad (4.6)$$

where we have assumed that

$$\xi = -(A/\Omega)\cos(kx + \Omega t) \qquad (4.7)$$

and similar Fourier mode expressions for the remaining disturbances. A denotes an amplitude that cannot be determined by the linear theory. Equation (4.6) is the *dissipative* harmonic oscillator equation. The damping coefficient may be positive, negative, or zero according to the sign and values given to β and $(\partial\sigma/\partial T)$.

Using the earlier defined capillary length as the space scale, l (to simplify notation we drop the subscript), and hence considering in practical terms that $d \gg l$, which is consistent with the fact that the viscous penetration depth of a wavy disturbance (in the high frequency limit approximation) is rather small, eq. (4.6) takes on a dimensionless form:

$$d^2\zeta/d\tau^2 + [4a^2 + \mathrm{Ma}/(2\mathrm{Pr}^3\omega^3)^{1/2}]d\zeta/d\tau + [a(1 + a^2)/\mathrm{PrCa}]\zeta = 0, \quad (4.8)$$

with $\zeta = \xi/l$, $\tau = tv/l^2$, and $\omega = \Omega/vk^2$. The groups Ma and Ca are now defined using l and not d as the length scale. For consistency, we have assumed that, in the *high-frequency* limit, $\mathrm{Ma} = O(\omega^{3/2})$, $\mathrm{Pr} = O(1)$, and $\mathrm{Ca} = O(1/\omega^2)$. The dissipative harmonic oscillator approximation to the oscillatory interfacial motion of the open surface in a Bénard layer contains all the relevant physics. Indeed, the damping coefficient vanishes whenever we set

$$\mathrm{Ma} = -4a^2(2\mathrm{Pr}^3\omega^3)^{1/2}, \qquad (4.9)$$

thus allowing a free oscillation of (dimensionless) frequency given by

$$\omega^2 = (1 + a^2)/a^3\mathrm{PrCa}. \qquad (4.10)$$

As a result of the preceding analysis, the lowest Marangoni number needed to sustain the oscillatory motion is

$$\mathrm{Ma}_c = -7.93(\mathrm{Pr}/\mathrm{Ca})^{3/4} \tag{4.11}$$

with a frequency

$$\omega_c = 6\sqrt{5}\,\mathrm{Pr}\mathrm{Ca} \tag{4.12}$$

for a wavenumber excitation

$$a_c = \sqrt{5/5}. \tag{4.13}$$

According to eq. (4.11) and the sign convention earlier established, this oscillatory interfacial instability is to be expected for *negative* Marangoni numbers only. Note, indeed, the very high values of the critical Marangoni number, in view of the very low capillary numbers in standard liquids. For standard liquids, this means that the heating is from above or the layer is cooled from below. However, for nonstandard liquids the predicted oscillatory instability can be seen when heating the liquid layer from below, i.e., from the liquid side. The argument given above was extended to the case of transverse waves in a two-liquid system with interface and mass transfer by Velarde and Chu (1988), and to the dilational waves in the same system but with negligible interface deformation by Velarde and Chu (1989b).

Earlier, Lucassen (1968; Lucassen-Reynders and Lucassen, 1969) had also described these modes of surface vibration in the wave dispersion relation for a layer subjected to transfer of a soluble surfactant. He derived a dispersion relation that describes the capillary-gravity waves as well as the dilational mode of frequency explicitly depending on the Marangoni number,

$$\omega^2 = -\mathrm{Ma}/\mathrm{Pr}^{3/2}. \tag{4.14}$$

However, he did not consider instability but rather damped motions. Instability in Lucassen's problem was studied by Hennenberg *et al.* (1979, 1980; see also Sanfeld *et al.*, 1979) for two semi-infinite fluid layers separated by an undeformable interface submitted to isothermal transfer of a surfactant. They predicted instability thresholds for viscous, longitudinal, or dilational waves in terms of the viscoelastic composition behavior of the interface. Furthermore, they considered the nonequilibrium surface desorption process, taking into account the different stages of the surfactant transfer. They generalized the earlier predictions by Sternling and Scriven (1959) and showed that when mass transport was controlled by the surface transfer, no longitudinal oscillation was possible. These authors studied in fact two fluids of almost equal densities and viscosities, which restricted the validity of their conclusions. Later on, Velarde and collaborators (Velarde and García-Ybarra, 1987; Velarde *et al.*, 1987; Chu and Velarde, 1988,

1989a,b,c, 1991; Hennenberg *et al.*, 1992; Velarde and Chu, 1988, 1989a,b) extended these studies to arbitrary density differences and provided a detailed analysis of the role played by deformation and surfactant transport, and by the adsorption–desorption process at the interface, particularly for highly insoluble surfactants.

In a series of papers, Chu and Velarde (1988, 1989a,b,c) discussed in detail the possible competition between the two oscillatory modes, the already mentioned transverse and the longitudinal or dilational, for a layer heated from the air side and for a two-liquid layer (Figure 9). They also discussed the possible competition between steady cellular Bénard instability and oscillatory convection in a layer heated from the liquid side. In particular, they showed that for low enough Prandtl numbers and high enough capillary numbers, oscillatory instability has the lowest threshold, hence it is the expected one in experiment rather than Bénard cells.

Nepomnyashchy and Simanovsky (1991) extended the study of transverse waves to the realistic case of liquid layers of finite depth (for a comprehensive account of their work see Simanovsky and Nepomnyashchy, 1993). In their numerical computations, the rigid boundaries of a two-liquid-layer system are considered to be good conductors. Then they found that for

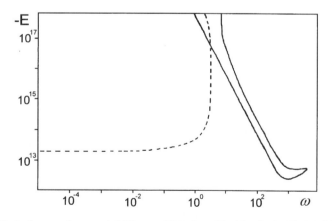

FIG. 9. Neutral curves for overstability providing the critical (surfactant/solutal) Marangoni number — also called elasticity number and denoted E — for capillary-gravity or transverse (solid line) and dilational or longitudinal (dashed line) waves as a function of frequency, ω, in a two-liquid system. $\rho_2/\rho_1 = 0.8$, $\eta_2/\eta_1 = 0.5$, $D_2/D_1 = 2$; $Bo = 1$, $Ca = 10^{-9}$; Schmidt and surface Schmidt numbers: $S = S_\Sigma = 10^3$. The surface excess solute number $H = \Gamma_0/b_1 l^2$ and the Langmuir adsorption number $H_z = k_1^l/l$ are defined using the concentration gradient and the Langmuir adsorption coefficient in the liquid at the bottom, 1, and the capillary length $l = (\sigma_0/g|\rho_1 - \rho_2|)^{1/2}$; $H = -10^{-9}$ and $H_z = 10^{-5}$. (Redrawn after Chu and Velarde, 1989a.)

small enough values of the capillary number, i.e., for large enough surface tension, the critical wavenumber is rather small but does not vanish. Depending on the physical parameters of the two liquids, two different kinds of neutral curves appear and they provided the corresponding critical Marangoni numbers, frequencies, and wavenumbers (Figures 10 and 11).

The oscillatory instability by heating from above can appear also in liquid mixtures when thermocapillarity is destabilizing and solutocapillarity is stabilizing, hence a consequence of the competition between the corresponding two Marangoni effects (see Joo, 1995, and references therein).

When the Marangoni effect is present but there is no surface deformation in a liquid layer, other competing agents leading to oscillatory instability are rotation of the layer, the action of electric or magnetic fields, or Soret

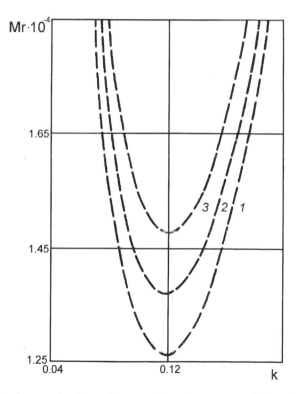

FIG. 10. Neutral curves for the capillary-gravity or transverse oscillatory instability in an air–water system heated from above ($d_1/d_2 = 1$) for $Ga_1 = 0$ (line 1), 3 (line 2), and 6 (line 3). Here $Mr = (Ma_1/P_1)(\eta_1/\eta_2)$ and $Ga_1 = ga_1^3/v_1^2$; subscript 1 (2) corresponds to the upper (lower) fluid. (Redrawn after Simanovsky and Nepomnyashchy, 1993.) Reprinted with permission of Gordon and Breach Publishers.

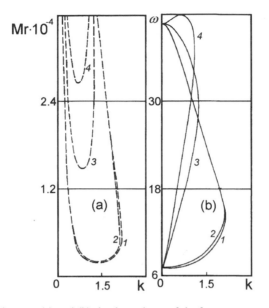

FIG. 11. Neutral curves (a) and (b) the dependence of the frequency ω on the wavenumber (dispersion relation) for the capillary-gravity or transverse oscillatory instability in a model system (all fluid parameters are equal except $\rho_1/\rho_2 = 0.999$ and $v_1/v_2 = 0.5$; $d_1/d_2 = 1$) for $Ga_1 = 0$ (lines 1), 10^4 (lines 2), 10^6 (lines 3), and 2.10^6 (lines 4). Definitions as in Fig. 10. (Redrawn after Simanovsky and Nepomnyaschchy, 1993.) Reprinted with permission of Gordon and Breach Publishers.

cross-transport etc. Results exist (Sarma, 1981) showing that, for low enough Prandtl number liquids at high enough angular velocities, overstability is possible. This is interesting because rotation, like lateral boundaries, tends to be a stabilizing agent for steady modes of instability. However, for oscillatory modes the local acceleration, inertia, partially offsets the constraining force of rotation. When, on the contrary, the Prandtl number is low enough, inertia dominates, hence having a destabilizing effect on time-dependent disturbances. In this case, as the Marangoni stress is at the surface, the analogy is appealing with near-surface Ekman layer flows in the ocean.

Gumerman and Homsy (1974) have studied the three-dimensional linear stability problem of Zeren and Reynolds (1972) with an imposed Couette flow due to both external rigid isothermal boundaries moving in opposite directions. In the application to the water–benzene case they found that criticality was achieved by buoyancy as the interfacial instability threshold is delayed by the Rayleigh number. These authors discussed the case of large surface tension, hence low two-liquid interfacial deformation, and large density contrast between the liquids that becomes the one-layer case where instability is not interfacial but buoyancy driven in a single layer. They also discussed the onset of long-wave disturbances, which are the modes appearing in a shearing flow.

Other Marangoni convection studies (Georis *et al.*, 1993; Georis and Legros, 1996; Funada, 1986; Birikh *et al.*, 1994, 1995; Maldarelli *et al.*, 1980; Oron *et al.*, 1995; Kats-Demyanets *et al.*, 1997; Liu *et al.*, 1998; Prakash and Koster, 1997; Steinchen *et al.*, 1982) targeted possibilities of microgravity experimentation, where the Galileo number becomes of order unity so that one can neglect buoyancy and focus now on purely surface tension gradient-driven flows. A curious case is a three-layer system (Georis *et al.*, 1993; Georis and Legros, 1996) submitted to a gradient normal to the interfaces, with both outer layers being identical liquids and differing only by their temperature gradient. The middle layer was a different immiscible liquid. This problem is interesting since the two liquid–liquid interfaces are opposedly oriented relative to the heat flux. For layers of equal depth, in the absence of surface deformation, the main parameter is the ratio of heat diffusivities. When it differs from unity, one has a monotonic instability and the critical Marangoni number is higher than that corresponding to both external layers being poor conducting inert gases (Funada, 1986). When the diffusivity ratio approaches unity, the coupling of both interfaces yields oscillatory motion.

The case where the external layers are different from one another has also been studied (Kats-Demyanets *et al.*, 1997) as well as instabilities in a layer with two deformable boundaries (Oron *et al.*, 1995). Another case (Birikh *et al.*, 1994, 1995) relevant to microgravity is that of a liquid layer, floating in space, partitioned midway by a porous solid divider of negligible width, characterized by two phenomenological coefficients expressing hydrostatic resistances across the divider. This divider can be used as a catalyzer causing internal heating or cooling of the two liquid-layer halves. It corresponds to a problem with two symmetric heat fluxes, at variance with a case considered by Georis *et al.* (1993) who considered an antisymmetric case, with the middle layer held at an intermediate temperature between the two outer ones. Birikh *et al.* (1994) dealt with undeformable surfaces and the onset of steady flow. The question is to know whether both halves of the layer behave as separate Bénard–Marangoni layers or if there is a common motion. Symmetric heating enables us to distinguish even and odd solutions. The even mode depends only on the characteristic coefficient of the divider linked to the jump in hydrostatic pressure across it. Their critical limiting value yields Pearson's (1958) result for a good heat conducting boundary. The odd modes depend on the parallel and transverse characteristics of the divider and their limiting behavior is that predicted by Pearson for a liquid layer placed on very poor heat conducting supports. Subsequently, Birikh *et al.* (1995), have extended the study to account for the interaction between the two outer deformable surfaces as a function of the two characteristics of the solid divider. This corresponds to having the whole system encapsulated in a gaseous enclosure of finite depth.

For monotonic instability, the even modes, whose bounding free surfaces were in phase, allow results in agreement with the predictions made by Sternling and Scriven (1964) and by Smith (1966) for a high resistance of the hotter divider and thus two different critical wavenumbers. For a low resistance of the divider, the transverse flow yields only a single critical wavenumber, which vanishes when heating from outside and is nonvanishing for inside heating. The odd modes are unstable at zero wavenumber only when internally heating with the solid device. Then the outer deformable surfaces are expected to move out of phase, hence the entire system behaves like a membrane showing flexural (stretching or antisymmetric) or squeezing (symmetric) modes of vibration (Maldarelli *et al.*, 1980; Steinchen *et al.*, 1982).

V. Some Recent Results on Patterned Convection

Let us provide now a short account of recent results concerning steady convective patterns generated by the (Bénard–Marangoni) surface tension gradient-driven instability (see Figure 1). Oscillatory instabilities are also expected but as secondary instabilities at high enough supercritical Marangoni numbers. For short-wavelength instability, the analysis, performed long ago by Scanlon and Segel (1967) for a semi-infinite layer, has been extended to a liquid layer of finite depth by Cloot and Lebon (1984). The latter authors established that a hexagonal planform appears in the subcritical region and that it is the only one near threshold (see also Bragard and Lebon, 1993), in agreement with experimental data (Schatz *et al.*, 1995). Two types of hexagons are possible, one with an ascending flow in the center of the cell for a liquid with high enough Prandtl number and the other with a descending flow in the center of the cell for low enough Prandtl number liquids. The hexagonal patterns are stable inside a certain region in the

(k, Ma) plane, like Busse's balloon in (Rayleigh–Bénard) buoyancy-driven convection (Busse, 1978) as suggested by Bestehorn (1993) and, in a more systematic approach, by Bragard and Velarde (1997, 1998). From eqs. (2.1)–(2.8), a suitable multiscale method provides equations for the (weakly nonlinear) amplitudes, A_j ($j = 0, 1, 2$), of the expected hexagonal thermoconvective mode at and near the threshold (Ma $\geqslant$ Ma$_c$). We have

$$\partial A_j/\partial t = s_0 A_j + D(\mathbf{n}_j \cdot \nabla)^2 A_j + \alpha A^*[j-1]A^*[j+1]$$

$$- A_j[\kappa(0)|A_j|^2 + \kappa(\pi/3)|A[j-1]|^2 + \kappa(\pi/3)|A[j+1]|^2)$$

$$+ iK_0(A^*[j+1]\mathbf{n}[j-1] \cdot \nabla A^*[j-1] + A^*[j-1]\mathbf{n}[j+1])$$

$$+ iK(A^*[j+1]\mathbf{t}[j-1] \cdot \nabla A^*[j-1] - A^*[j-1]\mathbf{t}[j+1]), \quad (5.1)$$

where $[j \pm 1] = (j \pm 1) \bmod.3$, $\mathbf{n}_j = \mathbf{k}_j/k_c$, $\mathbf{t}_j$ are unit vectors orthogonal to $\mathbf{n}_j$. The $\mathbf{k}_j$ are wavevectors satisfying the condition $\mathbf{k}_1 + \mathbf{k}_2 + \mathbf{k}_3 = 0$. The other parameters s_0, α, and κ are not needed explicitly for our purpose here and are given by Bragard and Velarde (1998). Note that the amplitude equations (5.1) contain terms that rule out the variational character of the problem, hence no potential exists as with the original, nonlinear and dissipative, Navier–Stokes equations, hence making complete (2.1). Accordingly, even near the instability threshold, these equations have not only steady solutions but the possibility also exists of space–time erratic solutions, i.e., spatio-temporal chaos, a kind of (interfacial) dissipative turbulence.

When the Marangoni number increases, other (square, rolls, or unsteady) patterns have been theoretically predicted (Bestehorn, 1993, 1996; Thess and Bestehorn, 1995; Thess and Orszag, 1995; Bragard and Velarde, 1997, 1998; Golovin et al., 1997a) and experimentally observed (Nitschke and Thess, 1995; Eckert et al., 1998). The transition between hexagons and other patterns, in the coupled buoyancy-driven and surface tension gradient-driven convection (the Rayleigh–Marangoni–Bénard case), was investigated theoretically by Parmentier et al. (1996) following the experiments by Cerisier et al. (1987). The influence of initially imposed constraints like a thermally drawn surface pattern was studied by Pantaloni et al. (1979), while the role of lateral boundaries of the layer on the formation of patterns was studied by Dauby et al. (1993). To be noted is the oscillatory alternating of hexagons and rolls, recently predicted by Bragard and Velarde (1997, 1998) in the absence of buoyancy.

For the limiting case of a long-wavelength instability, a nonlinear theory was developed by Funada (1987) who from the nonlinear extension of eqs.

(2.1)–(2.8), with (2.7) appropriately augmented to include surface deformation, $\xi(x, y, t)$, derived an asymptotic equation describing weakly nonlinear
surface deformations:

$$\partial\xi/\partial t + \Delta(a\xi + b\Delta\xi + c\xi^2/2) = 0, \qquad (5.2)$$

where a, b, and c denote some appropriate parameters whose explicit
expression is not needed for our purpose here. Funada's eq. (5.2) resembles
the Kuramoto–Sivashinsky equation, but the nonlinear term contains an
additional spatial derivative (see, e.g., Christov and Velarde, 1993). Stationary solutions describing solitary and periodic distortions of the surface can
be easily found (Funada and Kotani, 1986). However, all of these solutions
are unstable, and there is a subcritical nonlinear instability of the quiescent
state. The equation describing strongly nonlinear long-wavelength convection for a two-layer system is known but its properties are still poorly
investigated (Badratinova, 1985). A simplified description of the thermocapillary convection also in a two-layer system with a deformable
interface was suggested by Nepomnyashchy and Simanovsky (1990).

The earlier mentioned coexistence of long- and short-wavelength instability modes (Smith, 1966) can give rise through their interaction, at high
enough Marangoni numbers, to various convective patterns including wavy
regimes generated by a secondary instability of a stationary convective
pattern (Golovin *et al.*, 1994, 1997a,b). Recently, this interaction of long and
short waves has attracted experimentalists (Van Hook *et al.*, 1997) but
further experimental work is needed before final conclusions can be drawn.

As already mentioned, for low Biot numbers (poor relative heat conductivity of the liquid layer boundaries), the stationary instability is characterized by a long wavelength at onset. The weakly nonlinear analysis of the
pattern formation (Hadji *et al.*, 1991) shows the possibility of competition
between hexagonal and square patterns. For infinite Prandtl number fluids,
problem (5.1) can be reduced to a nonlinear evolution equation for the mean
temperature in the liquid layer. The spatially one-dimensional (1D) equation
of this kind (describing 2D patterns) was obtained by Sivashinsky (1982) in
the case of a nondeformable surface and by Oron and Rosenau (1989) for a
deformable surface. An equation describing 3D patterns was derived for an
infinite Prandtl number and nondeformable surface by Knobloch (1990).
This equation governs the two-dimensional field, $T(x, y)$, of the temperature
averaged across the layer, which is the only active variable because in the
high Prandtl number case the velocity is a slave to temperature, and has the

following structure (after suitable scaling):

$$\partial\Theta/\partial t = \alpha\Theta - \mu\Delta\Theta - \Delta^2\Theta + \nabla\cdot(|\nabla\Theta|^2\nabla\Theta) + \beta\nabla\cdot(\Delta\Theta\nabla\Theta) + \delta\Delta|\nabla\Theta|^2,$$

(5.3)

with Θ as the suitably scaled temperature field. The coefficient α is proportional to the Biot number, μ is the scaled bifurcation parameter depending on the Marangoni number, and the coefficients β and δ are constants of order unity whose explicit expressions are not needed here.

Shtilman and Sivashinsky (1991) generalized eq. (5.3) by taking into account the mean drift flow generated by distortions of convection patterns at finite Pr, in the same way as Pismen (1986) did for (Rayleigh–Bénard) buoyancy-driven convection. They showed that at finite Pr, the three-dimensional convective motions may generate a large-scale horizontal flow. The suitably scaled stream function, $\Psi(x, y)$, of such a flow satisfies the equation

$$\Delta\Psi = p^{-1}[\nabla\Delta\Theta, \nabla\Theta]_z,$$

(5.4)

where p is a coefficient proportional to Pr. Accordingly, an additional term $[\nabla\Theta, \nabla\Psi]_z$ describing the advection of the temperature by the horizontal flow should be added to the left-hand side of eq (5.3).

Pontes *et al.* (1996, 2000) performed a systematic numerical simulation of patterns governed by the Knobloch equation. They analyzed hexagonal and square patterns (see Figures 12 and 13 on p. 206), the transition between them, and unsteady regimes (spatiotemporal chaos). This study has been further extended by several groups of authors (Bestehorn, 1996; Gunaratne, 1993; Gunaratne *et al.*, 1994). For a detailed nonlinear analysis when deformation of the interface is present, see the work of Golovin *et al.* (1995).

VI. Further Discussion of Oscillatory Motions: Surface and Internal Waves, and Boundary-Layer Effects

Let us provide further insight about the expected oscillatory motions and waves in a Bénard layer. To do this we focus on the various time-scales involved in the problem. When the liquid layer is heated from the air side or open to suitable mass adsorption of a "light" component from a vapor phase above, with subsequent absorption in the bulk a (stabilizing) thermal gradient inside the liquid layer and, consequently, a stably stratified

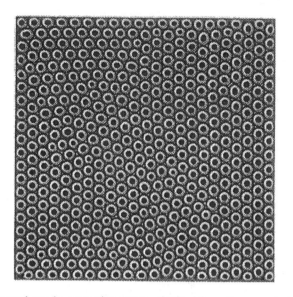

FIG. 12. Hexagonal steady convective pattern obtained from the numerical integration of Knobloch's eq. (31) with $\alpha = 0.8$, $\mu = 2.7$, $\beta = 0.125\sqrt{7}$, and $\delta = 0.75\sqrt{7}$. Compare with Fig. 1b. (Redrawn after Pontes *et al.*, 1996, with permission of World Scientific.)

FIG. 13. Square steady convective pattern obtained from the numerical integration of Knobloch's eq. (31) with $\alpha = 0.9$, $\mu = 2.5$, and $\beta = \delta = 0.25\sqrt{7}$. (Redrawn after Pontes *et al.*, 1996, with permission of World Scientific.)

layer is created. Then, the problem with Marangoni stresses, gravity and buoyancy, involves several timescales. On the one hand we have the viscous and thermal scales, $t_{vis} = d^2/v$, $t_{th} = d^2/\kappa$, respectively. There also exist two other timescales associated with gravity and surface tension (Laplace overpressure) that tend to suppress surface deformation: $t_{gr} = (d/g)^{1/2}$ and $t_{cap} = (\rho d^3/\sigma)^{1/2}$. The timescale related to the Marangoni effect is $t_{mar} = (\rho d^2/|\sigma_T \beta|)^{1/2}$. There is also another timescale related to buoyancy due to the stratification imposed by the temperature gradient, $t_{st} = (1/|\alpha\beta|g|)^{1/2}$. Here $\beta > 0$ when heating the liquid layer from below and that α is positive for standard liquids. The various ratios between timescales, and, accordingly, the ratios between forces involved in the dynamics, provide the earlier defined dimensionless groups [Prandtl, Marangoni, Rayleigh, Galileo, and (static) Bond numbers, respectively]. We have $\mathrm{Pr} = t_{th}/t_{vis} = v/\kappa$, $\mathrm{Ma} = -(d\sigma/dT)t_{th}t_{vis}/t_{mar}^2 = -(d\sigma/dT)\beta d^2/\eta\kappa$, $\mathrm{Ra} = t_{th}t_{vis}/t_{st}^2 = \alpha\beta g d^4/v\kappa$, $\mathrm{Ga} = t_{th}t_{vis}/t_{gr}^2 = gd^3/v\kappa$, and $\mathrm{Bo} = t_{cap}^2/t_{gr}^2 = gd^2/\sigma$.

These timescales are not always of the same quantitative order. For example, for the simplified problem treated by Pearson (1958) when dealing with a liquid layer with undeformable surface, we have $\mathrm{Ma} \approx 1$, but $\mathrm{Ga} \gg 1$ and $\mathrm{Ra} \ll \mathrm{Ma}$. Indeed, although Pearson neglected gravity his assumption of undeformability was practically equivalent to gravity being able to keep the surface level, whatever flows and thermal inhomogeneities exist. The characteristic timescale of the problem is $t_{th} \approx t_{vis} \approx t_{mar}$ (at $\mathrm{Pr} \approx 1$). For monotonic instability hence the case leading to Bénard cells when heating the liquid layer from the liquid side there exists a finite limit of the critical Marangoni number as $\mathrm{Ga} \to \infty$. No oscillatory instability appears in the one-layer problem with unedeformable surface. If such instability is possible the critical Marangoni numbers tends to infinity with $\mathrm{Ga} \to \infty$. Thus the critical Marangoni number should better be scaled with Ga, as Ga becomes very large, in agreement with earlier studies (Chu and Velarde, 1988; García-Ybarra and Velarde, 1987; Velarde and Chu, 1988, 1989a).

For high enough values of Ga the oscillatory mode is the capillary-gravity wave. The timescales t_{gr} and t_{cap} associated with this twofold wave are much smaller than the viscous and thermal timescales (at least for $\mathrm{Pr} \approx 1$, $\mathrm{Bo} \approx 1$). Then dissipative effects are relatively weak and the dispersion relation is [compare to eq. (4.10)]

$$\omega^2 = \mathrm{GaPr}\, k[1 + k^2/\mathrm{Bo}] \tanh(k) \qquad (6.1)$$

(to nondimensionalize ω the thermal timescale is used hereafter, k is the dimensionless wave number in units of d^{-1}). Clearly, the higher Ga is (and the wave frequency), the stronger the work of the Marangoni stresses (i.e., the higher the critical Marangoni number must be) to excite and sustain capillary-gravity waves, in agreement with a result given earlier. We indeed reobtain that for a standard liquid, $d\sigma/dT < 0$, this instability appears when heating the liquid layer from the air side (Ma < 0).

Let us now focus on the viscous, longitudinal, or dilational mode (Chu and Velarde, 1988, 1989a; Levchenko and Chernyakov, 1981; Lucassen, 1968; Reichenbach and Linde, 1981; Sternling and Scriven, 1959; Velarde and Chu, 1989b). Indeed when a liquid element rises to the surface, it creates a cold spot there. Then, the surface tension gradient acts toward this spot, pushing the element back to the bulk, hence overstability. High values of Ma ensure that the oscillations exist as earlier stated. Let their characteristic timescale also be t_{mar}. Calculation (Rednikov *et al.*, 1998a,b,c; see also Lucassen, 1968) yields the following expression for the frequency of the longitudinal wave (in the limit Ma $\rightarrow -\infty$):

$$\omega^2 = -\mathrm{Ma}[\mathrm{Pr}/(\mathrm{Pr}^{1/2} + 1)]k^2. \qquad (6.2)$$

Although this longitudinal wave is intrinsically dissipative, the damping rate is asymptotically smaller, $O(|\mathrm{Ma}|^{1/4})$, than its frequency. Up to some extent the flow field accompanying the dilational wave is qualitatively similar to that of the capillary-gravity wave. Potential flow can be assumed in the bulk of the layer, while vorticity is present only in boundary layers at the bottom rigid plate and at the upper free surface. The boundary layer thickness is of order of $O(|\mathrm{Ma}|^{-1/4})O(\mathrm{Ga}^{-1/4})$. For the longitudinal wave the horizontal velocity field in the surface boundary layer is stronger than the potential flow in the bulk [by $O(|\mathrm{Ma}|^{1/4})$] at variance with the capillary-gravity wave. Thus, the longitudinal motion is really concentrated near the surface. Noteworthy is that with an undeformable surface ($1 \ll |\mathrm{Ma}| \ll \mathrm{Ga}$), the dilational mode is always damped as stated earlier, thus justifying that no oscillatory instability was expected in the one-layer Marangoni problem without surface deformability (Pearson, 1958; Vidal and Acrivos, 1966; Takashima, 1981a,b). However, if the longitudinal wave is accompanied by non-negligible surface deformation ($|\mathrm{Ma}| \geqslant \mathrm{Ga}$), it can be amplified, a striking result. Accordingly, at Ga $\gg 1$, two tightly coupled thresholds for oscillatory Marangoni instability are expected with corresponding two (high-frequency) wave modes, capillary-gravity and dilational wave motions

(Lucassen, 1968). As already said, to sustain the longitudinal wave one needs surface deformability. Alternatively, to sustain a capillary-gravity wave one needs the Marangoni stress. The most dramatic manifestation of the tight coupling occurs at resonance, when the frequencies are equal to each other. Near resonance there is mode mixing. Namely, the capillary-gravity mode in the parameter half-space from one side of the resonance manifold is swiftly converted into the longitudinal one when crossing the manifold, and vice versa. Another feature of resonance is that the damping/amplification rates are drastically enhanced here, namely, $O(Ga^{3/5})$ versus $O(Ga^{1/4})$ far from resonance.

Recent work by Rednikov *et al.* (1998a,b,c) provides a thoroughly detailed analytical study of these oscillatory instabilities with consideration given to the role of the boundary layers near the solid bottom and the open surface or interface. Indeed, taking the Galileo number to be rather large, the layer can be divided into three portions: two boundary layers, one at the bottom where there is no slip and the other at the open surface, while the bulk of the layer can be well approximated by inviscid flow. Because the threshold for overstability is expected at relatively high values of the Marangoni number, Rednikov *et al.* (1998a) introduced the earlier mentioned *dynamic* Bond number, which can be written as the inverse of a modified Marangoni number, $1/m = Ga/Ma$. The marginal curves found delineate a bag or bubble-like region whose lower boundary corresponds to the onset of capillary-gravity waves while the upper one gives the dilational waves. These authors provided the frequencies of each of these two modes, as well as an analysis of their resonance and mode mixing, thus providing a most complete counterpart of the pioneering work by Lucassen on damped waves (Lucassen, 1968; Lucassen-Reynders and Lucassen, 1969). Some typical results obtained in the realistic case of two-fluid layers of finite depth are shown in Figures 14 and 15. At a critical value of *m*, the stable longitudinal mode becomes the stable capillary-gravity mode.

Now let us turn to yet another oscillatory mode in the liquid layer. Indeed, if the liquid layer is deep enough and hence stratified there is also the possibility of internal waves of frequency given by the Brunt–Väisälä frequency (Normand *et al.*, 1977):

$$\omega^2 = -\mathrm{Ra}\,\mathrm{Pr}\frac{k^2}{k^2 + \pi^2 n^2} \qquad (n = 1, 2, \ldots). \tag{6.3}$$

Disregarding capillary-gravity waves, hence surface deformation, the possibility exists of coupling longitudinal waves to internal waves with $|Ra| \ll Ga$.

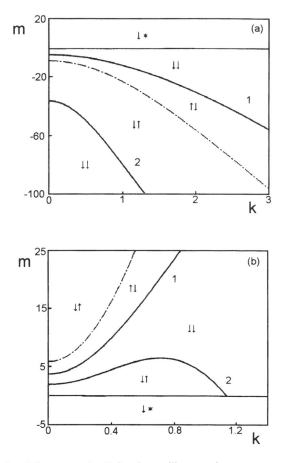

FIG. 14. Neutral stability curves (m, k) for the capillary-gravity or transverse (solid line 1) and dilational or longitudinal (solid line 2) waves, and the resonance curve (dot-dashed line) for (a) $d_2/d_1 = 0.2$, $\rho_2/\rho_1 = 0.5$, $v_2/v_1 = 4.5$, $\kappa_2/\kappa_1 = 1.0$, $\lambda_2/\lambda_1 = 0.5$, $Bo_1 = 1$, and $Pr_1 = 6.0$; (b) $d_2/d_1 = 0.5$, $\rho_2/\rho_1 = 0.5$, $v_2/v_1 = 0.4$, $\kappa_2/\kappa_1 = 2.0$, $\lambda_2/\lambda_1 = 0.5$, $Bo_1 = 0.17$, and $Pr_1 = 6.0$. In each region the arrows indicate whether the transverse (first arrow) and longitudinal (second arrow) modes are amplified (arrow pointing up) or damped (arrow pointing down). When the dilational mode is not oscillatory the second arrow is replaced by a star. (Redrawn after Rednikov *et al.*, 1998b).

This may be called the Rayleigh–Marangoni problem and it is the natural extension to overstable motions of Nield's (1964) study of monotonic instability. Rednikov *et al.* (1998c) have analyzed this combined buoyancy and surface tension gradient-driven problem with undeformable surface for high enough Rayleigh and Marangoni numbers. As expected, two unstable

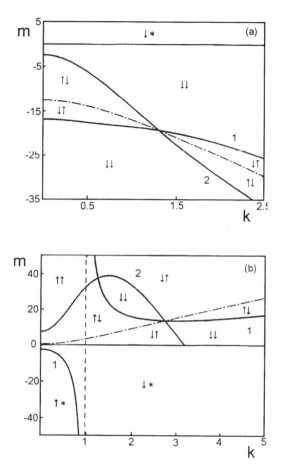

FIG. 15. Neutral stability curves (m, k) as in Fig. 14 for (a) $d_2/d_1 = 1.6$, $\rho_2/\rho_1 = 0.65$, $\nu_2/\nu_1 = 3.0$, $\kappa_2/\kappa_1 = 0.5$, $\lambda_2/\lambda_1 = 0.3$, $Bo_1 = 3.0$, and $Pr_1 = 6.0$; (b) $d_2/d_1 = 0.45$, $\rho_1/\rho_2 = 0.95$, $\nu_2/\nu_1 = 0.8$, $\kappa_2/\kappa_1 = 2.0$, $\lambda_2/\lambda_1 = 0.5$, $Bo_1 = 3.0$, and $Pr_1 = 2.0$. Note the resonant intersection of neutral stability curves. Compare with Fig. 7. (Redrawn after Rednikov *et al.*, 1998b.)

modes do exist in this case, the longitudinal wave and the Brunt–Väisälä internal wave. The marginal curves are again in the form of closed bags or bubbles with the upper boundary providing the threshold for internal waves and the lower one for dilational surface waves. The bubbles collapse when lowering the (absolute value) Rayleigh number. In the absence of the Marangoni effect, no unstable oscillatory motion persists undamped, which again stresses the crucial role played by the coupling of the two wave disturbances (Rednikov *et al.*, 1998c). Note that in the Rayleigh–Marangoni case we have a countable number ($n = 1, 2, \ldots$) of internal wave modes, and the dilational wave can be coupled to each of them and, hence, to a countable number of marginal stability conditions. Relative to the earlier case of coupling capillary-gravity to longitudinal waves the form of the marginal curves is qualitatively different. Furthermore, there exists the minimally possible Rayleigh number (in absolute value), below which there is no oscillatory instability. No such bound was found for the Galileo number in the other case (at least in the region where Ga remains high).

VII. Nonlinear Waves and Dissipative Solitons

Past threshold, Ma_c, the nonlinear evolution of either capillary-gravity or dilational waves poses formidable tasks. Velarde and collaborators (Chu and Velarde, 1991; Velarde *et al.*, 1991a,b; Garazo and Velarde, 1992; Nepomnyashchy and Velarde, 1994) have studied the former case in the drastically simplified approximation of a shallow Bénard layer. First, these authors neglected buoyancy, because a reasonable simplification for a shallow layer showed that provided the Marangoni number is *negative* (i.e., the heating is from the air side), capillary-gravity waves can be excited in the form studied by Boussinesq (1871, 1872, 1877), Lord Rayleigh (1876), and Korteweg and de Vries (1895), i.e., as (long) solitary or cnoidal waves. Chu and Velarde (1991; see also Velarde *et al.*, 1991a,b) obtained a Korteweg–de Vries–Burgers (KdV-B) equation (Johnson, 1972; Whitham, 1974) where the coefficient of the second-order space derivative (the Burgers term), which contains the viscosity and energy supply due to the Marangoni effect, vanishes at the threshold for instability. Later on Garazo and Velarde (1991), Rednikov *et al.* (1995), and, with greater generality, Nepomnyashchy and Velarde (1994) have developed the asymptotics leading to dissipation modified Korteweg and de Vries (KdV), Kadomtsev–Petsviashvili (KP), and 3D-Boussinesq equations for wave evolution (Boussinesq, 1872, 1877; Whitham, 1974; Drazin and Johnson, 1989; Wu and Zhang, 1996).

Nepomnyashchy and Velarde (1994) found as solvability conditions for the problem, first the linear (phase) wave velocity and the instability threshold $Ma_c = -12$. This particular value is not relevant to standard experiments with a rigid bottom and may only be considered of interest when dealing with a liquid overlying another liquid with a largely different (dynamic) shear viscosity. Nepomnyashchy and Velarde (1994) used the longwave shallow layer approximation (Gardner *et al.*, 1967) and a slippery bottom with the appropriate surface deformability and Ursell number limit (Benjamin, 1982; Christov *et al.*, 1996; Cole, 1968; Ursell, 1953; Wu, 1995, 1998; Yih and Wu, 1995). As with monotonic instability, i.e., Bénard cells (Normand *et al.*, 1977; Bragard and Velarde, 1997, 1998), the qualitative picture provided by the approximation of a stress-free bottom has some universal validity. Quantitatively, the consideration of a rigid bottom, and hence no slip, is expected to yield a higher critical Marangoni number in absolute value and to move the instability scale away from the long wavelength limit (Depassier, 1984; Rednikov *et al.*, 1995; Velarde *et al.*, 1999). Further below we come back to this question. In the simplest case of a 2D geometry and restricting consideration to left-to-right steadily propagating long waves, past the instability threshold the second-order solvability condition is the following dissipation-modified KdV equation (Garazo and Velarde, 1991; Nepomnyashchy and Velarde, 1994):

$$\xi_t + C_{\text{exp}}\xi_x + \alpha_1\xi\xi_x + \alpha_2\xi_{xxx} + \alpha_3\xi_{xx} + \alpha_4\xi_{xxxx} + \alpha_5(\xi\xi_x)_x + \alpha_6\xi = 0, \quad (7.1)$$

where, as earlier, ξ denotes the suitably scaled dimensionless surface deformation and the coefficients α_i ($i = 1, \ldots, 6$) are explicit functions of Ma, Ra, Pr, and Ga. The term $\alpha_6\xi$ accounts for bottom friction in an *ad hoc* way. A suitable transformation of variables and choice of the reference frame, brings eq. (7.1) to the more compact form

$$\eta_\tau + (\eta^2)_x + \eta_{xxx} + \delta[\eta_{xx} + \eta_{xxxx} + D(\eta^2)_{xx} + \alpha\eta] = 0, \quad (7.2)$$

where $\eta(x, \tau)$ is once more the suitably scaled surface deformation. The coefficient D can be either positive or negative, while α and δ are semidefinite positive. In particular, α_3 and, consequently, δ is a function of (Ma $-$ Ma$_c$). When δ vanishes, the ideal KdV equation is recovered (Whitham, 1974). The cylindrical case was, recently, considered by Huang *et al.* (1998). For a thoroughly detailed study of eq. (7.1) using the corresponding qualitative dynamical system approach, see Nekorkin and Velarde (1994).

Equation (7.1), or (7.2), contains a fourth-order derivative dissipating energy at short scales and the nonlinear term $D(\eta^2)_{xx}$, which helps redistribution, supercritically, of the energy supplied by the Burgers term ($\delta\eta_{xx}$ and the Marangoni effect). The factor C_{exp} in eq. (7.1) is a function of the actual Marangoni number in an experiment thus providing the measurable nonlinear wave velocity in the laboratory frame of reference. Its linear approximation is $C_0^2 = \text{Pr}(\text{Ga} - \text{Ma})$, which to the lowest order reduces to $C_0^2 = C_{\text{KdV}}^2(1 + 12/\text{Ga})$ where $C_{\text{KdV}}^2 = gd$ is the linear velocity of the ideal KdV solitary wave (Whitham, 1974). For the linear approximation, the correction due to the thermal constraint is already about 30% for a silicone oil layer of $d = 0.01$ mm. Subsequently, corrections are induced by the distance of the Marangoni number from the threshold value.

Equation (7.1) has also been obtained by Depassier and collaborators (Alfaro and Depassier, 1989; Aspe and Depassier, 1990) for buoyancy-driven convection with no Marangoni effect in shallow layers with open deformable surface. The physical relevance of their approach was discussed above. The same and some other related equations appear in a variety of problems, e.g., free-falling films, Eckhaus instability for traveling waves, and other wave phenomena (Castillo *et al.*, 1988; Chang and Demekhin, 1996; Christov and Velarde, 1993, 1995; Elphick *et al.*, 1991; García-Ybarra *et al.*, 1987; Janiaud *et al.*, 1992; Johnson, 1972; Kawahara, 1983; Kawahara and Toh, 1988; Koulago and Parséghian, 1996; Kuramoto and Tsuzuki, 1976; Nepomnyashchy, 1974, 1976; Shkadov, 1967; Shkadov *et al.*, 1976; Sivashinsky, 1977; Zeytounian, 1990, 1994, 1995, 1998). (Incidentally, eq. 283, footnote p. 360, of Boussinesq (1877) is the KdV equation (Korteweg and de Vries, 1895).

Note that the linear terms with second and fourth derivatives, on the one hand, provide a (phase) wave velocity selection for eq. (7.1) and, on the other hand, they are expected to add a wavy forerunner to the ideal bell-shaped KdV wave. Such a wavy forerunner, found indeed with the numerical integration of (7.1), acts like an infinitely long-ranged albeit exponentially weak attractive "potential," which at variance with the ideal, dissipation-free KdV equation, suggests the easiness in forming *bound* states (Nekorkin and Velarde, 1994) and the possibility of inelastic wave or particle-like collisions as in the imperfect (van der Waals) gas (Kac *et al.*, 1963). For particular combinations of the coefficients in eq. (7.1) or (7.2), some exact analytical, nonperturbative solutions in the form of solitary and periodic waves propagating with constant velocity have been found (Kudryashov, 1988, 1990; Lou *et al.*, 1991; Estevez and Gordoa, 1993; Porubov, 1993; Porubov and Samsonov, 1995; Huang *et al.*, 1998). Also some numerical solutions and stability studies exist (Christov and Velarde, 1995; Rednikov *et al.*, 1995; Toh and Kawahara, 1985; Topper and Kawahara, 1978; Trifonov, 1992; Tsvelodub and Trifonov, 1989; Velarde *et al.*, 1995; Bar and Nepomnyashchy, 1995; Oron and Rosenau, 1997).

Let us see how eq. (7.1) or (7.2) accommodates the energy balance at and past the instability threshold. By multiplying eq. (7.2) by η, and integrating over the full space or one wavelength in the variable x, the energy $E = \frac{1}{2}\int \eta^2$ is governed by the balance

$$\frac{dE}{dt} = \delta\left(\int \eta_x^2 \, dx - \int \eta_{xx}^2 \, dx + 2D \int \eta \eta_x^2 \, dx - \alpha \int \eta^2 \, dx\right), \qquad (7.3)$$

whose value vanishes at the steady state. Clearly, it now appears that the first term on the right-hand side of (7.3) describes the energy input at rather long wavelengths due to instability (the Burgers term; Chu and Velarde, 1991); the second and fourth terms describe energy dissipation on short and long wavelengths, respectively; and the third term accounts for the convective nonlinearity redistributing energy as a (feedback) correction to long-wave energy input (for η positive, positive if D is positive and negative otherwise). Recalling that the ideal KdV equation possesses a one-parameter family of solitary waves or cnoidal waves (periodic wave trains) thanks to the dispersion-nonlinearity balance, also existing in eq. (7.2), we note that here the input–output energy balance (7.3) selects indeed either a single wave or a periodic wavetrain or a bound state or a spatially chaotic wavetrain (Nekorkin and Velarde, 1994).

On the other hand, when considering the three-dimensional problem (Nepomnyashchy and Velarde, 1994) phase shifts following wave collisions or reflections at walls depend on the incident angle, α_i [e.g., measured front to front or twice the value front to wall, i.e., by $(\pi/2) - \alpha_i$; apparently, a reflection is like a collision with a mirror image wave]. At the approximate value of $\pi/2$ no phase shift is expected while for lower collision angles the phase shift has the sign of phase shifts on head-on collisions. Higher values than $\pi/2$ (or $\alpha_i < \pi/4$) lead to a change of sign in the phase shift and the formation of a *third* wave, or *stem*, evolving *phase locked* with the post collision or reflected front. This result for (Bénard–Marangoni) surface tension gradient-driven waves generalizes a result obtained by Miles (1977a,b) for a dissipation-free shallow liquid layer. These phenomena were discovered a century ago by Russell (1885; plate VII of the Appendix from his 1844 report) for water waves in a wave tank, where viscosity plays a negligible role, and by Mach and collaborators for shocks in gases, where dissipation is essential for the existence of the shock. It is known that surface waves in liquids (including hydraulic pumps) share some common kinematic features with shocks in compressible gases (Bazin, 1865; Bouasse, 1924; Courant and Friedrichs, 1948; Drazin and Johnson, 1989; Hornung, 1988;

Krehl and van der Geest, 1991; Lax, 1968; Maxworthy, 1976; Melville, 1980; Russell, 1844, 1885; Weidman and Maxworthy, 1978). The phase shift in overtaking collisions was discovered by Zabusky and Kruskal (1965) in their seminal paper where they introduced the *soliton* concept. Finally, starting with, e.g., an initial condition of two nearby "solitary" pulses the system is expected to evolve, according to eq. (7.1) or (7.2), to a *bound* state, i.e., a wavetrain with unequally spaced maxima but equal amplitude, and hence wave crests traveling with the same velocity dictated by the energy balance (7.3) (Nekorkin and Velarde, 1994; Nepomnyashchy and Velarde, 1994; Christov and Velarde, 1995; Velarde *et al.*, 1995b; Rednikov *et al.*, 1995). Different types of interactions between those *dissipative* solitons are shown in Figures 16, 17, and 18.

The consideration of a solid support, hence non-negligible bottom friction, demands a sophisticated analysis with inclusion of the boundary layers near the wall and the open surface or interface as recently done by Velarde and Rednikov (1998; Velarde *et al.*, 1999). These authors have considered two horizontally infinite layers, liquid and air both of finite depth, that for convenience we denote here by h_1 and h_2, respectively. Thus they have restricted consideration to the problem in two-dimensional geometry. The air layer above the liquid is covered by a rigid support and has no negligible role on the liquid placed on another flat rigid support, a geometry that fits well with experimental setups. At rest, there exists a linear vertical temperature distribution.

In the spirit of earlier work (Boussinesq, 1872, 1877; Ursell, 1953; Gardner *et al.*, 1967; Miles, 1976) for "long" enough waves in a shallow layer a smallness parameter, ε, can be defined as the ratio of the layer depth to a characteristic wavelength of disturbances. Then, there are two timescales in the problem. One of them is defined by heat diffusion, $t_{th} = h_1^2/\kappa_1$ (as the Prandtl number of the liquid is assumed of order unity, the viscous timescale, $t_{vis} = h_1^2/\nu_1$, is of order of the thermal one). The other timescale, $t_{gr} = \varepsilon h_1/g^{1/2}$, is associated with gravity waves. When $t_{th} \ll t_{gr}$ the heat and viscous effects are predominant, which, in practice, corresponds to very shallow liquid layers or microgravity conditions. In the opposite situation, $t_{th} \gg t_{gr}$, the dissipation is limited to the boundary layers at the bottom and, if the Marangoni effect is significant, at the air–liquid interface. In terms of the Galileo number the first case corresponds to $\mathrm{Ga}\varepsilon^2 \ll 1$, while in the second $\mathrm{Ga}\varepsilon^2 \gg 1$. Velarde and Rednikov (1998) restricted consideration to high enough Galileo numbers, $\mathrm{Ga} \gg 1$. As $t_{gr} \approx t_{th}$ where t_{gr} is as above while now $t_{th} = d^2/\kappa_1$, the thickness, d, of the boundary layers is such that

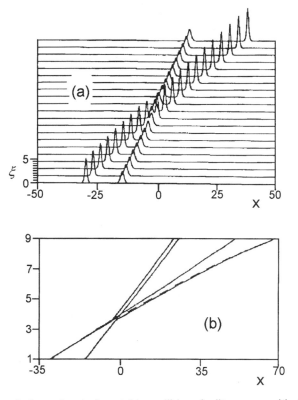

FIG. 16. Numerical experiment of overtaking collision of solitary waves with phase velocities $c_1 = 10$ and $c_2 = 5$, governed by eq. (7.1), in a weakly dissipative case ($\alpha_1 = 3$, $\alpha_2 = 1$, $\alpha_3 = \alpha_4 = 0.001$, $C_{exp} = \alpha_5 = \alpha_6 = 0$): (a) Evolution of bell shapes with time. (b) Trajectories of centers of solitary waves exhibiting negative phase shifts. Note that if we wait a much longer time lapse than shown in the figure (generally longer the lower the value of α_3 and α_4 is) the two unequal waves eventually radiate (or absorb) their excess (or missing) mass and energy to adjust to the energy balance eq. (7.3), and hence amplitudes become equal. Compare with Fig. 17. (Redrawn after Christov and Velarde, 1995, with permission of Elsevier Science.)

$$d/h_1 \approx \varepsilon^{-1/2} \mathrm{Ga}^{-1/4}. \tag{7.4}$$

With x and z denoting the original (dimensionless) horizontal and vertical coordinates, respectively, the bottom of the layer is taken at $z = -1$, the interface at $z = \eta(x, t)$, and the top of the air layer at $z = h$ where t is time and $\eta(x, t)$ describes once more the suitably scaled, dimensionless, surface deformation [here h_1 is the unit of length and $(h_1/g)^{1/2}$ is the unit of time].

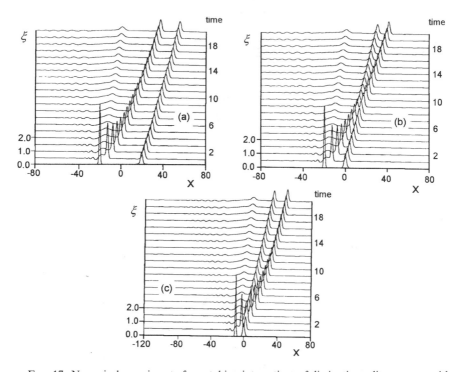

FIG. 17. Numerical experiment of overtaking interactions of dissipative solitary waves with phase velocities $c_1 = 10$ and $c_2 = 2$, governed by eq. (7.1), with $\alpha_1 = 3$, $\alpha_2 = 1$, $\alpha_3 = \alpha_4 = 0.1$ and $C_{exp} = \alpha_5 = \alpha_6 = 0$: (a) Large initial separation. (b) Moderate initial separation. (c) Small initial separation. Note the tendency of two unequal waves to form a *bound* state of crests of equal amplitude traveling together with the same velocity, in accordance with the energy balance eq. (7.3), particularly in case (a). Note also that the nearer the two waves start their collision, the sharper the definition becomes of the particle-like, solitonic event. (Redrawn after Christov and Velarde, 1995, with permission of Elsevier Science.)

Note that, for simplicity, in this section $\beta > 0$ corresponds to heating from the air side. Going to a moving frame the horizontal variable is $\xi = \varepsilon(x - Ct)$, with C the phase velocity to be determined. Then, the horizontal velocity, pressure, and surface deformation can be scaled with ε^2, the vertical velocity with ε^3, and the slow timescale is $\tau = \varepsilon^3 t$. For temperature the scale is determined by the leading convective contribution to the temperature field, which is of order ε^2.

For the case $\varepsilon^{-1/2}G^{-1/4} \ll 1$, i.e., when the liquid layer can be subdivided into its bulk where the flow is potential and the corresponding boundary layers at the solid supports and interface, the most interesting asymptotics

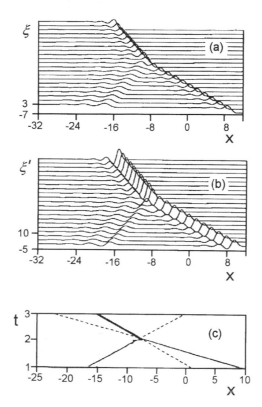

FIG. 18. Numerical experiment of head-on collision of two kinks [$\xi(-\infty) = 8/3$, $\xi(0) = 0$, $\xi(+\infty) = -16/3$], governed by eq. (7.1), with $\alpha_1 = 3$, $\alpha_2 = 3$, $\alpha_3 = \alpha_4 = 1$, $C_{exp} = \alpha_5 = \alpha_6 = 0$: (a) Shape of kinks. (b) Their derivative as bell-shaped solitary waves. (c) Trajectories of the centers of solitons. Note that dissipation adds *inelasticity* to the collision event. Accordingly, following collision two waves as particle-like objects may stick together as two colliding pieces of clay or soft, sticky balls. Ideal KdV–soliton collisions are purely *elastic*, and hence their collisions merely produce phase shifts in trajectories. Otherwise, ideal KdV–solitons behave like billiard balls or hard spheres. (Redrawn after Christov and Velarde, 1995, with permission of Elsevier Science.)

corresponds to the case when the boundary layer thickness and the deformation of the surface are of the same order. From (7.4), it follows then that $\varepsilon^{-1/2}Ga^{-1/4} \approx \varepsilon^2$, i.e., $\varepsilon = Ga^{-1/10}$ (this defines ε in terms of Ga). Then the effects of energy output (due to heat and viscous dissipation) and input (due to the Marangoni effect) will be of the same order as nonlinearity and dispersion. When the latter two are in appropriate balance we have the ideal KdV equation for long waves in a shallow layer.

For the air layer, due to the relatively large kinematic viscosity ($v_{air} > v_{liquid}$) and thermal diffusivity ($\kappa_{air} > \kappa_{liquid}$) ratios, inertia does not dominate over dissipative effects. Then weakly nonlinear, weakly dispersive long waves driven by the Marangoni effect in a shallow Bénard layer obey the following equation:

$$2\left(1 - \frac{m}{Pr^{1/2} + 1}\right)\left[\eta_\tau + \frac{3}{2}\eta\eta_\xi + \left(\frac{1}{6} - \frac{1}{2B}\right)\eta_{\xi\xi\xi}\right]$$

$$- \frac{2m}{Pr^{1/2}} \frac{1}{2\pi} \int_{-\infty}^{+\infty} Q_2(\xi - \xi')\left[\eta_\tau + \frac{3}{2}\eta\eta_{\xi'} + \left(\frac{1}{6} - \frac{1}{2B}\right)\eta_{\xi'\xi'\xi'}\right] d\xi'$$

$$+ \left(m\frac{2Pr^{1/2} + 1}{Pr + Pr^{1/2}} - 1\right)\frac{1}{\pi^{1/2}} \frac{d}{d\xi} \int_\xi^\infty \frac{\eta(\xi')}{(\xi' - \xi)^{1/2}} d\xi'$$

$$- \frac{m}{Pr^{1/2}} \frac{1}{2\pi} \int_{-\infty}^{+\infty} Q_1'(\xi - \xi')\eta(\xi') \, d\xi' = 0 \tag{7.5}$$

where $m = Ma\varepsilon^{10}/Pr = Ma/PrGa$ is a modified Marangoni number using the liquid layer parameters. Note that m, which is an inverse *dynamic* Bond number, is of order unity. This is the case of interest as viscous and thermocapillary stresses are of the same order and Ma is of the order of Ga. The coefficients Q_i ($i = 1, 2$) depend on the various parameters of the problem and their explicit values are not needed here (Velarde *et al.*, 1999). Equation (7.5) significantly generalizes the earlier discussed results of eq. (7.1), or (7.2) and provides the evolution equation for nonlinear dispersive–dissipative waves in a realistic situation with bottom friction. Under appropriate limiting conditions, eq. (7.5) reduces, on the one hand, to the ideal KdV equation and on the other hand to an earlier known dissipative KdV result with no Marangoni effect (Miles, 1976). The dissipation-modified KdV eq. (7.5) possesses the necessary ingredients to have solutions in the form of (solitary and cnoidal) stationary propagating waves: There is an unstable wavenumber interval, where the energy is brought by the Marangoni effect and dissipation occurs on the wavenumbers belonging to the stability interval. As for eq. (7.1) or (7.2), here again the convective nonlinearity redistributes the energy from long to short waves, making possible the dynamic equilibrium and the appropriate energy balance for the dissipative wave. Accordingly, sustained dissipative periodic wavetrains or solitary waves are expected at and past an instability threshold whose quantitative value is now relevant to standard experiments as we show below.

For the particular case of a thin air gap above the liquid layer ($h \ll 1$, however, h should remain larger than the surface deformation) where the air motion is dissipation dominated, eq. (7.5) reduces to

$$
2\left(1 - \frac{m}{\mathrm{Pr}^{1/2} + 1}\right)\left[\eta_\tau + \frac{3}{2}\eta\eta_\xi\left(\frac{1}{6} - \frac{1}{2B}\right) + \eta_{\xi\xi\xi}\right]
$$

$$
+ \left(m\frac{2\mathrm{Pr}^{1/2} + 1}{\mathrm{Pr} + \mathrm{Pr}^{1/2}} + m\frac{1}{h\mathrm{Pr}^{1/2}} - 1\right)\frac{1}{\pi^{1/2}}\frac{d}{d\xi}\int_\xi^\infty \frac{\eta(\xi')}{(\xi' - \xi)^{1/2}}\,d\xi' = 0
$$

$$(7.6)$$

Then the threshold value of the modified Marangoni number is

$$
m_b = \left(\frac{2\mathrm{Pr}^{1/2} + 1}{\mathrm{Pr} + \mathrm{Pr}^{1/2}} + \frac{1}{h\mathrm{Pr}^{1/2}}\right)^{-1} \tag{7.7}
$$

Noteworthy is that the threshold value considerably decreases with decreasing air gap to liquid depth ratio. Thus to observe sustained capillary-gravity waves as the result of (Bénard–Marangoni) surface tension gradient-driven instability, the thinner the air gap the better. Take, for example, a water-like liquid. For illustration choose in CGS units: $h_1 = 0.1$, $h_2 = 0.01$; hence, $h = 0.1$, $d\sigma/dT = -0.15$, $g = 10^3$, and $\mathrm{Pr} = 6$. Then, from eq. (7.7) it follows that the temperature difference for the onset of capillary-gravity waves driven by the Marangoni effect is 15 K or a 150 K/cm gradient, heating from the air side.

Finally, let us mention that the vanishing of the coefficient of the [bracketed] ideal KdV terms in eq. (7.5), or (7.6), at $m = \mathrm{Pr}^{1/2} + 1$, corresponds to the earlier discussed resonance between capillary-gravity and dilational waves (Section VI). This resonance value is always higher than m_c.

The available experimental data, although scarce, support the salient predictions of the theory. For instance, for the case of heat transfer with an octane liquid layer open to heated ambient air, gradients of 100–200 K/cm have been found by Linde and collaborators (Linde *et al.*, 1997; Weh and Linde, 1997). They have found similar values for other liquids. They have also found (long) solitary waves and wavetrains exhibiting the nonlinear solitonic behavior shared by both solitary waves in water canals and supersonic shocks, including the appearance of the earlier mentioned third wave or (Mach–Russell) *stem*. Indeed, for the case of *negative* Marangoni numbers, Linde and collaborators (Weidman *et al.*, 1992; Linde *et al.*, 1993a,b,c, 1997, 2000; Velarde *et al.*, 1995; see also Santiago-Rosanne *et al.*,

1997) have identified sequences of nearly planar waves that show waves with permanent form, moving at velocities of some 2 mm/sec, that preserve their form after experiencing *head-on* collisions with only phase shifts in their trajectories with, however, opposite sign to that of *overtaking* collisions (Zabusky and Kruskal, 1965; Christov and Velarde, 1995). When these waves have oblique reflections at walls, if the incident angle to the wall is small enough, a third wave appears analogous to the Mach–Russell *stem* (Figure 19). The possibility of no wave reflected exists as also reported by Russell for the solitary wave in water canals (Russell, 1844, 1885; Bazin, 1865; Bouasse, 1924). For oblique collisions, Linde *et al.* (1993a,b,c, 1997) have shown that the phase shifts depend on the collision angle, thus proceeding from negative to positive values as the angle goes from acute (and also head-on) to obtuse (overtaking collisions are expected to fall in this category) (Figure 20). There is a critical angle about $\pi/2$, in agreement with theory (Nepomnyashchy and Velarde, 1994), at which the collision is neutral as no phase shift occurs. The value predicted for ideal, dissipation-free flows is $\pi/3$ (Miles, 1997a, b, 1980), which agrees with the data reported by Russell and Bazin and with the more recent experiments by Maxworthy (1976) and Melville (1980).

Recent experimental work (Linde *et al.*, 1997, 2000; Wierschem *et al.*, (1999, 2000) has produced further interesting findings on the 2D and 3D

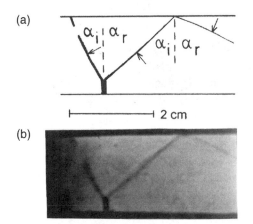

Fig. 19. Reflection pattern in an experiment of pentane vapor absorption on liquid toluene (layer depth, 2 mm): (a) Graphic reconstruction of the (b) shadow picture of two reflections, one with Mach-Russell stem and the other without, corresponding to *small* and *large* enough incident angles, respectively. (Redrawn after Linde *et al.*, 1993c.)

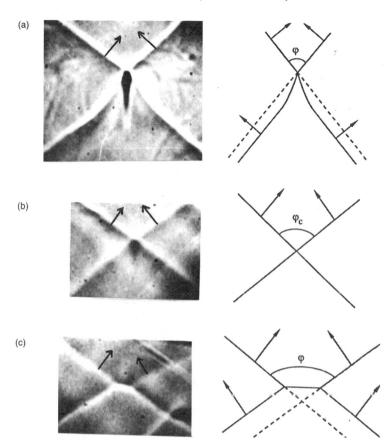

FIG. 20. Oblique collisions of two solitary waves in an experiment of pentane vapor absorption on liquid toluene (layer depth, 2 mm). For each event we provide on the left part the shadow picture and on the right its corresponding graphic reconstruction: (a) Acute collision angle, φ, with *negative* phase shift as for *overtaking* collisions. (b) Neutral collision with no phase shift (collision angle, φ_c, about $\pi/2$). (c) Obtuse angle, φ, with *positive* phase shift and corresponding Mach-Russell *stem* as for *head-on* collisions. (Redrawn after Linde *et al.*, 1993a.)

characteristics of the above mentioned nonlinear long waves (Figure 21). In experiments with absorption of pentane vapor on toluene, it appears that as soon as the surfactant vapor is absorbed a set of wave crests or wavetrains develops in the liquid layer with flow motions mostly confined near the surface of the layer, hence one is tempted to call them *surface* waves. Subsequently, as the surfactant penetrates deeper and deeper into the liquid

(a)

(b)

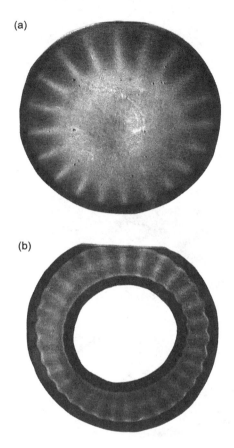

FIG. 21. Snapshots of traveling wavetrains in experiments of vapor absorption on liquid toluene: (a) Circular cylindrical container (hexane vapor; diameter, 43 mm; liquid depth, 18 mm). (b) Annular circular cylindrical container (pentane vapor; outer diameter, 57 mm; inner diameter, 37 mm; liquid depth, 4 mm). (Redrawn after Velarde *et al.*, 1995a, with permission of World Scientific.)

bulk, the first set of (surface) waves seems to act like "wind" on the stably stratified pentane-toluene layer (pentane is lighter than toluene) thus exciting waves and wavetrains of shorter wavelength that one is tempted to term *internal* waves. As with solitary waves, the collisions of either *surface* or *internal* wave crests and their reflections at walls show also the earlier mentioned (kinematic) *solitonic* signatures (Figure 22). Beautiful daisy-like and other dynamic patterns have been observed in multiple collision events (Figure 23) (Wierschem *et al.*, 1999; see also Santiago-Rosanne *et al.*, 1997).

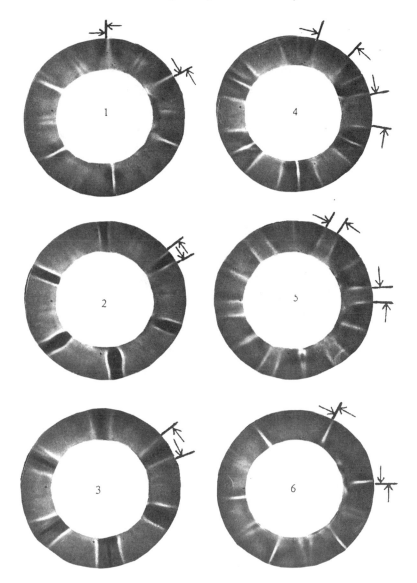

Fɪɢ. 22. Six snapshots to illustrate two consecutive (synchronic) head-on collisions of wave crests of two counter-rotating periodic wavetrains with six wave crests each, in an experiment of pentane vapor absorption on liquid toluene (outer diameter, 57 mm; inner diameter, 37 mm; liquid depth, 4 mm). Time step between two consecutive snapshots: 0.1 s. (Redrawn after Velarde *et al.*, 1995a, with permission of World Scientific.)

(a)

(b)

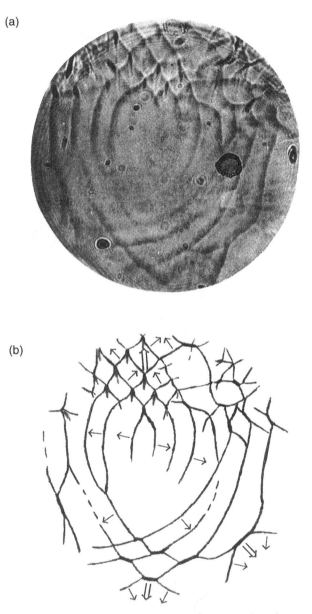

FIG. 23. (a) Snapshot and (b) corresponding graphic representation of a pattern appearing in an experiment of benzene desorption on nonane (diameter, 49 mm; liquid depth, 4 mm), with simultaneous multiple collisions at various places with varying collision angles thus forming occasionally Mach–Russell stems. Arrows ($\rightarrow$) indicate motion of *individual* wave crests while double arrows ($\Rightarrow$) indicate overall motion of the pattern as a whole. Note the similarity with pictures taken in some combustion processes. (Redrawn after Linde *et al.*, 2000.)

VIII. Concluding Remarks

Bénard convection when the liquid layer is heated from the air rather than from the liquid side offers a challenging problem, both theoretical and experimental. Although progress has been achieved since the pioneering, albeit unsystematic, experiments conducted by Block (1956), and the theory developed by Sternling and Scriven (1959), only a few properties have been firmly established. Certainly, a systematic exploration of the onset of (oscillatory) instability with heat transfer is needed to establish unambiguous results and a testing ground for theory. Mass transfer experiments have the added difficulty of being non-stationary by nature. On the other hand, almost nothing has been explored of the eventually time-dependent flows in the standard Bénard–Marangoni problem, heating the layer from the liquid side and proceeding well beyond patterned, steady Bénard convection. Thus, the surface tension gradient-driven instability offers a field of research with a high potential for development, both theoretical and experimental.

Acknowledgments

The authors acknowledge fruitful discussions with Dr. X.-L. Chu, Prof. C. I. Christov, Dr. A. N. Garazo, Prof. G. Z. Gershuni, Prof. H. Linde, Dr. A. Mendes-Tatsis, Prof. J. Pantaloni, Prof. Yu. S. Ryazantsev, Prof. S. Slavchev, Dr. A. Ye. Rednikov, Prof. A. Sanfeld, Dr. M. Santiago-Rosanne, Dr. M. Vignès-Adler, Prof. P. D. Weidman, and Dr. A. Wierschem. Two of the authors (AAN and MH) wish to express their appreciation to the Spanish Ministry of Education and to the Universidad Complutense de Madrid for sabbatical positions held at the Instituto Pluridisciplinar where research leading to this manuscript was carried out. This research has been sponsored by DGICYT (Spain) under grants PB90-264, PB93-81, and PB96-599, and by the European Union under network grant 960010.

References

Alfaro, C. M., and Depassier, M. C. (1989). Solitary waves in a shallow viscous fluid sustained by an adverse temperature gradient. *Phys. Rev. Lett.* **62**, 2597–2599.

Aspe, H., and Depassier, M. C. (1990). The evolution equation of surface waves in a convecting fluid. *Phys. Rev. A.* **41**, 3125–3128.

Badratinova, L. G. (1985). In *Dynamics of Fluids with Free Interfaces*, Institute for Hydrodynamics, Siberian Branch USSR Acad. Sci., No. 69, 3–18. (in Russian)

Bar, D., and Nepomnyashchy, A. A. (1995). Stability of periodic waves governed by the modified Kawahara equation. *Physica D* **86**, 586–602.

Bazin, H. (1865). Recherches experiméntales sur la propagation des ondes. *Mém. présentés par divers savants à l'Acad. Sci. Inst. France*, **19**, 495–644.

Bénard, H. (1900). Les tourbillons cellulaires dans une nappe liquide. Première partie: description générale des phénomènes. *Rev. Gén. Sci. Pures Appl.* **11**, 1261–1271.

Bénard, H. (1901). Les tourbillons céllulaires dans une nappe liquide transportant de le chaleur par convection en règime permanent. *Ann. Chim. Phys.* **23**, 62–143.

Benguria, R. D., and Depassier, M. C. (1987). Oscillatory instabilities in the Rayleigh-Bénard problem with a free surface. *Phys. Fluids* **30**, 1678–1682.

Benguria, R. D., and Depassier, M. C. (1989). On the linear stability theory of Bénard–Marangoni convection. *Phys. Fluids A* **1**, 1123–1127.

Benjamin, T. B. (1982). The solitary wave with surface tension. *Quart. Appl. Maths*, **40**, 231–234.

Berg, J. C., and Acrivos, A. (1965). The effect of surface active agents on convection cells induced by surface tension. *Chem. Engng. Sci.* **20**, 737–745.

Berg, J. C., Acrivos, A., and Boudart, M. (1966a). Natural convection in pools of evaporating liquids. *J. Fluid Mech.* **24**, 721–735.

Berg, J. C., Acrivos, A., and Boudart, M. (1966b). Evaporative convection. *Adv. Chem. Eng.* **6**, 61–123.

Bestehorn, M. (1993). Phase and amplitude instabilities for Bénard–Marangoni convection in fluid layers with large aspect ratio. *Phys. Rev. E* **48**, 3622–3634.

Bestehorn, M. (1996). Square patterns in Bénard–Marangoni convection. *Phys. Rev. Lett.* **76**, 46–49.

Birikh, R. V., and Rudakov, R. N. (1966). Thermocapillary instability of a deformable liquid film. *Izv. Rus. Akad. Nauk., Mekh. Zhid. Gaza* **5**, 30–36. [in Russian; English translation: *Fluid Dyn.* **31**(5), 655–660]

Birikh, R. V., Briskman, V. A., Rudakov, R. N., and Velarde, M. G. (1994). Marangoni–Bénard convective instability driven by a heated divider. *Int. J. Heat Mass Transfer* **37**, 493–498.

Birikh, R. V., Briskman, V. A., Rudakov, R. N., and Velarde, M. G. (1995). Marangoni–Bénard instability of a floating liquid layer with an internal, permeable, heated or cooled divider and two deformable open surfaces. *Int. J. Heat Mass Transfer* **38**, 2723–2731.

Block, M. J. (1956). Surface tension as the cause of Bénard cells and surface deformation in a liquid film. *Nature (Lond.)* **178**, 650–651.

Bouasse, H. (1924). *Houle, Rides, Seiches et Marées*, pp. 291–292. Delagrave, Paris.

Boussinesq, J. V. (1871). Théorie de l'intuméscence liquide appelée onde solitaire ou de translation se propageant dans un canal rectangulaire. *C. R. Hebd. Séances Acad. Sci.* **72**, 755–759.

Boussinesq, J. V. (1872). Théorie des ondes et des remous qui se propagent le long d'un canal rectangulaire horisontal en communiquant au liquid contenu dans ce canal des vitesses sensiblement pareilles de la surface au fond. *J. Math. Pures Appl.* **17**, 55–108.

Boussinesq, J. V. (1877). Essai sur la theorie des eaux courantes. *Mem. préséntes par divers savants à l'Acad. Sci. Inst. France* **23**, 1–680.

Bragard, J., and Lebon, G. (1993). Non-linear Marangoni convection in a layer of finite depth. *Europhys. Lett.* **21**, 831–836.

Bragard, J., and Velarde, M. G. (1997). Bénard convection flows. *J. Non-Equilib. Thermodyn.* **22**, 1–19.

Bragard, J., and Velarde, M. G. (1998). Bénard–Marangoni convection: Theoretical predictions about planforms and their relative stability. *J. Fluid Mech.* **368**, 165–194.

Brian, P. (1971). Effect of Gibbs adsorption on Marangoni instability. *A.I.Ch.E. J.* **17**, 765–772.

Busse, F. H. (1967). The stability of finite amplitude cellular convection and its relation to an extremum principle. *J. Fluid Mech.* **30**, 625–649.

Busse, F. H. (1978). Non-linear properties of thermal convection. *Rep. Prog. Phys.* **41**, 1929–1967.

Cardin, Ph., Nataf, H.-C., and Dewost, Ph. (1991). Thermal coupling in layered convection: Evidence for an interface viscosity control from mechanical experiments and marginal stability analysis. *J. Phys. II (Paris)* **1**, 599–622.

Castillo, J. L., García-Ybarra, P. L., and Velarde, M. G. (1988). In *Synergetics and Dynamic Instabilities* (G. Caglioti, H. Haken, and L. A. Lugiato, eds.), pp. 219–243. North-Holland, Amsterdam.

Cerisier, P., Jamond, C., Pantaloni, J., and Pérez-García, C. (1987). Stability of roll and hexagonal patterns in Bénard-Marangoni convection. *Phys. Fluids.* **30**, 954–959.

Chandrasekhar, S. (1961). *Hydrodynamic and Hydromagnetic Stability.* Clarendon Press, Oxford.

Chang, F.-P., and Chiang, K.-T. (1998). Oscillatory instability analysis of Bénard–Marangoni convection in a rotating fluid under a uniform magnetic field. *Int. J. Heat Mass Transfer* **41**, 2667–2675.

Chang, H.-C., and Demkhin, E. A. (1996). Solitary wave formation and dynamics of falling films. *Adv. Appl. Mech.* **32**, 1–58.

Char, M. I., and Chiang, K. T. (1996). Effect of interfacial deformation on the onset of convective instability in a doubly diffusive fluid layer. *Int. J. Heat Mass Transfer* **39**, 407–418.

Christov, C. I., and Velarde, M. G. (1993). On localized solutions in an equation in Bénard convection. *Appl. Math. Model* **17**, 311–318.

Christov, C. I., and Velarde, M. G. (1995). Dissipative solitons. *Physica D* **86**, 323–347.

Christov, C. I., Maugin, G. A., and Velarde, M. G. (1996). Well-posed Boussinesq paradigm with purely spatial higher-order derivatives. *Phys. Rev. E* **54**, 3621–3638.

Chu, X.-L., and Chen, L.-Y. (1987). Hydrodynamic stability of two-component fluid heated from below and with surface adsorption. *Commun. Theor. Phys. (Beijing)* **8**, 167–177.

Chu, X.-L., and Velarde, M. G. (1988). Sustained transverse and longitudinal waves at the open surface of a liquid. *Physicochem. Hydrodyn.* **10**, 727–737.

Chu, X.-L., and Velarde, M. G. (1989a). Transverse and longitudinal waves induced and sustained by surfactant gradients at liquid-liquid interfaces. *J. Colloid Interface Sci.* **131**, 471–484.

Chu, X.-L., and Velarde, M. G. (1989b). Dissipative hydrodynamic oscillators V Interfacial oscillations in a finite container.. *Il Nuovo Cimento D* **11**, 1615–1629.

Chu, X.-L., and Velarde, M. G. (1989c). Dissipative hydrodynamic oscillators. VI. Transverse interfacial motions not obeying the Laplace–Kelvin law. *Il Nuovo Cimento D* **11**, 1631–1643.

Chu, X.-L., and Velarde, M. G. (1991). Korteweg–de Vries soliton excitation in Bénard–Marangoni convection. *Phys. Rev. A* **43**, 1094–1096.

Chu, X.-L., Velarde, M. G., and Castellanos, A. (1989). Dissipative hydrodynamic oscillators. III. EHD interfacial phenomena. *Il Nuovo Cimento D* **11**, 723–737.

Cloot, A., and Lebon, G. V. (1984). A nonlinear stability analysis of the Bénard–Marangoni problem. *J. Fluid Mech.* **145**, 447–469.

Cole, J. D. (1968). *Perturbation Methods in Applied Mathematical Blaisdell*, Sec. 5.2.2. Waltham, MA.

Courant, R., and Friedrichs, K. O. (1948). *Supersonic Flow and Shock Waves.* Wiley-Interscience, New York.

Dauby, P. C., Lebon, G., Colinet, P., and Legros, J. C. (1993). Hexagonal Marangoni convection in a rectangular box with slippery walls. *Quart. J. Mech. Appl. Math.* **46**, 683–707.

Dauzère, C. (1908). Recherches sur la solidification céllulaire. *J. Phys. (Paris)* **7**, 930–934.

Dauzère, C. (1912). Sur les changements qu'éprouvent les tourbillons céllulaires lorsque la température s'élève. *C. R. Hebd. Séances Acad. Sci.* **155**, 394–398.

Davis, S. H. (1987). Thermocapillary instabilities. *Ann. Rev. Fluid Mech.* **19**, 403–435.

Davis, S. H., and Segel, L. A. (1968). Effects of surface curvature and property variation on cellular convection. *Phys. Fluids* **11**, 470–476.

de Boer, P. C. T. (1984). Thermally driven motion of strongly heated fluids. *Int. J. Heat Mass Transfer* **27**, 2239–2251.

de Boer, P. C. T. (1986). Thermally driven motion of highly viscous fluids. *Int. J. Heat Mass Transfer* **29**, 681–688.

Depassier, M. C. (1984). A note on the free boundary condition in Rayleigh–Bénard convection between insulating boundaries. *Phys. Lett. A* **102**, 359–361.

Drazin, P. G., and Johnson, R. S. (1989). *Solitons. An Introduction.* Cambridge University Press, Cambridge.

Drazin, P. G., and Reid, W. H. (1981). *Hydrodynamic Stability.* Cambridge University Press, Cambridge.

Eckert, K., Besterhorn, M., and Thess, A. (1998). Square cells in surface-tension-driven Bénard convection: Experiment and theory. *J. Fluid Mech.* **356**, 155–197.

Elphick, C., Ierley, G. R., Regev, O., and Spiegel, E. A. (1991). Interacting localized structures with Galilean invariance. *Phys. Rev. A* **44**, 1110–1122.

Estevez, P. G., and Gordoa, P. R., (1993). In *Applications of Analytic and Geometric Methods to Nonlinear Differential Equations* (P. A. Clarkson, ed.), pp. 287–288. Kluwer, Dordrecht.

Funada, T. (1986). Marangoni instability of thin liquid sheet. *J. Phys. Soc. Jpn.* **55**, 2191–2202.

Funada, T. (1987). Nonlinear surface waves driven by the Marangoni instability in a heat transfer system. *J. Phys. Soc. Jpn.* **56**, 2031–2038.

Funada, T., and Kotani, M. (1986). A numerical study of nonlinear diffusion equation governing surface deformation in the Marangoni convection. *J. Phys. Soc. Jpn.* **55**, 3857–3862.

Garazo, A. N., and Velarde, M. G. (1991). Dissipative Korteweg–de Vries description of Marangoni–Bénard convection. *Phys. Fluids A* **3**, 2295–2300.

Garazo, A. N., and Velarde, M. G. (1992). Marangoni-driven solitary waves. *ESA-SP-333 (Paris)* 711–715.

García-Ybarra, P. L., and Velarde, M. G. (1987). Oscillatory Marangoni–Bénard interfacial instability and capillary-gravity waves in single- and two-component liquid layers with or without Soret thermal diffusion. *Phys. Fluids* **30**, 1649–1655.

García-Ybarra, P. L., Castillo, J. L., and Velarde, M. G. (1987). Bénard–Marangoni convection with a deformable interface and poorly conducting boundaries. *Phys. Fluids* **30**, 2655–2661.

Gardner, P. L., Greene, J. M., Kruskal, M. D., and Miura, R. M. (1967). Method for solving Korteweg-de Vries equation. *Phys. Rev. Lett.* **19**, 1095–1097.

Georis, Ph., and Legros, J. (1996). In *Materials and Fluid under Low Gravity* (L. Ratke, H. Walter, and B. Feuerbacher, eds.), pp. 293–311. Springer, Berlin.

Georis, P., Hennenberg, M., Simanovski, I. B., Nepomnyashchy, A., Wertheim, I. B., and Legros, J. C. (1993). Thermocapillary convection in a multilayer system. *Phys. Fluids A* **5**, 1575–1582.

Gershuni, G. Z., and Zhukhovitsky, E. M. (1976). *Convective Stability of Incompressible Fluids.* Keter, Jerusalem.

Gershuni, G. Z., and Zhukhovitsky, E. M. (1982). Monotonic and oscillatory instabilities of a two-layer system of immiscible liquids heated from below. *Sov. Phys. Dokl.* **27**, 531–533.

Gilev, A. Yu, Nepomnyashchy, A. A., and Simanovsky, I. B. (1987). Convection in a two-layer system due to the combined action of the Rayleigh and thermocapillary instability mechanisms. *Izv. Akad. Nauk SSSR, Mekh. Zhidk. i. Gaza*, 166. (in Russian; English translation, *Fluid Dynam.* **22**, 142–145)

Golovin, A. A., Nepomnyashchy, A. A., and Pismen, L. M. (1994). Interaction between short-scale Marangoni convection and long-scale deformational instability. *Phys. Fluids* **6**, 34–48.

Golovin, A. A., Nepomnyashchy, A. A., and Pismen, L. M. (1995). Pattern formation in large-scale Marangoni convection with deformable interface. *Physica D* **81**, 117–147.

Golovin, A. A., Nepomnyashchy, A. A., and Pismen, L. M. (1997a). Nonlinear evolution and secondary instabilities of Marangoni convection in a liquid-gas system with deformable interface. *J. Fluid Mech.* **341**, 317–341.

Golovin, A. A., Nepomnyashchy, A. A., Pismen, L. M., and Riecke, H. (1997b). Steady and oscillatory side-band instabilities in Marangoni convection with deformable interface. *Physica D* **106**, 131–147.

Gouesbet, G., Maquet, J., Rozé, C., and Darrigo, R. (1990). Surface-tension and coupled buoyancy-driven instability in a horizontal liquid layer. Overstability and exchange of stability. *Phys. Fluids A* **2**, 903–911.

Goussis, D. A., and Kelly, R. E. (1990). On the thermocapillary instabilities in a liquid layer heated from below. *Int. J. Heat Mass Transfer* **33**, 2237–2245.

Gumerman, R. J., and Homsy, G. M. (1974). Convective instabilities in concurrent two phase flow. *A.I.Ch.E. J.* **20**, 981–1172; Part II. Global stability, *A.I.Ch.E. J.* **20**, 1161–1172.

Gunaratne, G. H. (1993). Complex spatial patterns on planar continua. *Phys. Rev. Lett.* **10**, 1367–1370.

Gunaratne, G. H., Ouyang, Q., and Swinney, H. L. (1994). Pattern formation in the presence of symmetries. *Phys. Rev. E* **50**, 2802–2820.

Hadji, L., Safar, J., and Schell, M. (1991). Analytical results on the coupled Bénard–Marangoni problem consistent with experiment. *J. Non-Equilib. Thermodyn.* **16**, 343–356.

Hennenberg, M., Bisch, P. M., Vignès-Adler, M., and Sanfeld, A. (1979). Mass transfer, Marangoni effect and instability of interfacial longitudinal waves, I. Diffusional exchanges. *J. Colloid Interface Sci.* **69**, 128–137.

Hennenberg, M., Bisch, P. M., Vignès-Adler, M., and Sanfeld, A. (1980). Mass transfer, Marangoni effect and instability of interfacial longitudinal waves, II. Diffusional exchanges and adsorption–desorption processes. *J. Colloid Interface Sci.* **74**, 495–508.

Hennenberg, M., Chu, X.-L., Sanfeld, A., and Velarde, M. G. (1992). Transverse and longitudinal waves at the air liquid interface in the presence of an adsorption barrier. *J. Colloid Interface Sci.* **150**, 721.

Hershey, A. V. (1939). Ridges in a liquid surface due to the temperature dependence of surface tension. *Phys. Rev.* **56**, 204.

Hoard, C. Q., Robertson, C. R., and Acrivos, A. (1970). Experiments on the cellular structure in Bénard convection. *Int. J. Heat Mass Transfer* **13**, 849–856.

Hornung, H. (1988). Regular and Mach reflection of shock waves. *Ann. Rev. Fluid Mech.* **18**, 33–58.

Huang, G.-H., Velarde, M. G., and Kurdiumov, V. (1998). Cylindrical solitary waves and their interaction in Bénard–Marangoni layers. *Phys. Rev. E* **57**, 5473–5482.

Imaishi, N., and Fujinawa, K. (1974a). Theoretical study of the stability of two-fluid layers. *J. Chem. Eng. Jpn.* **7**, 81–86.

Imaishi, N., and Fujinawa, K. (1974b). Thermal instability in two-fluid layers. *J. Chem. Eng. Jpn.* **7**, 87–92.

Imaishi, N., Hozawa, M., Fujinawa, K., and Suzuki, Y. (1983). Theoretical study of interfacial turbulence in gas–liquid mass transfer, applying Brian's linear-stability analysis and using numerical analysis of unsteady Marangoni convection. *Int. Chem. Eng.* **23**, 466–476.

Izakson, V. Kh. (1969). On the influence of the surface tension on the onset of convection in a fluid layer with free surface. *Zh. Prikl. Mekh. i. Tekhn. Fiz.* **3**, 89–92. (in Russian)

Izakson, V. Kh., and Yudovich, V. I. (1968). On the onset of convection in a fluid layer with free surface. *Izv. Akad. Nauk SSSR, Mekh. Zhidk. i Gaza* **4**, 23–28. (in Russian)

Janiaud, B., Pumir, A., Bensimon, D., Croquette, V., Richter, H., and Kramer, L. (1992). The Eckhaus instability for traveling waves. *Physica D* **55**, 269.

Johnson, R. S. (1972). Shallow water waves on a viscous fluid. The undular bore. *Phys. Fluids* **15**, 1693–1699.

Joo, S. W. (1995). Marangoni instabilities in liquid mixtures with Soret effects. *J. Fluid Mech.* **293**, 127–145.

Kac, M. Uhlenbeck, G. E., and Hemmer, P. C. (1963). On the van der Waals theory of the vapor–liquid equilibrium. I. Discussion of a one-dimensional model. *J. Math. Phys.* **4**, 216–28.

Kats-Demyanets, V., Oron, A., and Nepomnyashchy, A. A. (1997). Marangoni instability in tri-layer liquid system. *Eur. J. Mech. B/Fluids* **16**, 49–74.

Kawahara, T. (1983). Formation of saturated solutions in a nonlinear dispersive system with instability and dissipation. *Phys. Rev. Lett.* **51**, 381–383.

Kawahara, T., and Toh, S. (1988). Pulse interactionsa in an unstable dissipative-dispersive nonlinear system. *Phys. Fluids* **31**, 2103–2111.

Knobloch, E. (1990). Pattern selection in long-wavelength convection. *Physica D* **41**, 450–479.

Korteweg, D. J., and de Vries, G. (1895). On the change of form of long waves advancing in a rectangular channel, and on a new type of long stationary waves. *Phil. Mag.* **39**(5), 422–443.

Koschmieder, E. L. (1966). On convection on a uniformly heated plane. *Beitr. Phys. Atmos.* **39**, 1–11.

Koschmieder, E. L. (1967). On convection under an air surface. *J. Fluid Mech.* **30**, 9–15.

Koschmieder, E. L. (1988). In *Physicochemical Hydrodynamics: Interfacial Phenomena* (M. G. Velarde, ed.), pp. 189–198. Plenum, New York.

Koschmieder, E. L. (1993). *Bénard Cells and Taylor Vortices.* Cambridge University Press, Cambridge.

Koschmieder, E. I., and Biggerstaff, M. I. (1986). Onset of surface-tension-driven Bénard convection. *J. Fluid Mech.* **167**, 49–64.

Koulago, A. E., and Parséghian, D. (1996). A propos d'une équation de la dynamique ondulatoire dans les films liquides. *J. Phys. III (Paris)* **5**, 309–312.

Kozhoukharova, Z. D., and Slavchev, S. (1992). Influence of the surface deformability on Marangoni instability in a liquid layer with surface chemical reaction. *J. Colloid Interface Sci.* **148**, 42–55.

Kraenkel, R. A., Pereira, J. G., and Manna, M. A. (1992a). Nonlinear surface-wave excitations in the Bénard–Marangoni system. *Phys. Rev. A* **46**, 4786–4790.

Kraenkel, R. A., Pereira, J. G., and Manna, M. A. (1992b). Surface perturbations of a shallow viscous fluid heated from below and the (2 + 1)-dimensional Burgers equation. *Phys. Rev. A* **45**, 838–841.

Krehl, P., and van der Geest, M. (1991). The discovery of the Mach reflection effect and its demonstration in an auditorium. *Shock Waves* **1**, 3–15.

Kudryashov, N. A. (1988) Exact soliton solutions of the generalized evolution equation of wave dynamics, *J. Appl. Math. Mech.* **52**, 361–365.

Kudryashov, N. A. (1990). Exact solutions of the generalized Kuramoto–Sivashinsky equation. *Phys. Lett. A* **147**, 287–291.

Kuramoto, Y., and Tsuzuki, T. (1976). Persistent propagation of concentration waves in dissipative media from thermal equilibrium. *Prog. Theor. Phys.* **55**, 356–369.

Lamb, H. (1945). *Hydrodynamics*, 6th ed., Sec. 349. Dover, New York.

Langmuir, I., and Langmuir, D. B. (1927). The effect of monomolecular films on the evaporation of ether solutions. *J. Phys. Chem.* **31**, 1719–1731.

Lax, P. D. (1968). Integrals of nonlinear equations of evolution and solitary waves. *Comm. Pure Appl. Math.* **21**, 467–490.

Levchenko, E. B., and Chernyakov, A. L. (1981). Instability of surface waves in a nonuniformly heated liquid. *Soviet Phys. JETP* **54**, 102–105.

Limat, L. (1993). Instabilité d'un liquide suspendu sous un surplomb solide: Influence de l'épaisseur de la couche. *C. R. Hebd. Séances Acad. Sci. Ser. II* **317**, 563–568.

Linde, H., Pfaff, S., and Zirkel, C. (1964). Strömungsuntersuchungen zur hydrodynamischen Instabilität flüssig-gasförmiger-Phasengrenzen mit Hilfa der Kapillarspaltmethode. *Z. Phys. Chem.* **225**, 72–100.

Linde, H., Schwartz, P., and Wilke, H. (1979). In *Dynamics and Instability of Fluid Interfaces* (T. S. Sørensen, ed.), pp. 75–119. Springer-Verlag, Berlin.

Linde, H., Chu, X.-L., and Velarde, M. G. (1993a). Oblique and head-on collisions of solitary waves in Marangoni–Bénard convection. *Phys. Fluids A* **5**, 1068–1070.

Linde, H., Chu, X.-L., and Velarde, M. G. (1993b). Solitary waves driven by Marangoni stresses. *Adv. Space Res.* **13**, 109–117.

Linde, H., Chu, X.-L., Velarde, M. G., and Waldhelm, W. (1993c). Wall reflection of solitary waves in Marangoni–Bénard convection. *Phys. Fluids A* **5**, 3162–3166.

Linde, H., Velarde, M. G., Wierschem, A., Waldhelm, W., Loeschcke, K., and Rednikov, A. Ye. (1997). Interfacial wave motions due to Marangoni instability. I. Traveling periodic wave trains in square and annular containers. *J. Colloid Interface Sci.* **188**, 16–26.

Linde, H., Velarde, M. G., Waldhelm, W., and Wierschem, A. (2000). Interfacial wave motions due to Marangoni instability. III. Solitary waves and (periodic) wavetrains and their collisions and reflections leading to dynamic network patterns. *J. Colloid Interface Sci.* (to be published).

Liu, Q. S., Roux, B., and Velarde, M. G. (1998). Thermocapillary convection in two-layer systems. *Int. J. Heat Mass Transfer* **41**, 1499–1511.

Lou, S.-Y., Huang, G.-X., and Yuan, H.-R. (1991). Exact solitary waves in a convecting fluid, *J. Phys. A: Math. Gen.* **24**, L587–L590.

Low, A. R. (1929). On the criterion for stability of a layer of viscous fluid heated from below. *Proc. Soc. London A* **125**, 180–195.

Lucassen, J. (1968). Longitudinal capillary waves, Part 1. Theory. *Trans. Faraday Soc.* **64**, 2221–2229; Part 2 Experiments, *Trans. Faraday Soc.* **64**, 2230–2235.

Lucassen-Reynders, E. H., and Lucassen, J. (1969). Properties of capillary waves. *Adv. Colloid Interface Sci.* **2**, 347–395.

Maldarelli, C., Jain, R. K., Ivanov, I. B., and Ruckenstein, E. (1980). Stability of symmetric and unsymmetric thin liquid films to short and long wavelength perturbations. *J. Colloid Interface Sci.* **78**, 118–143.

Maxworthy, T. (1976). Experiments on collisions between solitary waves. *J. Fluid Mech.* **76**, 177–185.

Melville, W. K. (1980). On the Mach reflection of a solitary wave. *J. Fluid Mech.* **98**, 285–297.

Mihaljan, J. (1962). A rigorous exposition of the Boussinesq approximations applicable to a thin layer of fluid. *Astrophys. J.* **136**, 1126–1133.

Miles, J. W. (1976). Korteweg–de Vries equation modified by viscosity. *Phys. Fluids* **19**, 1063.

Miles, J. W. (1977a). Obliquely interacting solitary waves. *J. Fluid Mech.* **79**, 157–169.

Miles, J. W. (1977b). Resonantly interacting solitary waves. *J. Fluid Mech.* **79**, 171–179.

Miles, J. W. (1980). Solitary waves. *Ann. Rev. Fluid Mech.* **12**, 11–43.

Nekorkin, V. I., and Velarde, M. G. (1994). Solitarty waves of a dissipative Korteweg–de Vries equation describing Marangoni–Bénard convection and other thermoconvective instabilities. *Int. J. Bifurc. Chaos* **4**, 1135–1146.

Nepomnyashchy, A. A. (1974). Stability of the wavy regimes in the film flowing down and inclined plane. *Izv. AN SSSR, Fluid Gas Mech. Mekh. Zhidk. Gaza* **3**, 28. (in Russian; English translation, *Fluid Dynam.* **9**, 354–359.

Nepomnyashchy, A. A. (1976). Wavy motions in the layer of viscous fluid flowing down the inclined plane. *Fluid Dynam.* **9**; *Trans. Perm State Univ.* **362**, 114–124. (in Russian)

Nepomnyashchy, A. A. (1983). On the longwave convective instability in the horizontal layer with deformable boundary. In *Convective Flows*, E. M. Zhukovitsky, ed., Perm Pedagogical Institute, pp. 25–32.

Nepomnyashchy, A. A., and Simanovsky, I. B. (1983). Thermocapillary convection in a two-layer system, *Sov. Phys. Dokl.* **28**, 838–839.

Nepomnyashchy, A. A., and Simanovsky, I. B. (1986). Thermocapillary convection in two-layer systems in the presence of surfactant on the interface. *Izv. Akad. Nauk SSSR, Mekh. Zhidk. i Gaza* 3–8. (in Russian; English translation, *Fluid Dynam.* **21**, 469–471)

Nepomnyashchy, A. A., and Simanovsky, I. B. (1990). Long-wave thermocapillary convection in layers with deformable interfaces. *J. Appl. Math. Mech.* **54**, 490–496.

Nepomnyashchy, A. A., and Simanovsky, I. B. (1991). Onset of oscillatory thermocapillary convection in systems with deformable interface. *Izv. Akad. Nauk SSSR, Mekh. Zhidk. i Gaza* **4**, 11–16. (in Russian)

Nepomnyashchy, A. A., and Velarde, M. G. (1994). A three-dimensional description of solitary waves and their interaction in Marangoni–Bénard layers. *Phys. Fluids* **6**, 187–198.

Nield, D. A. (1964). Surface tension and buoyancy effects in cellular convection. *J. Fluid Mech.* **19**, 341–352.

Nitschke, K., and Thess, A. (1995). Secondary instability in surface tension driven Bénard convection. *Phys. Rev. E* **52**, R 5772–5775.

Normand, C., Pomeau, Y., and Velarde, M. G. (1977). Convective instability: A physicist's approach. *Rev. Mod. Phys.* **49**, 581–624.

Oron, A., and Rosenau, Ph. (1989). Evolution of the coupled Bénard–Marangoni convection, *Phys. Rev. A* **39**, 2063–2069.

Oron, A., Deissler, R. J., and Duh, J. C. (1995). Marangoni convection in a liquid sheet. *Adv. Space Res.* **16**, 83–86.

Oron, A., Davis, S. H., and Bankoff, G. (1997). Long-scale evolution of thin liquid films. *Rev. Mod. Phys.* **69**, 931–980.

Ostrach, S. (1982). Low-gravity fluid flows. *Ann. Rev. Fluid Mech.* **14**, 313–345.

Palm, E. (1960). On the tendency towards hexagonal cells in steady convection. *J. Fluid Mech.* **8**, 183–192.

Palmer, H. J., and Berg, J. C. (1972). Hydrodynamic stability of surfactant solutions heated from below. *J. Fluid Mech.* **51**, 385–402.

Pantaloni, J., Bailleux, R., Salan, J., and Velarde, M. G. (1979). Rayleigh–Bénard–Marangoni convective instability: New experimental results. *J. Non-Equilib. Thermodyn.* **4**, 201–218.

Parmentier, P. M., Regnier, V. C., and Lebon, G. (1996). Nonlinear analysis of coupled gravitational and capillary thermoconvection in thin fluid layers. *Phys. Rev. E* **54**, 411–423.

Pearson, J. R. A. (1958). On convection cells induced by surface tension. *J. Fluid Mech.* **4**, 489–500.

Pérez-Cordón, R., and Velarde, M. G. (1975). On the (nonlinear) foundations of Boussinesq approximation applicable to a thin layer of fluid. *J. Phys. (Paris)* **36**, 591–601.

Pérez-García, C., and Carneiro, G. (1991). Linear stability analysis of Bénard–Marangoni convection in fluids with a deformable free surface. *Phys. Fluids A* **3**, 292–298.

Pismen, L. M. (1986). Inertial effects in long-scale thermal convection. *Phys. Lett. A* **116**, 241–244.

Pontes, J., Christov, C. I., and Velarde, M. G. (1996). Numerical study of patterns and their evolution in finite geometries. *Int. J. Bifurc. Chaos* **6**, 1883–1890.

Pontes, J., Christov, C. I., and Velarde, M. G. (2000). Numerical approach to pattern selection in a model problem for Bénard convection in a finite fluid layer, *Annu. Univ. Sofia* **93** (to be published).

Porubov, A. V. (1993). Exact traveling wave solutions of nonlinear evolution equation of surface waves in a convecting fluid. *J. Phys. Math. Gen. A* **26**, L797–L800.

Porubov, A. V., and Samsonov, A. M. (1995). Solitary waves on the surface of a thin layer of viscous inhomogeneous liquid. *Tech. Phys.* **40**(3), 232–237.

Prakash, A., and Koster, J. N. (1997). Steady natural convection in a two-layer system of immiscible liquids. *Int. J. Heat Mass Transfer* **40**, 2799–2812.

Rasenat, S., Busse, F. H., and Rehberg, I. (1989). A theoretical and experimental study of double-layer convection. *J. Fluid Mech.* **199**, 519–540.

Rayleigh, Lord (1876). On waves. *Phil. Mag.* **1**, 257–279.

Rayleigh, Lord (1916). On convection currents in a horizontal layer of fluid, when the higher temperature is on the under side. *Phil. Mag.* **32**, 529–536.

Rednikov, A. Ye., Colinet, P., Velarde, M. G., and Legros, J. C. (1998a). Oscillatory thermocapillary instability in a liquid layer with deformable open surface: Capillary-gravity waves, longitudinal waves and mode-mixing (submitted).

Rednikov, A. Ye., Colinet, P., Velarde, M. G., and Legros, J. C. (1998b). Two-layer Bénard–Marangoni instability and the limit of transverse and longitudinal waves. *Phys. Rev. E* **57**, 2872–2884.

Rednikov, A. Ye., Colinet, P., Velarde, M. G., and Legros, J. C. (2000). Rayleigh Marangoni oscillatory instability in a horizontal liquid layer heated from above: Coupling between internal and surface waves. *J. Fluid Mech.* **405**, 57–77.

Rednikov, A. Ye., Velarde, M. G., Ryazantsev, Yu. S., Nepomnyashchy, A. A., and Kurdyumov, V. (1995). Cnoidal wave trains and solitary waves in a dissipation-modified Korteweg–de Vries equation. *Acta Appl. Math.* **39**, 457–475.

Reichenbach, J., and Linde, H. (1981). Linear perturbation analysis of surface-tension-driven convection at a plane interface (Marangoni instability). *J. Colloid Interface Sci.* **84**, 433–443.

Renardy, M., and Renardy, Y. (1988). Bifurcation solutions at the onset of convection in the Bénard problem for two fluids. *Physica D* **32**, 227–252.

Renardy, Y. (1986). Interfacial stability in a two-layer Bénard problem. *Phys. Fluids* **29**, 356–363.

Renardy, Y., and Joseph, D. D. (1985). Oscillatory instability in a Bénard problem of two fluids. *Phys Fluids* **28**, 788–793.

Renardy, Y., and Renardy, M. (1985). Perturbation analysis of steady and oscillatory onset in a Bénard problem with two similar liquids. *Phys. Fluids* **28**, 2699–2708.

Russell, J. S. (1844). Report on waves. *Rep. 14th Meet. British Ass. Adv. Sci.*, York, pp. 311–390, J. Murray, London; reprinted as appendix in Russell (1885).

Russell, J. S. (1885). *The Wave of Translation in the Oceans of Water, Air and Ether*. Trübner, London.

Sanfeld, A., Steinchen, A., Hennenberg, M., Bisch, P. M., Van Lamsweerde-Gallez, D., and Dalle-Vedove, W. (1979). In *Dynamics and Instability of Fluid Interface* (T. S. Sørensen, ed.), pp. 168–204. Springer-Verlag, Berlin.

Santiago-Rosanne, M., Vignès-Adler, M., and Velarde, M. G. (1997). Dissolution of a drop on a liquid surface leading to surface waves and interfacial turbulence. *J. Colloid Interface Sci.* **191**, 65–80.

Sarma, G. S. R. (1981). On oscillatory modes of thermocapillary instability in a liquid layer rotating about a transverse axis. *Physicochem. Hydrodyn.* **2**, 143–151.

Scanlon, J. W., and Segel, L. A. (1967). Finite amplitude cellular convection induced by surface tension. *J. Fluid Mech.* **30**, 149–162.

Schatz, M. F., Van Hook, S., McComick, W., Swift, J. B., and Swinney, H. D. (1995). Onset of surface-tension-driven Bénard convection. *Phys. Rev. Lett.* **75**, 1938–1940.

Scriven, L. E., and Sternling, C. V. (1960). The Marangoni effects. *Nature* **187**, 186–188.

Scriven, L. E., and Sternling, C. V. (1964). On cellular convection driven by surface-tension gradients: Effects of mean surface tension and surface viscosity. *J. Fluid Mech.* **19**, 321–340.

Shkadov, V. (1967). Wave flow regimes of a thin layer of viscous fluid subject to gravity. *Fluid Dynam.* **2**, 29–34.

Shkadov, V., Kholpanov, L., and Mochalova, N. (1976). Hydrodynamics and heat-transfer in a liquid layer on an rotating surface taking into account interaction with a gas flow. *J. Eng. Phys.* **4**(31), 684–690.

Shtilman, L., and Sivashinsky, G. (1991). Hexagonal structure of large-scale Marangoni convection. *Physica D* **52**, 477–488.

Simanovsky, I. B., and Nepomnyashchy, A. A. (1993). *Convective Instabilities in Systems with Interfaces.* Gordon and Breach, New York.

Sivashinsky, G. (1977). Nonlinear analysis of hydrodynamic instability of laminar flames, Part 1, Derivation of basic equations. *Acta Astronautica* **4**, 1177–1206.

Sivashinsky, G. (1982). Large cells in nonlinear Marangoni convection. *Physica D* **4**, 227–235.

Smith, K. A. (1966). On convective instability induced by surface-tension gradients. *J. Fluid Mech.* **24**, 401–414.

Steinchen, A., Gallez, D., and Sanfeld, A. (1982). A viscoelastic approach to the hydrodynamic stability of membranes. *J. Colloid Interface Sci.* **85**, 5–15.

Sternling, C. V., and Scriven, L. E. (1959). Interfacial turbulence: Hydrodynamic instability and the Marangoni effect. *A.I.Ch.E.J.* **5**, 514–523.

Takashima, M. (1981a). Surface tension driven instability in a horizontal liquid layer with a deformable free surface, I. Stationary convection. *J. Phys. Soc. Jpn.* **50**, 2745–2750.

Takashima, M. (1981b). Surface tension driven instability in a horizontal liquid layer with a deformable free surface. II. Overstability. *J. Phys. Soc. Jpn.* **50**, 2751–2756.

Tanford, Ch. (1989). *Ben Franklin Stilled the Waves. An Informal History of Pouring Oil on Water with Reflections on the Ups and Downs of Scientific Life in General.* Duke Univ. Press, London.

Thess, A., and Bestehorn, M. (1995). Planform selection in Bénard–Marangoni convection: l-hexagons versus g-hexagons. *Phys. Rev. E* **52**, 6358–6367.

Thess, A., and Orszag, S. A. (1995). Surface tension driven Bénard–convection at infinite-hexagons, Prandtl numbers. *J. Fluid Mech.* **283**, 201–230.

Toh, S., and Kawahara, T. (1985). On the stability of soliton-like pulses in a nonlinear dispersive system with instability and dissipation. *J. Phys. Soc. Jpn.* **54**, 1257–1269.

Topper, J., and Kawahara, T. (1978). Approximate equations for long nonlinear waves on a viscous film. *J. Phys. Soc. Jpn.* **44**, 663–666.

Trifonov, Yu. Ya. (1992). Two-periodical and quasi-periodical wave solutions of the Kuramoto–Sivashinsky equation and their stability and bifurcations. *Physica D.* **54**, 311–330.

Tsvelodub, O. Yu., and Trifonov, Yu. Ya. (1989). On steady-state traveling solutions of an evolution equation describing the behavior of disturbances in active dissipative media. *Physica D* **39**, 336–351.

Ursell, F. (1953). The long-wave paradox in the theory of gravity waves. *Proc. Cambridge Phil. Soc.* **49**, 685–694.

Van Hook, S., Schatz, M. F., McCornick, W., Swift, J. B., and Swinney, H. D. (1995). Long-wavelength instability in surface-tension-driven Bénard convection. *Phys. Rev. Lett.* **75**, 4397–4400.

Van Hook, S., Schatz, M. F., McCornick, W., Swift, J. B., and Swinney, H. D. (1997). Long-wavelength surface-tension-driven Bénard convection: Experiment and theory. *J. Fluid Mech.* **345**, 45–78.

Velarde, M. G., and Chu, X.-L. (1988). The harmonic oscillator approximation to sustained gravity-capillary (Laplace) waves at liquid interfaces. *Phys. Lett. A* **131**, 430–432.

Velarde, M. G., and Chu, X.-L. (1989a). Waves and turbulence at interfaces. *Phys. Scr. T* **25**, 231–237.

Velarde, M. G., and Chu, X.-L. (1989b). Dissipative hydrodynamic oscillators. I. Marangoni effect and sustained longitudinal waves at the interface of two liquids. *Il Nuovo Cimento D* **11**, 707–716.

Velarde, M. G., and Chu, X.-L. (1992). Dissipative thermohydrodynamic oscillators. *Adv. Thermodyn.* **6**, 110–145.

Velarde, M. G., and García-Ybarra, P. L. (1987). Oscillatory and other Bénard–Marangoni instabilities in fluid layers under microgravity conditions. *European Space Agency–Special Pub.* **256**, 45–48.

Velarde, M. G., and Pérez-Cordón, R. (1976). On the (non-linear) foundations of Boussinesq approximation applicable to a thin layer of fluid. II. Viscous dissipation and large cell gap effects. *J. Phys. (Paris)* **37**, 177.

Velarde, M. G., and Rednikov, A. (1998). Time-dependent Bénard–Marangoni instability and waves. In *Time-Dependent Nonlinear Convection* (P. A. Tyvand, ed.), pp. 177–218. Computational Mechanics Publications, Southampton.

Velarde, M. G., Chu, X.-L., and Garazo, A. N. (1991a). Onset of possible solitons in surface tension-driven convection. *Physica Scr. T* **35**, 71–74.

Velarde, M. G., Garazo, A. N., and Christov, C. I. (1991b). In *Spontaneous Formation of Space–Time Structures and Criticality* (T. Riste and D. Sherrington, eds.), pp. 263–272. Kluwer, Dordrecht.

Velarde, M. G., García-Ybarra, P. L., and Castillo, J. L. (1987). Interfacial oscilations in Bénard–Marangoni layers. *Physicochem. Hydrodyn.* **9**, 387–392.

Velarde, M. G., Linde, H., Nepomnyashchy, A. A., and Waldhelm, W. (1995a). Further evidence of solitonic behavior in Bénard Marangoni convection: Periodic wave trains. In *Fluid Physics* (M. G. Velarde and C. I. Christov, eds.), pp. 433–441. World Scientific, Singapore.

Velarde, M. G., Nekorkin, V. I., and Maksimov, A. G. (1995b). Further results on the evolution of solitary waves and their bound states of a dissipative Korteweg–de Vries equation. *Int. J. Bifurc. Chaos* **5**, 831–839.

Velarde, M. G., Rednikov, A. Ye., and Linde, H. (1999). Waves generated by surface tension gradients and instability. In *Fluid Dynamics at Interfaces* (W. Shyy, ed.), pp. 43–56. Cambridge University Press, Cambridge.

Vidal, A., and Acrivos, A. (1966). Nature of the neutral state in surface-tension driven convection. *Phys. Fluids* **9**, 615–616.

Wahal, S., and Bose, A. (1988). Rayleigh–Bénard and interfacial instabilities in two immiscible liquid layers. *Phys. Fluids* **31**, 3502–3510.

Weh, L., and Linde, H. (1997). Marangoni-stress-driven "solitonic" (periodic) wave trains rotating in an annular container during heat transfer. *J. Colloid Interface Sci.* **187**, 159–165.

Weidman, P., and Maxworthy, T. (1978). Experiments on strong interactions between solitary waves. *J. Fluid Mech.* **85**, 417–431.

Weidman, P., Linde, H., and Velarde, M. G. (1992). Evidence for solitary waves in Marangoni-driven unstable liquid layers. *Phys. Fluids A* **4**, 921–926.

Whitham, G. B. (1974). *Linear and Nonlinear Waves.* John Wiley, New York.

Wierschem, A., Velarde, M. G., Linde, H., and Waldhelm, W. (1999). Interfacial wave motions due to Marangoni instability. II. Three dimensional characteristics of surface waves. *J. Colloid Interface Sci.* **212**, 365–383.

Wierschem, A., Linde, H., and Velarde, M. G. (2000). Internal waves excited by the Marangoni effect. *Phys. Rev. E* (to be published).

Wilson, S. K., and Thess, A. (1997). On the linear growth rates of the long-wave modes in Bénard–Marangoni convection. *Phys. Fluids* **9**, 2455–2457.

Wu, T. Y. (1995). Bidirectional soliton street. *Acta Mech. Sinica* **11**, 289–306.

Wu, T. Y. (1998). Nonlinear waves and solitons in water. *Physica D* **123**, 48–63.

Wu, T. Y., and Zhang, J. E. (1996). On modeling nonlinear waves. In *Mathematics is for Solving Problems* (L. P. Cook, V. Roytburd, and M. Tulin, eds.) pp. 233–247. SIAM, Philadelphia.

Yih, C.-S., and Wu, T. Y. (1995). General solution for interaction of solitary waves including head-on collisions. *Acta Mech. Simica* **11**, 193–199.

Zabusky, N., and Kruskal, M. D. (1965). Interaction of "solitons" in collisionless plasma and the recurrence of initial states. *Phys. Rev. Lett.* **15**, 57–62.

Zeren, R. W., and Reynolds, W. C. (1972). Thermal instabilities in two-fluid horizontal layers. *J. Fluid Mech.* **53**, 305–327.

Zeytounian, R. Kh. (1990). *Asymptotic Modeling of Atmospheric Flows.* Springer-Verlag, Berlin.

Zeytounian, R. Kh. (1994). A quasi-one-dimensional asymptotic theory for non-linear water waves. *J. Eng. Maths.* **28**, 261–296.

Zeytounian, R. Kh. (1995). Nonlinear long waves on water and solitons. *Uspekhi Fiz. Nauk* **165**, 1403–1456. (English translation: *Physics-Uspekhi* **38**, 1333–1381.)

Zeytounian, R. Kh. (1998). The Bénard–Marangoni thermocapillary instability problem. *Uspekhi Fiz. Nauk* **168**, 259–286. (In Russian; English translation: *Physics-Uspekhi* **41**, 241–267.)

Role of Cryogenic Helium in Classical Fluid Dynamics: Basic Research and Model Testing

KATEPALLI R. SREENIVASAN

Mason Laboratory, Yale University
New Haven, Connecticut

and

RUSSELL J. DONNELLY

Cryogenic Helium Turbulence Laboratory, Department of Physics, University of Oregon
Eugene, Oregon

ADVANCES IN APPLIED MECHANICS, VOL. 37
ISBN 0-12-002037-8 .
ISSN 0065-2165/01 $35.00

I. Introduction

A. ADVANTAGES OF CRYOGENIC HELIUM AS A TEST FLUID

Only dimensionless similarity parameters such as Reynolds and Rayleigh numbers matter for the scaling properties of incompressible fluid flows. Natural phenomena in geophysical and astrophysical flows occur at very high ("ultra-high") values of these parameters. Ultra-high Reynolds numbers also occur in navy and aerospace applications. A fundamental understanding of turbulence, often described as the most important unsolved problem of classical physics, requires the study of high-Reynolds-number fluid flows. The generation of such flows in laboratory-scale devices is facilitated by the use of fluids with small kinematic viscosity and thermal conductivity. The smallest kinematic viscosity of any substance belongs to liquid and gaseous helium. For instance, liquid helium at $2.2\,K$ has a kinematic viscosity coefficient of about $1.8 \times 10^{-4}\,cm^2\,s^{-1}$, so that a Reynolds number of 100 million can be generated with a modest flow velocity of $4\,ms^{-1}$ and a wing chord of $50\,cm$. For the same speed of water, the corresponding facility would have to be about 55 times larger. The size for air at atmospheric pressure would be about 830 times as large. Thus, using helium, one can create even in modest-scale facilities ultra-high Reynolds numbers suitable for fundamental work and model testing. Similarly, astronomically relevant Rayleigh numbers of the order of 10^{20} can be reached in an apparatus that is of the order of $10\,m$ in height.

Helium has additional advantages as well. The physical properties of its gaseous state near the critical point vary rather sensitively with pressure, so that one can attain a vast *range* of Reynolds and (especially) Rayleigh numbers in an apparatus of *fixed* size and design. By mixing helium II with its lighter isotope, namely, helium 3, it is possible to vary the Prandtl number between about 0.01 and 1. Helium lends itself very well for combined heat transfer and fluid dynamical studies in which both Reynolds and Rayleigh numbers play an essential role, and large ranges of parameters are required. The use of helium allows dynamic similarity to be attained simultaneously in more than one parameter, for example in Reynolds and Froude numbers. In general, this would not be possible with familiar fluids such as water. An array of instrumentation based on superconducting technology is available at cryogenic temperatures. In recent years, high-energy particle accelerators with superconducting magnets have demanded reliable and large refrigerators. As a consequence, a technology for large-scale helium refrigeration is readily available for pushing the limits in the space of similarity parameters. Finally, safety of operation is not a critical issue with helium.

B. QUESTIONS ABOUT HELIUM

These attractive attributes of helium have been exploited in a few instances (for a summary, see Donnelly, 1991a, and Donnelly and Sreenivasan, 1998). Yet, much remains to be done before the benefits of helium in classical fluid dynamics can be realized in full. Even though the low viscosity of helium is a familiar fact, some genuine questions continue to be asked about the fluid in the context of classical hydrodynamics. To what degree do helium flows in navy and aerodynamic testing correspond truthfully to air and water flows? Is is possible to make all the crucial fluid dynamical measurements? In particular, can one acquire turbulence data with adequate spatial and temporal resolution? Is the case for helium compelling enough to override the inflexibility of *large* cryogenic operations? These questions need some discussion. Finally, not every fluid phenomenon of practical interest can be replicated in helium — for example, compressibility effects characteristic of high-speed flight (because they depend on the ratio of specific heats of the fluid medium).

There are alternative methods for attaining ultra-high Reynolds numbers. For example, Smits and Zagarola (1997) have argued that air can be pressurized enough to lower the kinematic viscosity to approach that of helium. Ashkenazi and Steinberg (1999) have made convection measurements at very high Rayleigh numbers using SF_6 gas near its critical point. The relative advantages and disadvantages of these schemes need to be assessed in any given context. Helium is not the answer in every situation.

C. SCOPE AND ORGANIZATION OF THE ARTICLE

Thus, on the one hand, there are frontier opportunities to explore with respect to cryogenic helium; on the other hand, there are questions to be answered regarding the suitability of helium as a hydrodynamic test fluid. In this article, we discuss both opportunities and challenges, and argue on balance that the former considerations outweigh the latter. While in some instances the difficulty of working at cryogenic temperatures is not worth the trouble, there are several others for which it may be the only option — for instance, if dynamical similarity in Reynolds and Froude numbers is simultaneously desired. These views are neither new nor revolutionary. They have surfaced with different emphases at different times, starting perhaps with Smelt (1945) who articulated the advantages of helium. New at present is the confluence of interests in nonlinear physics, turbulence, model testing, instrumentation, and the technology of large-scale refrigeration, as summarized in the books edited by Donnelly (1991a) and Donnelly and Sreenivasan (1998). Over time, experience has been gained in small helium facilities typical of a laboratory-scale situation. The realization is now at hand that, if the use of helium can be pushed to the next level in size and versatility, a broad class of fluid dynamics problems can be addressed far more adequately that before. It is to the articulation of this theme that this article is devoted.

The rest of the article is organized as follows. Section II is a brief primer on the properties of helium. Section III contains examples of ultra-high Rayleigh and Reynolds numbers found in nature and technology, and Section IV makes the case that there are basic and applied problems that could benefit immeasurably by creating and studying flows at such ultra-high parameter values. In Section V, we remark briefly on a few specific examples for which helium offers unique opportunities. Some remarks will be made on the refrigeration opportunities available. Those refrigeration units have been developed at Brookhaven National Laboratory (BNL) and other such institutions for purposes of cooling superconducting magnets. The discussion in these sections is concerned largely with helium as a desirable fluid for hydrodynamic testing. In reality, this desirability is intertwined with promises that helium offers for fundamental new physics as well. The two aspects cannot be separated easily, and we shall specifically consider in Section VI an example of the latter by discussing the use of helium II. A summary of available instrumentation is given in Section VII. In Section VIII, some limitations of helium vis-à-vis our stated purposes will be discussed. We conclude the article with a brief summary in Section IX.

II. Brief Note on Helium

Cryogenic helium is one of the well-studied fluids, probably next in detail only to water and air. It offers essentially three fluids of dynamical interest, each with its own special and attractive properties. It is convenient to discuss these fluids with respect to the phase diagram of helium (Figure 1). The different states of the fluid are marked on it.

To the right of the λ line and above the liquid–vapor coexistence line is liquid helium I. This liquid obeys Navier–Stokes equations and satisfies

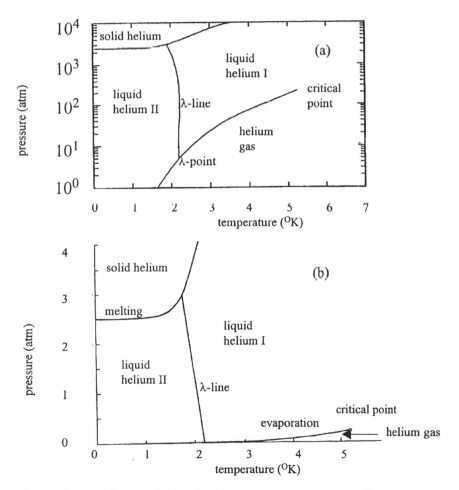

FIG. 1. The phase diagram of helium in the pressure–temperature plane. Pressure is given in the logarithmic scale in (a) and in linear scale in (b).

viscous boundary conditions. The viscosity of helium I is small but finite. If we reduce the pressure, helium I boils like any other liquid and enters the vapor state. The normal boiling point of liquid helium is 4.2 K. Gaseous helium is also a Navier–Stokes fluid. The critical temperature and pressure of helium are 5.2 K and 2.26 bar, respectively. In the vicinity of the critical point* the viscosity and thermal conductivity of gaseous helium depend sensitively on the temperature and, especially, on the pressure (see Figure 2). This latter dependence is not unlike the inverse power law for air, say, except that, even at 1 or 2 atm of pressure, very low viscosities can be attained. Thermal conductivity changes similarly, so one can easily attain a vast *range* of Rayleigh numbers (see Section III).

The point of intersection of the λ line with the coexistence line is the so-called "λ point." The corresponding temperature is about 2.17 K. The liquid state to the left of the λ line is helium II. Helium II exists as a liquid down to absolute zero and does not solidify except at pressures of the order of 25 atm. Roughly speaking, helium II is a mixture of a superfluid component and a normal component. The former has no viscosity or

---

*We consider conditions far enough away from the critical point for the effects of finite correlation length to matter.

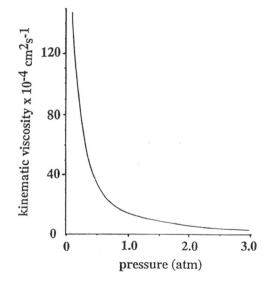

FIG. 2. The variation with pressure of the kinematic viscosity of the critical gas at 5.4 K. Similar curves exist for other temperatures.

TABLE 1
FLUID DYNAMICAL PROPERTIES OF HELIUM COMPARED WITH AIR AND WATER

Fluid	T (Pressure)	$\nu \, (\text{cm}^2/\text{sec})$	$\rho \, (\text{gm/cm}^3)$	$\alpha \, (\text{K}^{-1})$	$\kappa \, (\text{cm}^2/\text{sec})$	$\alpha/\nu\kappa$
Air	20°C	0.150	1.21×10^{-3}	3.67×10^{-3}	0.200	0.122
Water	20°C	1.004×10^{-2}	0.998	2.07×10^{-4}	1.43×10^{-3}	14.4
Helium I	2.2 K (SVP)	1.78×10^{-4}	0.146	1.03×10^{-2}	2.49×10^{-4}	2.32×10^{5}
Helium II	1.8 K (SVP)	8.94×10^{-5}	0.145	X[a]	X	X
Helium gas	5.5 K (2.8 Bar)	3.21×10^{-4}	0.0685	2.86	6.31×10^{-5}	1.41×10^{8}

[a]Properties are not defined because the superfluid component has infinite thermal conductivity.

entropy, while the latter, in addition to having a small viscosity, carries the entire heat content of the fluid. The two fluids are, of course, not separable because all helium atoms are identical. The hydrodynamics of the two fluids can be described at very low velocities by the Tisza–Landau two-fluid equations (see, for example, Roberts and Donnelly, 1974). We argue in Section VI that the interaction between normal and superfluid components of helium II can be used to advantage for hydrodynamic studies. A special feature of helium II is that variations in temperature do not propagate according to the Fourier law of heat conduction, but as a true wave motion. This longitudinal wave in temperature (or entropy) is called the *second sound*. Its speed depends on the temperature but a typical value is on the order of 20 ms^{-1}. We shall see in Section VII that second sound is an important tool for studies of turbulence in helium II.

Finally, the lighter isotope of helium, namely, helium 3, also serves useful fluid dynamical purposes (e.g., Metcalfe and Behringer, 1990). The Prandtl number of a mixture of helium II and helium 3 can be tuned by varying the mean temperature of the mixture. A major distinction between helium II and helium 3 is that the latter is not a superfluid. We shall not concern ourselves with helium 3 any further.

Table 1 lists a few representative properties of helium compared to those of water and air.

III. Some Examples of Flows at Very High Rayleigh and Reynolds Numbers

Recall that the Reynolds number, Re, and the Rayleigh number, Ra, are defined as $\text{Re} = UL/\nu$ and $\text{Ra} = \alpha g \Delta T L^3/\nu\kappa$, where U and L are characteristic velocity and length scales of the flow; ΔT is a characteristic temperature difference; v, κ and α are, respectively, the kinematic viscosity, thermal conductivity, and isobaric thermal expansion coefficient of the fluid; and g is the gravitational acceleration. How high are the Reynolds and Rayleigh numbers in physical situations of interest? We consider some specific examples next.

A. Geophysical Flows

On the average, earth's atmosphere and oceans are stably stratified, so that the motion is generally a combination of three-dimensional turbulence and waves. Under special conditions, however, parts of these systems do become unstable. The large-scale overturns of water masses in Mediterranean and Polar seas are two examples. For water, the combination $\alpha g/v\kappa \sim 1.5 \times 10^4$ in CGS units. For a length scale of a little over a kilometer and for temperature differences of the order of 1 K, one obtains a Rayleigh number of the order 10^{19}. For the atmosphere, unstable conditions obtain if the lapse rate is greater than the adiabatic value of about $10°C\,km^{-1}$. The most unstable conditions are observed off the coasts of Africa and Brazil in the Atlantic, and off California and Honolulu in the Pacific (Krishnamurti, 1975). Rayleigh numbers of the order of 10^7 are typical, and are smaller than those encountered in the ocean. Furthermore, because the wind shear is relatively pronounced, a more useful indicator of unstable conditions in the atmosphere is the Richardson number (see, for example, Monin and Yaglom, 1971).

Hurricanes, tornadoes, and other large-scale geophysical disturbances are sources of high-Reynolds-number phenomena in geophysical flows.

B. Solar Convection

The computation of Rayleigh and Reynolds numbers for the sun is nontrivial. We are content with rough estimates appropriate to the convection zone (the outer 30% of the solar radius except toward the surface where the fractional ionization is very low). From the knowledge of the temperature in that region and hence of the mean free path (Gray, 1988), one estimates the kinematic viscosity and computes the Reynolds number to be of the order of 10^{13}, and the Rayleigh numbers in the rough vicinity of 10^{22}.

TABLE 2
SOME EXAMPLES OF HIGH RAYLEIGH AND REYNOLDS
NUMBERS

Example	Ra	Re
Sun	10^{22}	10^{13}
Ocean	10^{19}	10^9
Atmosphere	10^7	10^9
Naval applications	—	10^9
Aerospace applications	—	5×10^8

C. AEROSPACE AND NAVY APPLICATIONS

The Rayleigh number is irrelevant in these instances. For reasonable operating conditions, the Reynolds number on the fuselage of a Boeing 747 could be of the order of 5×10^8. A modern torpedo (MK48) operates at a Reynolds number (based on length) of about 1.6×10^8. For an attack submarine (SSN688), the length-based Reynolds number could be as high as 10^9. An aircraft carrier (CVN68) produces a Reynolds number that is about five times higher; other ships on sea have comparably high Reynolds numbers.

Table 2 lists some of these numbers. Needless to say, they should be interpreted generously in a rough order-of-magnitude sense.

IV. Need for Studies at Conditions Approaching Ultra-High Parameter Values

Granted that there exist flow phenomena at these ultra-high Reynolds or Rayleigh numbers, it is not obvious that one should necessarily make tests under such extreme conditions. What reasons could compel such studies? Why is a straightforward extrapolation from a lower parameter range not adequate? In a brief attempt to address these questions, we shall consider both applications and basic research, although there are qualitative differences between the two instances. Basic research usually requires a "one of a kind" experiment, made with important questions in mind; this uniqueness renders irrelevant, or at least less pressing, the many considerations — such as the ease of repeated operation, minimum operating and turnaround times, and low operating costs — that are paramount in applications.

A. MODEL TESTING AND DIFFICULTIES WITH EXTRAPOLATION

Even in these days of supercomputers and advanced methods of computational fluid dynamics, there is still a place for a good experimental flow facility. Recent instances in which wind tunnel testing has played an essential role are for transonic wing and gas turbine development. An immense variety of problems could similarly benefit from model testing, but we shall consider instances where large gaps exist between the test and operating conditions. This gap often leads to "almost unmanageable risks" (Bushnell, 1998), a situation one would wish to avoid.

As an example, consider a submarine moving at some angle to its longitudinal axis. The vortices shed from the frontal fin will interact with the rear fin and the propeller, which in turn affects the latter's performance tremendously. The sound generated from these regions of intense interaction radiates outward and can be detected in the far field. Thus, one would not only like to understand the development of the boundary layer on the submarine body, but also the entire flow field including far-field acoustics and cavitation on the propeller blade. What makes navy and aerodynamic testing at ultra-high Reynolds numbers especially important is that the complexity of flow fields and the multiplicity of interactions among their various elements render uncertain the extrapolation to higher Reynolds numbers. Needless to say, new designs and development cannot occur without the ability to make rapid tests under realistic conditions.

Figure 3 shows some of the water facilities available in the world and the maximum Reynolds numbers attainable in them. The Reynolds numbers are based on the length of the largest model that can be tested in the facility. (Given the length-to-diameter ratio of the model, the maximum allowable blockage determines the maximum testable length.) For submarine-like bodies, Reynolds numbers of the order of 10^8 can be attained in these facilities, and it is sensible to ask why the knowledge acquired at these (or even lower) Reynolds numbers cannot be extrapolated adequately. Two superficial arguments might suggest that an extrapolation is possible. First, no surprising and qualitatively new physical phenomena may occur once a "sufficiently high" Reynolds number is reached. If so, the returns for working at these ultra-high Reynolds numbers are meager in relation to the costs. Second, such quantitative changes as might occur beyond this "sufficiently high" Reynolds number are slow, and so extrapolations for a decade or two in Reynolds number should be reasonably adequate. A plausible strategy for understanding the flow at Reynolds numbers of 10^9 would then be to

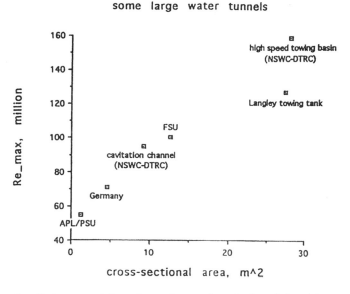

FIG. 3. A representation of some large-scale water tunnels in existence.

acquire solid information for Reynolds numbers up to, say, 10^7 and extrapolate it.

Unfortunately this is not always possible. While certain types of changes are indeed slow with respect to Reynolds number (Section IV.C.2), some are not — especially when several types of interactions occur. As an example, one does not know how to extrapolate the interaction of the intense vorticity field with the propeller by a scale factor of 10, let alone 100. One does not know how to calculate the far-field pressure reliably from the knowledge at low Reynolds numbers. The practice in the U.S. Navy is to build quarter-scale models and operate them in lakes with radio control. Even these enormously expensive and roughly realistic tests do not yield results that can be satisfactorily extrapolated for full-scale submarines. Here is a case where almost nothing but a full-scale test can produce satisfactory answers.

Similar circumstances exist for aerodynamic testing. In Figure 4, we show on the left ordinate the chord Reynolds numbers estimated for various aircraft, while on the right we show Reynolds numbers attainable in a few available test facilities. By convention, chord length is taken as a tenth of the square root of the cross-sectional area. Again, the gap in the Reynolds

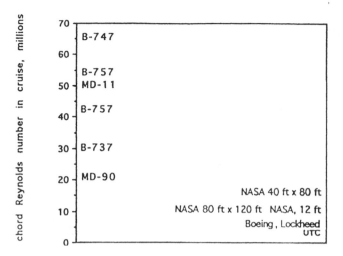

FIG. 4. Some aircraft Reynolds numbers and existing subsonic wind tunnels.

number between flight conditions and wind tunnel tests is an order of magnitude or larger. More thorough accounts of the wind tunnel situation can be found in various NASA documents of limited availability and AIAA information papers, but our purpose here is to point out that full-scale testing cannot be done in any existing facilities.

The one important exception to this statement is the National Transonic Facility (NTF) at NASA Langley (e.g., Kilgore, 1991, 1998); as is well known, NTF operates at cryogenic conditions of liquid nitrogen. The value of cryogenic testing has long been appreciated and used to advantage in NASA. The use of helium is the natural next step.

B. OTHER APPLIED ISSUES

A few Reynolds-number-dependent issues of interest to the navy and air force are the dynamic response to nonlinear maneuvers, transition to turbulence, effects of tripping the boundary layers (which could be significantly different between low and high Reynolds numbers), scaling of submarine propellers, and so forth. Bushnell and Greene (1991) cite other practical instances of low-speed aerodynamics where research at ultra-high Reynolds numbers is critical. Their first example is the hazard to lighter aircraft encountering strong wing-tip vortices left behind larger aircraft. The distance needed for the natural dissipation of these vortices is unacceptably large (probably proportional to their Reynolds number) in a modern airport. A useful strategy would be to "control" the vortices so as to ameliorate their effects. Wind tunnel tests made for the purpose have been at Reynolds numbers that are smaller by about two orders of magnitude. This mismatch in Reynolds numbers is thought to be responsible for the observed discrepancy between the flight and laboratory data. A second example is the enhancement of the maneuverability of jet fighters by particular use of vortex generation techniques. Here again, Reynolds number effects are known to be critical. The third example is the development and evaluation of high-lift devices where, for instance, one cannot predict the position of separation. In general, the interaction between vortices and solid body can be understood only by controlled studies at very high Reynolds numbers. Helium flows offer tremendous opportunities here.

C. BASIC PROBLEMS IN HIGH-REYNOLDS-NUMBER TURBULENCE

Turbulence is intrinsically a high-Reynolds-number phenomenon. Much is known about it (e.g., Monin and Yaglom, 1971, 1975, and the hundreds of papers that occupy journal pages year after year) and yet, only a small part of that knowledge is impeccable. The theory is very hard for good reasons, and is still a long way from being satisfactory (e.g., L'vov and Procaccia, 1996). The so-called "universal aspects" of turbulence can be found (if at all) only at very high Reynolds numbers. One does study high-Reynolds-number turbulence in atmospheric and oceanic flows, but they are not well controlled.

What, specifically, are the types of questions that one supposes will be answered by studying turbulence at "high enough" Reynolds numbers? How high is "high enough"? Finally, at the risk of inviting ridicule, we might as well ask: What constitutes a solution of the turbulence problem, and how does one recognize it when it appears on the horizon? These questions are considered below.

1. *Large-Scale Phenomena*

Historically, one has always learned something valuable and unexpected when the Reynolds number boundary has been pushed behind by one or two orders of magnitude. The drag crisis for the sphere (see Schlichting,

1956), Roshko's (1961) work on the drag coefficient for the circular cylinder, Kistler and Vrebalovich's (1966) data in grid turbulence, spectral measurements of Grant *et al.* (1962) in the ocean, and Saddoughi and Veeravalli's (1994) experiments in the NASA AMES wind tunnel are a few examples worth citing. The Nusselt number measurements of Castaing *et al.* (1989), Chevanne *et al.* (1997), and Niemela *et al.* (2000a) have revealed unexpected features. The mean velocity measurements in pipe flow at very high Reynolds numbers (Zagarola and Smits, 1997) possess unsuspected elements. In all of these instances, new global aspects have come to surface; even if some findings confirmed what was previously expected, the value of these high-Reynolds-number studies cannot be exaggerated.

It may be worth emphasizing that one does not yet know for certain basic quantities such as the asymptotic value of the drag coefficient of a smooth sphere. One does not yet know at what Rayleigh numbers, if any, the Nusselt number begins to vary as the half power of the Rayleigh number (Kraichnan, 1962; Howard, 1972). What is the effect of the aspect ratio of the apparatus on the observed power law, and what is its form for large aspect ratio? What is the effect of surface roughness on convective heat transport, and how does one conveniently parameterize it? Do thermal "plumes" survive at ultra-high Rayleigh numbers? More generally, how much of the coherent structure observed at low and moderate Reynolds numbers survives at ultra-high Reynolds numbers? What is the effect of the large-scale motion on various power law exponents? How much does the large-scale motion itself depend on initial conditions? Beyond what Rayleigh numbers does the average value of the energy dissipation rate follow the Kolmogorov form? An empirical observation of Niemela *et al.* (2000a,b) is that the variation of the Nusselt number with the Rayleigh number has a *functional form* that is very close — over 11 orders of magnitude in Ra — to the prediction of a weakly nonlinear theory. Whether this is a happenstance or has a profound theoretical basis of some generality remains to be understood.

Instead of listing more questions, we emphasize that a sound theory of turbulence will not be possible without putting such basic issues on firm footing: One well-executed experiment at very high Reynolds numbers may be superior to a host of repeats at low to modest Reynolds numbers.

2. *Small-Scale Turbulence*

Let us now turn our attention to the scaling properties of turbulence in inertial and dissipative ranges. This is an area of active research, propelled not the least by the extraordinary success that has occurred in critical phenomena during the recent two or so decades (e.g., Goldenfeld, 1992); that success is in no small measure due to simultaneous progress in theoretical and experimental work. A reasonable goal of research in small-scale turbulence is to reach a comparable state of certainty with respect to scaling. Some typical problems are mentioned below.

As has been well known now for almost 60 years, Kolmogorov's (1941) ideas have ruled the horizons of research in turbulence physics (see, e.g., Hunt *et al.*, 1991). Experiments have revealed deviations from Kolmogorov's theory (e.g., Frish, 1995), and these deviations are attributed to the intermittency of small scales. The role of intermittency is not fully understood. Its importance for certain aspects of turbulence such as energy spectral density may be small, but could be rather large for turbulent combustion, breakage of liquid droplets, particulate aggregation, and the like. The finiteness of the Reynolds numbers renders the observed effects of intermittency to varied interpretations (e.g., Nelkin, 1994; Sreenivasan and Antonia, 1997). The kinematic and dynamic effects of the sweep of small scales by the large scale are not fully understood (Tennekes and Lumley, 1972). The much simpler problem of passive scalars mixed by turbulence contains partially understood aspects: Some examples are the fractal character of isotherms (Constantin *et al.*, 1991) and the effect of shear on it, the asymptotic shape of the probability density functions of temperature increments and derivatives (e.g., Sinai and Yakhot, 1989; Shraiman and Siggia, 1995), and the limitations of cascade-type ideas for describing the interscale transfer of scalar variance. Looming large are potential limitations of the very notion of universality of small scales — a supposition that has been at the core of the subject.

To improve the state of these long-standing uncertainties, first and foremost, one needs solid experimental data. However, for experiments to be compelling in this respect, one needs to have a large scaling range (say, three decades) and the information extracted from them should not be subject to dubious artifacts of data processing. (The situation becomes less stringent if a plausible theory emerges.) An important fact about turbulence is that the scaling range increases only logarithmically with Reynolds number. So do the number of steps in the spectral cascade (Onsager, 1949); the number of effectively independent layers in wall-bounded flows (Tennekes and Lumley, 1972); plausible corrections for finite Reynolds number effects (Barenblatt and Chorin, 1996); the Reynolds number dependence of the volume occupied by the dissipation field and of the fine-scale vortex structures (Sreenivasan and Meneveau, 1988), and so forth. If we ignore flows that are very close to solid boundaries, and those driven by extremely large shear, the number of decades of inertial scaling varies with the Reynolds number roughly as

$$\text{number of decades} = \log_{10} R_\lambda - 1.75 \text{ for } R_\lambda > 200,$$

where R_λ is the microscale Reynolds number based on the root-mean-square velocity fluctuation and the Taylor microscale (see, e.g., Batchelor, 1953). This suggests that one needs an R_λ of the order of 50,000 to obtain three decades of scaling. Translating R_λ to the bulk Reynolds number, Re, is not precise but the equivalent Re is roughly on the order 3×10^8. (It must be said, however, that the scaling range shrinks if the forcing is very strong or spectrally broad.) Theory will then have a stronger foothold.

3. The "Turbulence Problem"

The so-called "turbulence problem" has been with us for substantially longer than a century. Given the multifaceted complexity of turbulent flows, a minimal set of conditions that ought to be satisfied before we can declare that the problem is "solved" is worth some consideration. This issue is not central to our thesis about helium here, but is marginally relevant because we are advocating very high Reynolds number studies as a means to this end. One possible scenario is that every flow of interest can be computed away fast enough to be useful in practice. Though computing is not "understanding," the problem will then assume less urgency. Given the large Reynolds numbers in some cases of interest, this scenario does not seem realistic despite rapid advances that continue to occur in computing power. There is thus a need to "model" turbulence. Modeling involves many issues; among them are functional relationships between different quantities as the Reynolds number is varied, and the numerical coefficients involved in these relationships. Scaling studies mentioned earlier focus on the interrelationships. Detailed numbers and coefficients have to come from measurements, theory, or simulations. A program of basic research in turbulence would thus involve the experimental determination of scaling relationships over the entire range of Reynolds number of interest, theoretical advances that understand such relationships, and a detailed exploration through simulations as well as experiment at a few Reynolds numbers. Helium offers an excellent opportunity in terms of the experiment.

V. Some Considerations about Large-Scale Helium Flows

We have so far argued that there is a need for ultra-high Reynolds number experiments from both practical and fundamental perspectives. Some at least of these needs cannot be met by existing wind and water facilities. Even those that can be met in principle by existing facilities cannot always be explored in them because of their meager availability and large operating expenses. In particular, it is nearly impossible to procure for fundamental research in turbulence the services of NTF at NASA Langley. As for high-Rayleigh-number research, no large-scale facilities are in existence. Alternatives are clearly needed.

A. Examples of Large-Scale Facilities

After making a careful study of these factors, Donnelly *et al.* (1994) recommended the following three large-scale helium facilities to be built:

1. A 10-m-high convection cell to generate a Rayleigh number as high as 10^{20} using cryogenic helium gas. This Rayleigh number is of direct interest in geophysical and astrophysical problems (Table 2).

2. A helium tunnel with a cross section of side 125 cm. A submarine model of 25 cm in diameter and 250 cm in length can be operated at a Reynolds number of the order 10^9, comparable to that of a full-scale submarine in operation.

3. A liquid helium tow tank, about 8 m wide and 40 m long, that could match both Reynolds and Froude numbers at the same time. The run time would be of the order of 10 s.

It is useful to keep these concrete examples in mind as we proceed with the following discussion.

B. Refrigeration

In building helium flow facilities that exceed a modest scale, one concern is the availability of refrigeration. Helium volume of more than a few thousand liters is difficult to manage in a university research laboratory. None of the facilities mentioned above could be operated in this mode, and so require special considerations. Fortunately, demands placed on the cooling of superconducting magnets used for high-energy particle accelerators have perfected techniques of large-scale refrigeration. For instance, the refrigeration available at BNL, as part of its Relativistic Heavy Ion Collider (RHIC) project, is about 25 kW (Sondericker, 1998). An inventive suggestion made in Donnelly (1994) was that such refrigeration facilities could be used off-line for the special helium facilities of the sort just discussed. For instance, the 10-m convection cell would require about 60,000 liters of helium and about a kW of refrigeration. The helium tunnel with the 125-cm test section would require about a million liters of helium and about 1 kW of cooling power at 1.6 K. The towing tank would also use about a million liters of helium and need comparable cooling power. All of them are, in principle, well within the BNL capacity (and those of Fermilab, Livermore, and CEBAF in the United States, and a few places such as at CERN in Europe; see Quack, 1998).

This granted, one may still question whether helium offers the right choice; we have already noted highly compressed air and SF_6 near the critical point as possible alternatives. There is less experience with SF_6, and so we will consider a brief comparison between helium and compressed air. Helium I at 2.5 K and 1 atm, helium gas at a temperature of 5.4 K and 2.9 atm, and compressed air at room temperature and about 200 atm all possess roughly the same kinematic viscosities. Thus, to obtain the same model Reynolds numbers, the same flow velocity is required in all three cases. To test an ellipsoid of aspect ratio 12 at a length-based Reynolds number of 10^9, assuming that the blockage permitted is of the order of 3%, one requires a test section of about 1-m diameter and a flow velocity of about $50 \, \mathrm{ms}^{-1}$. Thus, issues such as allowable surface roughness of the model and the resolution required of the instrumentation will be of equal import in all three cases. What would render one of the fluids more, or less, desirable are issues such as the dynamic pressure (which directly determines the forces on the model), the flexibility of use, possibilities for research and further development, sophistication of available instrumentation, and so forth. Some of these issues are discussed below. For comparisons to be useful in practice, reliable cost estimates and power requirements should be made to the same degree of detail in all the cases. We have not made such detailed studies (nor has anymore else). Without the benefit of such studies, we have to be content with discussing some general considerations with respect to helium.

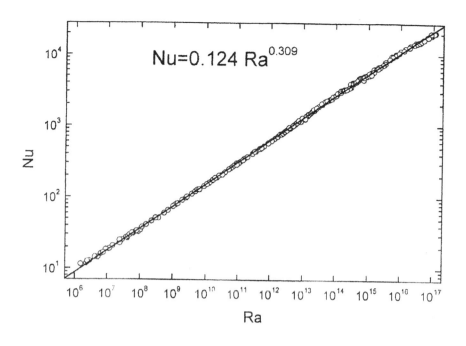

Fig. 5. A plot of the Nusselt number, Nu, versus Rayleigh number in the 1-m convection apparatus at the University of Oregon. (Nusselt number is the ratio of the convective heat transport to the conductive transport for the same temperature difference between bottom and top plates.) The Rayleigh number covers about 11 orders of magnitude. Over this range, the power law has an exponent of about 0.3. (From Niemela *et al.*, 2000a.)

C. THERMAL CONVECTION EXPERIMENT

The suitability of helium for the convection studies has been recognized for more than 25 years now, and studies exploiting its special features have been made in various laboratories around the world (e.g., Threlfall, 1975; Castaing *et al.*, 1989; Chevanne *et al.*, 1997; Niemela *et al.*, 2000a,b). Presently, the largest cell in operation is at the University of Oregon; it is 1 m high and 0.5 m in diameter and produces Rayleigh numbers between 10^6 and 10^{17}. At the upper end of its use, it uses 60 liters of helium and requires 10 W of refrigeration.

The uniqueness of this cell is demonstrated by Figure 5, which plots the Nusselt number data over 11 orders of magnitude of Rayleigh number — all acquired in the same apparatus by controlling the gas pressure and the temperature difference between top and bottom plates. Rayleigh number is roughly proportional to the square of the Reynolds number, so one has in the same facility 5 or more orders of magnitude variation in Reynolds number, or about 2.5 orders of magnitude in the Taylor microscale Reynolds number. It is extremely rare that one can obtain such a range in one apparatus. One can also observe an extensive scale range, transition from Bolgiano scaling (1962) to Kolmogorov scaling (1941), and so forth. More details can be found in Niemela *et al.* (2000a,b).

The Oregon apparatus is nearly at the limit of what is possible in a university laboratory. The large convection cell proposed in the previous subsection cannot be built and operated in this same way. It is also expensive. Keeping this latter in mind, Donnelly (1994) designed the same cryogenic cell to house other flows such as turbulence behind a towed grid, Taylor–Couette flow, a medium-sized helium tunnel, and basic drag experiments of falling bodies. These other purposes to which the convection call can be put are illustrated in Figure 6. Perhaps the most basic of them is the towed grid experiment. (A small-scale version of this flow is in operation at Yale University.)

D. Helium Flow Tunnel

The operation of a helium tunnel is no different in principle from that of a water tunnel, but the advantage of a helium I tunnel is the small size. Although the smallness could also be true of highly compressed air — as has been noted already — some features make helium additionally desirable. First, the forces on the model are smaller. We show in Figure 7 the ratios of the dynamic head of compressed air at 300 K, as functions of the air pressure, to that of liquid helium at two temperatures, keeping the Reynolds number fixed. This ratio is appreciably larger than unity even for air pressures of a few hundred atmospheres. Second, an array of instrumentation is available at low temperature, as we shall see briefly in Section VII. A superconducting magnetic suspension could be used to support the model in the test section (Goodyer, 1991; Lawing, 1991; Britcher, 1991, 1998).

The limiting parameter in the design and operation of a helium tunnel is the refrigeration available. In circumstances where refrigeration is limited, transient operation is a plausible alternative. This is discussed in Donnelly (1991a). As a precursor to the 125-cm tunnel, a 6-cm tunnel has been built at the University of Oregon.

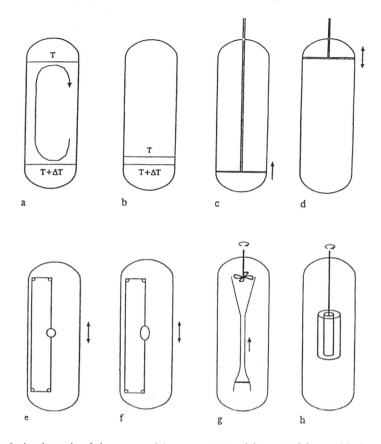

Fɪɢ. 6. A schematic of the proposed large cryostat and its potential uses: (a) ultra-high-Rayleigh number convection, (b) convection with variable aspect ratio, (c) towed grid, (d) oscillating grid, (e) towed sphere, (f) towed ellipsoid, (g) a tunnel insert, and (h) a Taylor–Couette insert. (From Donnelly, 1994.)

E. Tow Tanks Using Liquid Helium

The properties of liquid helium, especially helium II, vary rapidly with temperature. These variations can be used to advantage in modeling the motion of ships or waves on a free surface. In these instances, we need to simulate the Froude number $Fr = U/(gL)^{1/2}$, where L is the length of the vessel. The difficulty in model studies is that the matching of the Froude number leads to a drastic mismatch of the Reynolds numbers.

As an illustration, consider the motion of a surface ship 100 m long, 10 m in lateral size, moving at 30 knots ($\sim 16\,\mathrm{ms}^{-1}$). For water temperature of

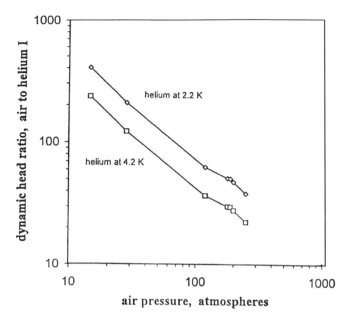

FIG. 7. The ratio of the dynamic head in compressed air flow at 300 K to that in helium I at 2.2 and 4.2 K, as a function of the air pressure in atmospheres.

15°C and a scale ratio of 25, a 4-m-long model in a water tow tank will match the Froude number of 0.52 if towed at 3.3 ms^{-1}. However, under these conditions the Reynolds number of the model is 1.3×10^7 compared to the full-scale Reynolds number of 1.7×10^9, smaller by a factor of about 130.

Operating in helium gives greater flexibility for matching similarity parameters. For instance, using helium I at 2.2 K, a scale ratio of 25 can match the Froude number exactly and the Reynolds numbers to within a factor of 2. Donnelly (1994) has provided additional details of design for an 8-m × 40-m facility (see Figure 8). Helium II offers even greater flexibility but needs a separate discussion because of its superfluid part. This is the subject of the next section.

VI. Superfluid Helium and the Hypothesis of Vortex-Coupled Superfluidity

We remarked in Section I that helium offers flexibility for exploring uncharted territories in hydrodynamics. Here we advance one such possibility with respect to helium II.

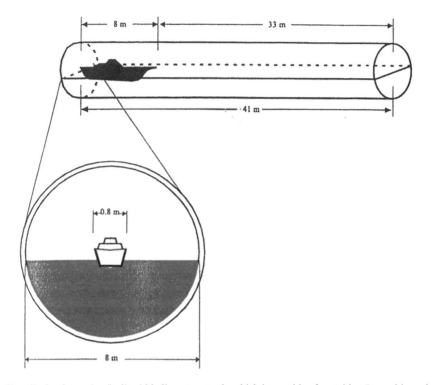

FIG. 8. A schematic of a liquid helium tow tank, which is capable of matching Reynolds and Froude numbers at the same time.

Helium II has several attractive attributes. One of them is the saving of refrigeration. The kinematic viscosity of liquid helium at 4.2 K is $2.6 \times 10^{-4}\,\text{cm}^2\,\text{s}^{-1}$; it falls steadily with temperature, reaching $9 \times 10^{-5}\,\text{cm}^2\,\text{s}^{-1}$ at 1.75 K, below which it begins to rise again. Since the pressure drop in a pipe flow, for example, depends on the cube of the kinematic viscosity, operation at 1.75 K instead of 4.2 K needs a mere 4% of the refrigeration. Another advantage is the ease of stable temperature control, with microdegree stability being routine in small cryostats. Additionally, the high heat conductivity of helium II suppresses cavitation that frequently plagues normal saturated liquid testing: the small heat of vaporization and surface tension of helium I forces one to operate under pressure to minimize the possibility of two-phase flow conditions.

However, there is a price to pay: helium II contains not only a normal fluid component but also a superfluid component with no viscosity, and one has to understand the interaction between the two fluids before helium II can be considered a hydrodynamic test fluid. We limit ourselves here to a few elementary comments. The relative fraction of the two fluids varies with temperature, from being entirely normal at the lambda point and entirely superfluid at absolute zero. The two fluids independently obey the hydrodynamic equations (Euler equations in the case of superfluid) until quantized vortex lines appear, as they do beyond a critical flow speed. Quantized vortex lines are similar to classical vortex lines except that their core radius is of the order of an Ångstrom, and the circulation κ around them is fixed by quantum mechanical considerations as $h/m \approx 9.97 \times 10^{-4}\,\mathrm{cm^2\,s^{-1}}$, where h is the Planck's constant and m is the mass of the helium atom. In the presence of quantized vortices, the interaction between the normal and superfluid components is described through a mutual friction coefficient. These facets of helium II have been known for about 60 years (see Vinen, 1961, and Donnelly, 1991b, for a historical account).

This interaction between the two fluids after the appearance of quantized vortices is simple in certain circumstances and complex in others. In the case of a rotating bucket containing helium II, quantized vortex lines run parallel to the axis of rotation, and the density of these vortex lines is such as to match the vorticity of a fluid in solid body rotation. Except in such simple situations, the quantized vortices occur in random orientations ("tangled mass") with respect to the main flow, and the interaction becomes more complex. It is relatively easy to do a kinematic simulation of the evolution of this tangled mass. From such calculations (e.g., Schwartz, 1985, 1988; Aarts and DeWaele, 1994), a great deal has been learned regarding the generation of quantized vortices at the boundary, their reconnection, their density, and the mutual friction. Barenghi and Samuels (1999) have incorporated in a simple example the back reaction of the superfluid vortex tangle on the normal fluid. More generally, experimental observations beginning with Donnelly and Hollis Hallett (1958) have suggested that the interaction between the classical vorticity and superfluid vorticity is strong beyond a Reynolds number (for the normal fluid) of the order of 100.

The question of present interest is the interaction of the superfluid tangle ("superfluid turbulence") with the classical turbulence generated by the flow of normal fluid: One supposes that classical turbulence can be generated when the Reynolds number of the normal component exceeds an appropriately large value. It is plausible that the two random fields of vorticity are coupled in some strong way. The coupling has been designated (Donnelly, 1991a) *vortex-coupled superfluidity* (VCS). VCS has been invoked in the interpretation of the turbulence experiments of Smith (1993) and Stalp *et al.* (1999) behind towed grids in helium II. Vinen (2000) has recently examined this issue theoretically. He has shown that the coupling is strong for classical turbulent eddies of size larger than the Kolmogorov scale. The differences that may arise in the two types of turbulent motion are thought to disappear in a timescale that is set by mutual friction, this being typically smaller than the characteristic small-scale eddy time of classical turbulence.

Even though our understanding of these interactions is tentative, it is helpful to state the VCS hypothesis explicitly as follows.

When forced sufficiently strongly, superfluid helium behaves as a Navier–Stokes fluid with respect to measurements made on a length scale large compared to the characteristic spacing between quantized vortices, and small compared to the length scale associated with the forcing. For those conditions, the quantum vortex lines track the high-vorticity filaments in the normal fluid. The inter-vortex spacing of the quantized vortices is of the order of the Kolmogorov scale.

Qualitatively, the hypothesis imagines that a turbulent flow will be threaded with a complex tangle of superfluid vortices. In a completely random tangle of quantized vortices, no large-scale superfluid vorticity can be present, because it will average out to zero. Therefore, superfluid vorticity field attains a macroscopic value only when superfluid vortex lines come together in the form of a tight bundle near the core of the normal fluid vorticity, and on scales that are small compared to that of external forcing.

The VCS hypothesis derives support not only from the numerical work of Barenghi *et al.* (1997) and Barenghi and Samuels (1999), but also from the sphere-drag experiments of Laing and Rorschach (1961) and Smith *et al.* (1999). See also Van Sciver (1991). Experiments, to the extent they are available, support the notion that helium I and helium II flows are very similar. The pipe friction data of Walstrom *et al.* (1988) also suggest that there is no perceptible difference between helium I and helium II data. Given the experimental difficulties, the absolute quality of results in all these cases is not as good as one desires, so there is a need to reaffirm the conclusions taking due account of features such as tunnel blockage and entry length. Finally, Maurer and Tabeling (1998) have shown that the spectral density of turbulence, measured at different temperatures—on the one end dominated by the superfluid component and on the other end by the normal component—display the same power-law scaling.

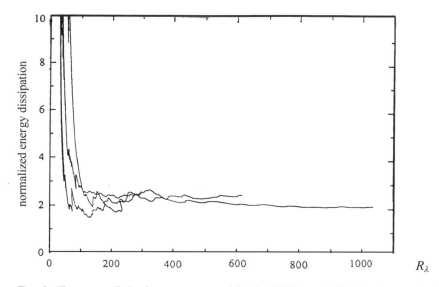

FIG. 9. The energy dissipation rate, measured by the VCS hypothesis behind a towed grid in helium II contained in a channel of a square centimeter cross section. The three curves correspond to the three different grid speeds of 10, 50, and 100 cm s^{-1}. The energy dissipation is normalized by the velocity and length scales characteristic of large-scale motion. The Taylor microscale Reynolds numbers, against which the normalized energy dissipation is plotted, are larger than those obtained in one of the largest wind tunnels (Kistler and Vrebalovich, 1966).

If the VCS hypothesis holds up under more intense scrutiny, it provides an important tool for studying classical problems using this nonclassical fluid (whose advantages have been mentioned earlier). In particular, the second sound attenuation in helium II is a sensitive tool for measuring the line density (line length per unit volume) of quantized vortex lines. When all the vortices are perpendicular to the second sound wave, as in the case of a rotating bucket, the attenuation is proportional to the vorticity (in addition to the speed of the second sound wave and the so-called mutual friction coefficient). No attenuation occurs along the axis of the quantized vortex lines. This fact can be used to estimate the attenuation of second sound in an isotropic and homogeneous distribution of quantized vortices. Under plausible assumptions, which need closer scrutiny, this attenuation can be related to the root-mean-square superfluid vorticity. Root-mean-square values as small as 10^{-2} Hz have been measured behind a towed grid (Smith, 1993; Stalp, 1998). By relating the second sound attenuation to classical vorticity, and thence to the turbulent energy dissipation rate, Stalp (1998) deduced the scaling of the latter as a function of Reynolds numbers. One of Stalp's graphs, reproduced here as Figure 9, shows that at high enough

Reynolds numbers, the energy dissipation scales on the large-scale velocity and length scales, just as in classical turbulence (Sreenivasan, 1984). The difference is that much higher Reynolds numbers have been achieved in a channel of square-centimeter cross section than in the largest wind tunnels. The observed decay laws are essentially identical to those of classical turbulence behind grids (Skrbek and Stalp, 1999), as are the inertial-range energy spectra. Though there are gaps in the arguments leading to the interpretation of the data, these results suggest that the VCS hypothesis is likely to be correct essentially as stated above. These are exciting possibilities worthy of further exploration.

VII. Summary of Instrumentation Development for Helium Turbulence

Creating ultra-high Reynolds number flows is merely a part of the challenge, the measurement of dynamically interesting quantities being the other. A summary of the present state of instrumentation is therefore worthwhile. Our focus here is on helium flows, but the problems of resolving small scales of turbulence at high Reynolds numbers is not peculiar to them by any means.

A. Average Quantities

1. *Mean Flow Velocity*

Several methods are available for the measurement of average velocity in cryogenic engineering. They have been reviewed by Van Sciver *et al.* (1991), and include turbine flow meters, venturis, fluidic flow meters based on vortex shedding, acoustic transit time flow meters, and flow meters based on the propagation of temperature or pressure pulses. These methods are applicable to experiments where there is a mean velocity, and not to counterflow experiments (where the classical and superfluid components of helium II move in opposite directions). Two examples of acoustic flow meters are given in Swanson and Donnelly (1998).

2. *Mean Temperature and Its Gradient*

Instrumentation for measuring absolute temperature and temperature gradient has been developed to an exquisite state for liquid helium studies. High-resolution thermometers now reach nanokelvin resolution (e.g., Lipa,

1998). Germanium resistance thermometers are stable and highly sensitive (10^{-4}–10^{-6} K). Their response times are of the order of 10 ms. More rapid responses of the order of a microsecond at 4.2 K belong to silicon diodes in unencapsulated form. Thus, mean characteristics such as the Nusselt number, requiring only average temperature measurements, are feasible even for the most demanding circumstances considered so far, and the available instrumentation has been used successfully (e.g., Wu, 1991; Chevanne *et al.*, 1997; Niemela *et al.*, 2000a,b).

3. *Differential Pressure*

Measurement of absolute pressure and pressure gradients has also been developed to a high degree of sensitivity, principally by using capacitance manometers consisting of light diaphragms near a fixed electrode. This sensitivity has been made extremely high by means of bridges or resonance circuits, and can easily detect Angstrom-range movements of the diaphragm. Typical resolution is 1 part in 10^8 of the capacitance under study. These devices maintain their calibration after multiple cycles to room temperature. Swanson *et al.* (1998) report details of a recent capacitance manometer.

4. *Aerodynamic Forces*

NASA has for a long time promoted the development of magnetic suspension and balance systems. Several authors in Donnelly (1991a) and Donnelly and Sreenivasan (1998) discuss various important details. Superconducting magnets can replace conventional ones at helium temperatures. These systems can measure lift and drag on a suspended body (Goodyer, 1991; Kilgore, 1991), and also execute sophisticated maneuvers. These devices are most useful when there is a substantial mean flow. Britcher (1998) summarizes the current status of this topic.

5. *Wall Stress Gauges*

Wall stress gauges were first developed for fluid mechanics by Lambert *et al.* (1965). They consist of a very sensitive bolometer that is heated above the environmental temperature so as to give information about the shear at the location of the bolometer. In Taylor–Couette flow, the wall stress gauge gives information equivalent to torque measurements.

Wall stress measurement in Taylor–Couette flow of helium I have been reported by Perrin (1982). Perrin used an aquadag painted layer

TABLE 3

ROUGH ORDERS OF MAGNITUDE OF SCALES AND FREQUENCIES IN A FEW PROPOSED EXPERIMENTS

Flow	R_λ	η (μm)	f_η(Hz)
Convection (Ra = 10^{19})	30,000	3	10^7
(Ra = 10^{16})	3,000	100	10^6
Towed grid	4,000	5–20	10^5–5×10^5
	600	85–350	6,000–30,000

Here, η is the Kolmogorov length scale and f_η is the corresponding frequency. The larger R_λ numbers are for the convection facility considered in Section V.A., and the towed grid experiment for which it can be used. The smaller R_λ numbers are for their counterparts presently in operation.

$\sim 1 \times 5$ mm carrying a dc power of typically 10^{-5} W. However, these devices are unlikely to be successful for experiments in helium II.

B. TURBULENCE MEASUREMENTS

Turbulence measurements with needed resolution are hard to make. Very small scales that also oscillate rapidly are generated at very high Reynolds numbers. If the apparatus is large, the problem is alleviated to some degree because all length and timescales increase correspondingly. The problem is serious in helium because, by design, the apparatus for a given Reynolds number tends to be smaller than for water or air. (This same observation applies also to compressed air flows at very high pressures.) These circumstances demand substantial upgrading of instrumentation capability. The resolution requirements for the few instances considered in Section V are illustrated in Table 3.

1. *Flow Visualization*

An important aspect of research in fluid dynamics is the ability to visualize flows. This is a nontrivial matter in helium because of its low density and the relatively low speeds of the fluid involved. Investigators working on very high resolution Brillouin scattering experiments in liquid helium find that, unless prefiltering at helium temperatures is successful, liquid helium consists of microscopic particles (probably dissolved solid air) that remain suspended for hours. Various tracer particles have also been introduced in a controlled manner with reasonable success. For instance, when a mixture of helium and deuterium is injected into a test section containing helium, the mixture solidifies on contact with helium. These particles can be arranged to have matched density with sizes of the order of microns (Murakami *et al.*, 1991; Ichikawa and Murakami, 1991). Hollow glass spheres and monodisperse silicon dioxide particles have also been used to visualize thermal counterflow jets of helium II (Murakami *et al.*, 1998), small-scale and large-scale vortex rings (Murakami *et al.*, 1987; Stamm *et al.*, 1994), and turbulent Taylor–Couette flow (Bielert and Stamm, 1993). Although these techniques appear highly promising, significant development work will be needed in terms of the uniformity of size distribution of the particles, avoidance of adhesion, etc.

Conventional optical techniques such as shadowgraphs are usually not attempted in helium because the refractive index is close to unity ($n \approx 1.02$). However, Woodcraft *et al.* (1998) point out that, under properly engineered circumstances, thermal convection in helium is only an order of magnitude more difficult to visualize than in water. They substantiated the claim by calculating the so-called "visualizability" parameter, and obtained working shadowgraphs of dilute mixtures of helium 3 in superfluid helium.

2. Temperature Fluctuations

Single-point and multipoint temperature measurements have already been made by fine bolometers (e.g., Wu, 1991; Niemela *et al.*, 2000a). Their response has to improve significantly before the smallest scales in the large convection cell of Section V.C can be resolved. For an example of the limitations imposed by poor probe response, see Grossmann and Lohse (1993). Continuing efforts suggest that an improvement by an order of magnitude is well within reach.

3. Velocity Fluctuations

Taking advantage of a "mean wind" present in his convection apparatus, Wu (1991) was able to use two arsenic-doped silicon bolometers, mounted closely together, to measure a mean flow velocity by means of delayed correlation of temperature fluctuations. This technique, used also by Niemela *et al.* (2000a,b), is crude at best and beset with many limitations. Hot wire methods work well at low temperatures, provided there is a substantial mean velocity. Castaing *et al.* (1992, 1994) have developed micron-sized hot wires working near 4 K, and using both resistive and superconducting thin films. Single-component velocity measurements with superconducting hotwire probes have also been made by Tabeling *et al.* (1996) in a helium apparatus with counter-rotating disks. No one has yet attempted cross-wire measurements. Even for single-wire probes, there are some difficult issues relating to temporal response (Emsellem *et al.*, 1997) but, on the whole, single-point single-component velocity measurements in helium flows have been successfully carried out. Whether the dissipative scales can be resolved in flows listed in Table 3 is still an open question, but further development ideas do exist (Wybourne and Smith, 1998).

If it is possible to seed helium gas or liquid helium for flow visualization studies, one can in principle use laser doppler velocimetry (LDV) and particle image velocimetry (PIV) methods. The development work necessary to obtain controlled seeding is currently under way.

Success in PIV depends on being able to seed the flow with particles that faithfully follow the fluid motion. The cross-correlation methods currently in use in water and air require a minimum of about five or ten particle pairs in the interrogation area. To resolve the finest scales, we need to have the interrogation volume no bigger than the Kolmogorov scale. The two requirements put serious constraints on the particle loading. Too high a particle loading (say, more than about 10^{-5} in volume fraction) will alter flow properties. Furthermore, the ability to image the particles depends on the laser power available and the index of refraction of the particles chosen. There are also issues related to the spatial resolution demands set by flow scales and constraints imposed by optics (such as fringe spacing).

It is thus clear that the ability to obtain a reliable PIV image depends on a combination of many specific issues. One will have to tailor a special PIV system for a given helium apparatus. Measurements (including the use of the so-called super-resolution PIV) are in progress at this time at Yale, where the first-generation measurements have already been made.

4. *Measurement of Vorticity*

Direct measurement of vorticity is still a challenging problem in classical turbulence. Even a qualitative method of visualizing the turbulent vorticity field would be of immense value. In helium II, ion measurements have been used for marking quantized vortices. Since ions are trapped on the cores of quantized vortices in helium II, they could, in principle, be used to obtain quantitative features such as root-mean-square vorticity. Complications arise because the ions can move along the cores of the vortices. At the present state of development, ions can locate vorticity spatially in a turbulent flow, but cannot easily determine the magnitude.

We have mentioned in Section VI a possible connection between second sound attenuation and root-mean-square vorticity. Studies at Oregon (Smith, 1993; Stalp, 1998) behind towed grids in a 1-cm^2 tube have shown the versatility of these measurements.

VIII. Limitations of Helium as a Hydrodynamic Test Fluid

We have cited a conceptual flow facility of 125 cm in diameter that can generate Reynolds numbers of the order 10^9. In Donnelly (1991a), several alternatives have also been considered to show that full-scale operation Reynolds numbers can be attained with helium. Although this satisfies the high Reynolds number requirement, it does not guarantee that satisfactory answers about *all* aspects of the overall field can be obtained. For instance, one does not know the nature of interaction between vorticity and the acoustic field reflected off the tunnel boundary, or the cavitation properties of helium in a turbulent environment. Some worries have also been expressed that turbulent motion at such high Reynolds numbers may not be the same in every respect as that for water flows. For example, local heating due to focused energy dissipation may affect the constitutive properties of helium (especially because of the extreme sensitivity of these properties to temperature changes); these local sources of heat may act as randomly distributed pressure sources. The smallness of velocity scales in ultra-high Reynolds number helium flows may render Navier–Stokes equations irrelevant to aspects of helium turbulence. These questions are often phrased, somewhat awkwardly but succinctly, as "Is helium a Navier–Stokes fluid?" The worry is not that some unknown stress-strain behavior is needed to describe flows of helium I; it is that faithful similarities between water and helium I flows may break down when the *total* environment, such as the interaction between sound and vorticity, sound propagation through the medium, its far-field properties, reflection from boundaries due to differences in acoustic impedance, cavitation effects, and so forth, are all considered. Some of these questions are relevant only to model testing, but others to basic research in turbulence. We have not pursued these questions to great depth but examined them via back-of-the-envelope calculations. These calculations suggest that no "show-stoppers" of principle exist, but more detailed calculations are called for.

Turning now to aerodynamics, usually ultra-high Reynolds numbers occur simultaneously with sizable compressibility effects, and there may even be regions of the flow where shocks are formed. Given that the ratio of specific heats for air ($\gamma = 1.4$) is different from that of helium gas ($\gamma = 1.67$), the shock structure will be undoubtedly different. The position of shocks depends to some degree on γ, as does the nature of shock boundary layer interaction. Thus, one has to be concerned about the degree to which the flow field observed in helium corresponds to that in air. In particular, this makes a transonic tunnel using helium gas less practical for aerodynamic testing.

Finally, one should be mindful of the fact that both the cool-down and warm-up phases of operation of any sizable helium facility would be significant.

IX. Concluding Remarks

This article has tried to provide a perspective on the use of helium as a test fluid for research and applications in classical fluid dynamics. Helium offers tremendous opportunities and advantages. At the moment, there is a convergence of interests from diverse fields such as turbulence, physics of helium, wind tunnel and water tunnel testing, instrumentation, and the technology of large-scale refrigeration plants. One should not lose sight of the uniqueness of this opportunity. The uniqueness of data that can be acquired by this means would amply justify the effort — on both fundamental and applied fronts.

Even well-known technologies, when applied to a different domain, pose unforeseen problems; this is not different with helium. However, the advantages of helium should not be buried under the cloud of uncertainties. Instead, some of the remaining questions should be given careful attention. Our discussion in this article — admittedly cursory on most fronts — shows that helium is an excellent option for some purposes. A large-scale experimental facility housed at BNL, say, can go a long way toward addressing some important problems of classical fluid dynamics. Already, some interim experiments in smaller scale facilities are under way. Their objectives are as follows:

1. To make turbulence measurements in helium flows with meaningful accuracy and resolution (using hot-wires, PIV and LDV), and make satisfactory comparisons with equivalent water or air flows;

2. To gain experience about vortex-coupled superfluidity, and the transmission of pressure waves generated by an oscillating body in still helium; and

3. To obtain aerodynamic data such as the drag coefficient on a cylinder or a sphere as a function of the Reynolds number through the "drag crisis" value.

The combined experience, which is now developing fast, will no doubt facilitate further decisions on the larger facilities. The outlook at the present is quite optimistic.

Acknowledgments

We are grateful to many colleagues for their contributions to our thinking on this problem over the years. Among them are Dennis Bushnell, Nigel Goldenfeld (especially for the statement of the vortex-coupled superfluidity), Adonios Karpetis, Richard Nadolink, Joseph Niemela, Ladik Skrbek, Steve Stalp, and Christopher White. We have been helped by some perceptive remarks made by Theodore Wu. This research was supported by the National Science Foundation under the grant DMR-95-29609.

References

Aarts, R. G. K. M., and DeWaele, A. T. A. M. (1994). Numerical simulation of superfluid turbulence near the critical velocity. *Physica B* **194–196**, 725.

Ashkenazi, S., and Steinberg, V. (1999). High Rayleigh number turbulent convection in a gas near the gas-liquid critical point. *Phys. Rev. Lett.* **83**, 3641.

Barenblatt, G. I., and Chorin, A. J. (1996). Scaling laws and zero viscosity limits for wall-bounded shear flows and the local structure in developed turbulence. *Proc. Nat. Acad. Sci.* **93**, 6749.

Barenghi, C. F., and Samuels, D. C. (1999). Self-consistent decay of superfluid turbulence. *Phys. Rev. B.* **60**, 1252.

Barenghi, C. F., Samuels, D. C., Bauer, G. H., and Donnelly, R. J. (1997). Superfluid vortex lines in a model of turbulent flow. *Phys. Fluids* **9**, 2631.

Batchelor, G. K. (1953). *The Theory of Homogeneous Turbulence*. Cambridge University Press, Cambridge.

Bielert, F., and Stamm, G. (1993). Visualization of Taylor–Couette flow in superfluid helium. *Cryogenics* **33**, 938.

Bolgiano, R. (1962). Structure of turbulence in stratified media. *J. Geophys. Res.* **67**, 3105.

Britcher, C. P. (1991). Recent aerodynamic measurements with magnetic suspension systems. In *High Reynolds Number Flows Using Liquid and Gaseous Helium* (R. J. Donnelly, ed.), p. 165. Springer-Verlag, New York.

Britcher, C. P. (1998). Application of magnetic suspension and balance systems to ultra-high Reynolds number facilities. In *Flow at Ultra-High Reynolds and Rayleigh Numbers: A Status Report* (R. J. Donnelly and K. R. Sreenivasan, eds.), p. 165. Springer-Verlag, New York.

Bushnell, D. M. (1998). High Reynolds number testing requirements in (civilian) aeronautics. In *Flow at Ultra-High Reynolds and Rayleigh Numbers: A Status Report* (R. J. Donnelly and K. R. Sreenivasan, eds.), p. 323. Springer-Verlag, New York.

Bushnell, D. M., and Greene, G. C. (1991). High Reynolds number test requirements in low speed aerodynamics. In *High Reynolds Number Flows Using Liquid and Gaseous Helium* (R. J. Donnelly, ed.), p. 19. Springer-Verlag, New York.

Castaing, B., Chabaud, B., Chilla, F., Hebral, B., Naert, A., and Peinke, J. (1994). Anemometry in gaseous ^{4}He around 4 K, *J. de Phys III (France)* **4**, 671.

Castaing, B., Chabaud, B., and Hebral, B. (1992). Hot wire anemometer operating at cryogenic temperatures. *Rev. Sci. Instr.* **63**, 4168.

Castaing, B., Gunaratne, G. H., Heslot, F., Kadanoff, L., Libchaber, A., Thomae, S., Wu, X.-Z., Zaleski, S., and Zanetti, G. (1989). Scaling of hard thermal turbulence in Raleigh-Bénard convection. *J. Fluid Mech.* **204**, 1.

Chevanne, X., Chilla, F., Castaing, B., Hebral, B., Chabaud, B., and Chaussy, J. (1997). Observations of the ultimate region in Rayleigh-Bénard convection. *Phys. Rev. Lett.* **79**, 3648.

Constantin, P., Procaccia, I., and Sreenivasan, K. R. (1991). Fractal geometry of isoscalar surfaces in turbulence: Theory and experiment. *Phys. Rev. Lett.* **67**, 1739.

Donnelly, R. J. (ed.). (1991a). *High Reynolds Number Flows Using Liquid and Gaseous Helium.* Springer-Verlag, New York.

Donnelly, R. J. (1991b). *Quantized Vortices in Helium II.* Cambridge University Press, Cambridge.

Donnelly, R. J. (ed.) (1994). Cryogenic helium gas convection research: A discussion of opportunities for using the cryogenic facilities of the SSC laboratories for high Rayleigh number and high Reynolds number turbulence research. Department of Physics, University of Oregon.

Donnelly, R. J., and Hollis Hallett, A. C. (1958). Periodic boundary layer experiments in liquid helium. *Ann. Phys.* **3**, 320.

Donnelly, R. J., and Sreenivasan, K. R. (eds.). (1998). *Flow at Ultra-High Reynolds and Rayleigh Numbers: A Status Report.* Springer-Verlag, New York.

Emsellem, V., Kadanoff, L. P., Lohse, D., Tabeling, P., and Wang, Z. J. (1997). Transitions and probes in turbulent helium. *Phys. Rev. E* **55**, 2672.

Frisch, U. (1995). *Turbulence: The Legacy of A. N. Kolmogorov.* Cambridge University Press, Cambridge.

Goldenfeld, N. (1992). *Lectures on Phase Transitions and the Renormalization Group.* Addison-Wesley, Reading, MA.

Goodyer, M. J. (1991). The six component magnetic suspension system for wind tunnel testing. In *High Reynolds Number Flows Using Liquid and Gaseous Helium* (R. J. Donnelly, ed.), p. 131. Springer-Verlag, New York.

Grant, H. L., Stewart, R. W., and Moilliet, A. (1962). Turbulent spectrum in a tidal channel. *J. Fluid Mech.* **12**, 241.

Gray, D. R. (1988). *Lectures on Spectral Line Analysis: F, G and K Stars.* Aylmer Express Ltd., Ontario.

Grossmann, S., and Lohse, D. (1993). Characteristic scales in Rayleigh-Bénard convection. *Phys. Lett. A*, **173**, 58.

Howard, L. N. (1972). Bounds on flow quantities. *Annu. Rev. Fluid Mech.* **4**, 473.

Hunt, J. C. R., Phillips, O. M., and Williams, D. (eds.). (1991). *Turbulence and Stochastic Processes: Kolmogorov's Idea 50 Years On.* Royal Society of London.

Ichikawa, N., and Murakami, M. (1991). Application of flow visualization technique to superflow experiment. In *High Reynolds Number Flows Using Liquid and Gaseous Helium* (R. J. Donnelly, ed.), p. 209. Springer-Verlag, New York.

Kilgore, R. A. (1991). Cryogenic wind tunnels. In *High Reynolds Number Flows Using Liquid and Gaseous Helium* (R. J. Donnelly, ed.), p. 53. Springer-Verlag, New York.

Kilgore, R. A. (1998). Cryogenic wind tunnels for aerodynamic testing. In *Flow at Ultra-High Reynolds and Rayleigh Numbers: A Status Report* (R. J. Donnelly and K. R. Sreenivasan, eds.), p. 66. Springer-Verlag, New York.

Kistler, A. L., and Vrebalovich, T. (1966). Grid turbulence at large Reynolds numbers. *J. Fluid Mech.* **26**, 37.

Kolmogorov, A. N. (1941). Local structure of turbulence in an incompressible fluid at very high Reynolds numbers. *Dokl. Akad. Nauk. SSSR,* **30**, 299; Energy dissipation in locally isotropic turbulence. *Dokl. Akad. Nauk, SSSR,* **32**, 19.

Kraichnan, R. H. (1962). Turbulent thermal convection at arbitrary Prandtl number. *Phys. Fluids* **5**, 1374.

Krishnamurti, R. (1975). On cellular cloud patterns. Part 1: Mathematical model, *J. Atm. Sci.* **32**, 1353.

Laing, R. A., and Rorschach, Jr., H. E. (1961). Hydrodynamic drag on spheres moving in liquid helium. *Phys. Fluids* **4**, 564.

Lambert, R. B., Snyder, H. A., and Karlsson, S. K. F. (1965). Hot thermistor anemometer for finite amplitude stability measurements. *Rev. Sci. Inst.* **36**, 924.

Lawing, P. L. (1991). Magnetic suspension—Today's marvel, tomorrow's tool. In *High Reynolds Number Flows Using Liquid and Gaseous Helium* (R. J. Donnelly, ed.), p. 153. Springer-Verlag, New York.

Lipa, J. (1998). Cryogenic thermometry for turbulence research: An overview. In *Flow at Ultra-High Reynolds and Rayleigh Numbers: A Status Report* (R. J. Donnelly and K. R. Sreenivasan, eds.), p. 179. Springer-Verlag, New York.

L'vov, V., and Procaccia, I. (1996). Hydrodynamic turbulence: A 19th century problem with a challenge for the 21st century. *Phys. World* **9**, 35.

Maurer, J., and Tabeling, P. (1998). Local investigation of superfluid turbulence. *Europhys. Lett.* **43**, 29.

Metcalfe, G. P., and Behringer, R. P. (1990). Convection in ^{3}He-superfluid-^{4}He mixtures: measurement of the superfluid effects. *Phys. Rev. A.* **41**, 5735.

Monin, A. S., and Yaglom, A. M. (1971). *Statistical Fluid Mechanics*, Vol. 1. The MIT Press, Cambridge, MA.

Monin, A. S., and Yaglom, A. M. (1975). *Statistical Fluid Mechanics*, Vol. 2. The MIT Press, Cambridge, MA.

Murakami, M., Hanada, M., and Yamazaki, T. (1987). Flow visualization study of large-scale vortex ring in He II. *Japan J. Appl. Phys.* **26**, 107.

Murakami, M., Nakano, A., and Iida, T. (1998) Applications of a laser Doppler velocimeter and some visualization methods to the measurement of He II thermo-fluid dynamic phenomena. In *Flow at Ultra-High Reynolds and Rayleigh Numbers: A Status Report* (R. J. Donnelly and K. R. Sreenivasan, eds.), p. 159. Springer-Verlag, New York.

Murakami, M., Yamazaki, T., Nakano, D. A., and Nakai, H. (1991). Laser Doppler velocimeter applied to superflow measurement. In *High Reynolds Number Flows Using Liquid and Gaseous Helium* (R. J. Donnelly, ed.), p. 215. Springer-Verlag, New York.

Nelkin, M. (1994). University and scaling in fully developed turbulence. *Adv. Phys.* **43**, 143.

Niemela, J., Skrbek, L., Sreenivasan, K. R., and Donnelly, R. J. (2000a). Turbulent convection at very high Rayleigh numbers. *Nature* **404**, 837–840.

Niemela, J., Skrbek, L., Sreenivasan, K. R., and Donnelly, R. J. (2000b). Comments on high Rayleigh number convection. To appear in *Proc. IUTAM Symposium on the Geometry and Statistics of Turbulence* (T. Kambe, ed.).

Onsager, L. (1949). Statistical hydrodynamics. *Nuovo Cimento Suppl.* **VI** (ser. IX), 279.

Perrin, B. (1982). Emergence of a periodic mode in the so-called turbulent region in a circular Couette flow. *J. Phys. Lett.* **43**, L-5.

Quack, H. H. (1998). European large scale helium refrigeration. In *Flow at Ultra-High Reynolds and Rayleigh Numbers: A Status Report* (R. J. Donnelly and K. R. Sreenivasan, eds.), p. 52. Springer-Verlag, New York.

Roberts, P. H., and Donnelly, R. J. (1974). Superfluid mechanics. *Annu. Rev. Fluid Mech.* **6**, 179.

Roshko, A. (1961). Experiments in the flow past a circular cylinder at very high Reynolds numbers. *J. Fluid Mech.* **1**, 345.

Saddoughi, S., and Veeravalli, S. (1994). Local isotropy in turbulent boundary layers at high Reynolds numbers. *J. Fluid Mech.* **268**, 333.

Schlichting, H. (1956). *Boundary-Layer Theory*. McGraw-Hill, New York.

Schwarz, K. W. (1985). Three-dimensional vortex-dynamics in superfluid ^{4}He: Line-line and line-boundary interactions. *Phys. Rev. B*, **31**, 5782.

Schwarz, K. W. (1988). Three-dimensional vortex dynamics in superfluid ^{4}He: Homogeneous superfluid turbulence. *Phys. Rev. B* **38**, 2398.

Shraiman, B., and Siggia, E. (1995). Anomalous scaling of a passive scalar in turbulent flow. *C. R. Acad. Sci. Paris*, **321**, Series IIb, 279.

Sinai, Y. G., and Yakhot, V. (1989). Limiting probability distributions of a passive scalar in a random velocity field. *Phys. Rev. Lett.* **63**, 1962.

Skrbek, L., and Stalp, S. (1999). On the decay of homogeneous isotropic turbulence (submitted for publication).

Smelt, R. (1945). Power economy in high-speed wind tunnel by choice of working fluid and temperature. In *High Reynolds Number Flows Using Liquid and Gaseous Helium* (R. J. Donnelly, ed.), p. 265. Springer-Verlag, New York.

Smith, M. R. (1993). Evolution and propagation of turbulence in helium II. Ph.D. thesis, University of Oregon, Eugene.

Smith, M. R., Hilton, D. K., and Van Sciver, S. W. (1999). Observed drag crisis on a sphere in flowing He I and He II. *Phys. Fluids*, **11**, 751.

Smits, A. J., and Zagarola, M. V. (1997). Design of a high Reynolds number testing facility using compressed air, AIAA Paper 97-1917, IV Shear Flow Conference, Snowmass, CO.

Sondericker, J. H. (1988). A brief overview of the RHIC cryogenic system. In *Flow at Ultra-High Reynolds and Rayleigh Numbers: A Status Report* (R. J. Donnelly and K. R. Sreenivasan, eds.), p. 436. Springer-Verlag, New York.

Sreenivasan, K. R. (1984). On the scaling of the turbulent energy dissipation rate. *Phys. Fluids* **27**, 1048.

Sreenivasan, K. R., and Antonia, R. A. (1997). The phenomenology of small-scale turbulence. *Annu. Rev. Fluid Mech.* **29**, 435.

Sreenivasan, K. R., and Meneveau, C. (1988). Singularities of the equations of fluid motion. *Phys. Rev. A* **38**, 6287.

Stalp, S. R. (1998). Decay of grid turbulence in superfluid turbulence. Ph.D. thesis, University of Oregon, Eugene.

Stalp, S. R., Skrbek, S., and Donnelly, R. J. (1999). Decay of grid turbulence in a finite channel. *Phys. Rev. Lett.* **82**, 4831.

Stamm, G., Bielert, F., Fisdon, W., and Piechna, J. (1994). Counterflow-induced macroscopic vortex rings in superfluid helium: Visualization and numerical simulation. *Physica B* **193**, 188.

Swanson, C., and Donnelly, R. J. (1998). Instrument development for high Reynolds number flows in liquid helium. In *Flow at Ultra-High Reynolds and Rayleigh Numbers: A Status Report* (R. J. Donnelly and K. R. Sreenivasan, eds.), p. 206. Springer-Verlag, New York.

Swanson, C. J., Johnson, K., and Donnelly, R. J. (1988). An accurate differential pressure gauge for use in liquid and gaseous helium. *Cryogenics* **38**, 673.

Tabeling, P., Zocchi, G., Belin, F., Maurer, J., and Willaime, H. (1996). Probability density function, skewness and flatness in large Reynolds number turbulence. *Phys. Rev. E* **53**, 1613.

Tennekes, H., and Lumley, J. L. (1972). *A First Course in Turbulence*. The MIT Press, Cambridge, MA.

Threlfall, C. (1975). Free convection in low-temperature gaseous helium. *J. Fluid Mech.* **67**, 17.

Van Sciver, S. W. (1991). Experimental investigations of He II flows at high Reynolds number. In *High Reynolds Number Flows Using Liquid and Gaseous Helium* (R. J. Donnelly, ed.), p. 223. Springer-Verlag, New York.

Van Sciver, S. W., Holmes, D. S., Huang, X., and Weisend, II, J. G. (1991). He II flowmetering. *Cryogenics* **31**, 75.

Vinen, W. F. (1961). Vortex lines in liquid helium II. In *Prog. Low Temp. Phys.*, Vol. III (C. J. Gorter, ed.), p. 1, North-Holland Publishing Co., Amsterdam.

Vinen, W. F. (2000). Why is turbulence in a quantum liquid often similar to that in a classical liquid? *Phys. Rev. B.* **61**, 1410.

Walstrom, P. L., Weisend II, J. G., Maddocks, J. R., and Van Sciver, S. W. (1988). Turbulent flow pressure drop in various He II transfer system components. *Cryogenics* **28**, 101.

Woodcraft, A. L., Lucas, P. G. J., Matley, R. G., and Wong, W. Y. T. (1998). First images of controlled convection in liquid helium. In *Flow at Ultra-High Reynolds and Rayleigh Numbers: A Status Report* (R. J. Donnelly and K. R. Sreenivasan, eds.), p. 436. Springer-Verlag, New York.

Wu, X.-Z. (1991). Along the road to developed turbulence: Free thermal convection in low temperature helium gas. Ph.D thesis, University of Chicago.

Wybourne, M., and Smith, J. (1998). Considerations for small detectors in high Reynolds number experiments. In *Flow at Ultra-High Reynolds and Rayleigh Numbers: A Status Report* (R. J. Donnelly and K. R. Sreenivasan, eds.), p. 329. Springer-Verlag, New York.

Zagarola, M., and Smits, A. J. (1997). Scaling of the mean velocity profiles for turbulent pipe flow. *Phys. Rev. Lett.* **78**, 239.

Recent Advances in Applications of Tensor Functions in Continuum Mechanics

JOSEF BETTEN

Department of Mathematical Models in Materials Science
Technical University of Aachen
Aachen, Germany

ADVANCES IN APPLIED MECHANICS, VOL. 37
ISSN 0065-2165/01 $35.00
ISBN 0-12-002037-8

I. Introduction

Continuum mechanics is concerned with the mechanical behavior of solids and fluids on the macroscopic scale. It ignores the discrete nature of matter, and treats material as uniformly distributed throughout regions of space. It is then possible to define quantities such as density, displacement, velocity, etc., as continuous (or at least piecewise continuous) functions of position. This procedure is found to be satisfactory provided that we deal with bodies whose dimensions are large in comparison with the characteristic lengths (e.g., interatomic spacings in a crystal or mean free paths in a gas) on the microscopic scale.

Continuum mechanics can also be applied to a granular material such as sand, concrete, or soil, provided that the dimensions of the regions considered are large compared with those of an individual grain.

The equations of continuum mechanics are of two main kinds. First, there are equations that apply equally to all materials. They describe universal physical laws, such as conservation of mass and energy. Second, there are equations characterizing the individual material and its reaction to applied loads; such equations are called *constitutive equations*, since they describe the macroscopic behavior resulting from internal constitution of the particular materials. But materials, especially in the solid state, behave in such complex ways when the entire range of possible temperatures and deformations is considered that it is not feasible to write down one equation or set of equations to describe accurately a real material over its entire range of behavior. Instead, we formulate separate equations describing various kinds of *ideal material response*, each of which is a mathematical formulation designed to approximate physical observations of a real material's response over a suitably restricted range.

Physical laws should be independent of the position and orientation of the observer, i.e., if two scientists using different coordinate systems observe the same physical event, it should be possible to state a physical law governing the event in such a way that if the law is true for one observer, it is also true for the other. For this reason, the equations of physical laws are *vector functions* or *tensor functions*, since vectors and tensors transform from one coordinate system to another in such a way that if the vector or tensor equation holds in one coordinate system, it holds in any other coordinate system not moving relative to the first one, i.e., in any other coordinate system in the same reference frame. Invariance of the form of the physical law referred to two frames of references in accelerated motion relative to each other is more difficult and requires the apparatus of general relativity theory, tensors in four-dimensional space–time. For simplicity, we limit ourselves in this article to tensors in three-dimensional *Euclidean space*. Furthermore, we use only rectangular Cartesian coordinates for the components of vectors and tensors.

Constitutive equations must be invariant under changes of frame of reference, i.e., two observers, even if in relative motion with respect to each other, observe the same stress in a loaded material. The *principle of material frame-indifference* is also called the *principle of material objectivity* (Betten, 1985b, 1993).

The formulation of constitutive equations is essentially a matter of experimental determination, but a theoretical framework is needed in order to devise suitable experiments and to interpret experimental results.

As has been pointed out in more detail by Avula (1987):

> The validity of a model should not be judged by mathematical rationality alone; nor should it be judged purely by empirical validation at the cost of mathematical and scientific principles. A combination of rationality and empiricism (logic and pragmatism) should be used in the validation.
>
> Experimental observations and measurements are generally accepted to constitute the backbone of physical sciences and engineering because of the physical insight they offer to the scientist for formulating the theory. The concepts that are developed from observations are used as guides for the design of new experiments, which in turn are used for validation of the theory. Thus, experiments and theory have a hand-in-hand relationship.

However, it must be noted, that experimental results can differ greatly from reality just like a bad mathematical model (Betten, 1973).

During the last two decades much effort has been devoted to the elaboration of phenomenological theories describing the relationship between force and deformation in bodies of materials that do not obey either the linear laws of the classical theories of elasticity or the hydrodynamics of viscous fluids. Such problems will play a central role for mathematicians, physicists, and engineers in the future (Astarita, 1979).

Material laws and constitutive theories are the fundamental bases for describing the mechanical behavior of materials under multiaxial states of stress involving actual boundary conditions. In solving such complex problems, the tensor function theory has become a powerful tool (Betten, 1987b,c, 1989a,b).

This article provides a short survey of some recent advances in the mathematical modeling of materials behavior including anisotropy and damage.

The mechanical behavior of anisotropic solids (materials with orientated internal structures, produced by forming processes and manufacturing procedures, or induced by permanent deformation) requires suitable mathematical modeling. The properties of tensor functions with several argument tensors constitute a rational basis for a consistent mathematical modeling of complex material behavior (Betten, 1987b,c, 1989a,b; Boehler, 1987).

This article presents certain principles, methods, and recent successful applications of tensor functions in solid mechanics. The rules of specifying irreducible sets of tensor invariants and tensor generators of material tensors of rank two and four are also discussed.

Furthermore, it is very important to determine the scalar coefficients in constitutive and evolutional equations as functions of the integrity basis and experimental data. It is explained in detail that these coefficients can be determined by using tensorial interpolation methods. Some examples for practical use are discussed.

Finally, we have carried out own experiments in order to examine the validity of the mathematical modeling.

Like applications in solid mechanics, tensor functions also play a significant role in mathematical modeling in fluid mechanics. This article, however, is restricted to the mechanical behavior of solids.

II. Nonlinear Constitutive Equations for Anisotropic Materials

It will be convenient in this text to use the compact notation often referred to as *indicial* or *index notation*. It allows a strong reduction in the number of terms in an equation and is commonly used in the current literature when stress, strain, and constitutive equations are discussed. Therefore, a basic knowledge of the index notation is helpful in studying continuum mechanics, especially constitutive modeling of materials. With such a notation, the various stress-strain relationships for materials under multiaxial states of stress can be expressed in compact form. Thus, greater attention can be paid to physical principles rather than to the equations themselves.

We consider vectors and tensors in three-dimensional *Euclidean space*. For simplicity, rectangular Cartesian coordinates x_i, $i = 1, 2, 3$, are used throughout. Results may, if desired, be expressed in terms of curvilinear coordinate systems by standard techniques of tensor analysis (Betten, 1987c). A matrix of tensor components will be denoted by a boldface letter.

The following text is based on lectures in continuum mechanics, especially *elasticity*, *plasticity*, and *creep mechanics* (Betten, 1985b, 1993), given at the Technical University of Aachen, Germany. Furthermore, some results are discussed that the author has presented at international conferences or published in international journals in recent years. A lot of results discussed in the following have not yet been published.

A. Elastic Behavior

It has been pointed out in more detail, for instance, by Betten (1985b, 1993) that an *isotropic material* can be characterized by a constitutive equation of the form

$$\sigma_{ij} = f_{ij}(V_{pq}) \tag{2.1}$$

relating the Cauchy stress tensor $\boldsymbol{\sigma}$ to the *left stretch tensor* $\mathbf{V}$, which is a positive definite symmetric tensor. It is closely related to the left Cauchy–Green tensor $\mathbf{B}$ and to the *deformation gradient* $\mathbf{F}$, as follows:

$$V_{ij}^{(2)} \equiv V_{ik}V_{kj} = B_{ij} = F_{ik}F_{jk}, \tag{2.2a}$$

or in matrix notation as:

$$\mathbf{V}^2 = \mathbf{B} = \mathbf{F}\mathbf{F}^t. \tag{2.2b}$$

In (2.2a) and in the following text Einstein's summation convention is introduced. The *deformation gradient* $\mathbf{F}$ can be interpreted as a linear transformation or a second-rank tensor: An initial line element vector $d\vec{s}_0$ is mapped onto the corresponding vector $d\vec{s} = \mathbf{F}\,d\vec{s}_0$ at time t, where a *translation*, a *rigid rotation*, and a *stretching* of the line element are produced. This decomposition of the total deformation corresponds with the *polar decomposition theorem*, which states that the nonsingular deformation gradient $\mathbf{F}$ can be decomposed, uniquely, in either of the products $\mathbf{F} = \mathbf{R}\mathbf{U}$ and $\mathbf{F} = \mathbf{V}\mathbf{R}$, where $\mathbf{R}$ is an orthogonal rotation tensor, and $\mathbf{U}$, $\mathbf{V}$ are positive definite symmetric tensors, which are called *right* ($\mathbf{U}$) and *left* ($\mathbf{V}$) *tensors*. However, the *translation part* of the total deformation is not involved in the polar decomposition since a parallel transport does not change the Cartesian components of a vector, here the line element vector. The *shifter*, which shifts a vector from one coordinate system to another, is here merely the unit tensor δ due to Kronecker. But in curvilinear coordinates, parallel transport does change the *covariant* and *contravariant* components of a vector (Betten, 1987c).

Although the deformation gradient tensor $\mathbf{F}$ plays a central role in the analysis of deformation, it is not itself a suitable (direct) *measure of strain*, since a *measure of strain* must be unchanged in a *rigid-body motion*. Thus, the difference $(ds)^2 - (ds_0)^2$ for two neighboring particles of a continuum can be taken for a suitable measure of strain, which occurs in the neighborhood of the particles between the initial and current configurations. This *strain measure* can be formulated with spatial coordinates x_i as independent variables (*spatial description*):

$$(ds)^2 - (ds_0)^2 = 2\eta_{ij}\, dx_i\, dx_j,$$

where η_{ij} are the cartesian components of the *Eulerian finite strain tensor* $\boldsymbol{\eta} = \frac{1}{2}(\delta - \mathbf{B}^{-1})$. In this expression $\mathbf{B}^{-1}$ is the inverse tensor of the *left Cauchy–Green tensor* $\mathbf{B} = \mathbf{FF}^t$ in (2.2a) and (2.2b).

Equation (2.1) is a tensor-valued function with one argument tensor and can be represented in three terms:

$$\sigma_{ij} = f_{ij}(V_{pq}) = \varphi_0 \delta_{ij} + \varphi_1 V_{ij} + \varphi_2 V_{ij}^{(2)}, \tag{2.3}$$

which are the contributions of zero, first, and second orders in the stretch tensor $\mathbf{V}$. Further terms on the right-hand side of (2.3) are redundant because of the *Hamilton–Cayley theorem*. It states that a second-order tensor of power three and all higher powers can be expressed in terms of lower powers than three:

$$V_{ij}^{(p)} = \sum_{v=1}^{3} {}^{p}Q_{3-v} V_{ij}^{(3-v)}, \qquad p \geqslant 3. \tag{2.4}$$

Thus, the representation (2.3) is *irreducible and complete*. The coefficients ${}^{p}Q_{3-v}$ in (2.4) are scalar polynomials of degree $p - 3 + v$ in the irreducible invariants

$$J_1 := V_{kk}, \quad J_2 := \tfrac{1}{2}(V_{ij}V_{ji} - V_{ii}V_{jj}), \quad J_3 := \det(V_{ij}) \tag{2.5a,b,c}$$

of the argument tensor $\mathbf{V}$ and can be determined by using the following recursion relation:

$$^{p}Q_{3-v} = J_{p-3+v} + \sum_{\mu=1}^{p-3} {}^{p-\mu}Q_{3-v} J_\mu, \tag{2.6}$$

which can be generalized to arbitrary dimension n as has been pointed out in detail by Betten (1987c, 1993).

The coefficients φ_0, φ_1, φ_2 in (2.3) are scalar-valued functions of the *integrity basis* and experimental data. The third term on the right-hand side in (2.3) expresses the *tensorial nonlinearity*, which is responsible for *second-order effects*, a representative sample of which is the *Poynting effect* (Betten, 1981a). The 1909 experiments of Poynting showed that wires of steel, copper, brass, and rubber lengthen when twisted. Poynting interpreted his results by means of a fragile analogy to shear, using a special nonlinear theory of three-dimensional elasticity.

When investigating the elastic behavior of *isotropic* materials one can presume the existence of an *elastic potential* (*strain-energy function*) W, which is a scalar-valued function of an appropriately defined strain tensor, e.g., $W = W(V_{ij})$. This function is said to be isotropic if the condition

$$W(a_{ip}a_{jq}V_{pq}) = W(V_{ij}) \qquad (2.7)$$

is fulfilled under any orthogonal transformation $\mathbf{a}$. It is evident from the theory of isotropic tensor functions that the elastic potential can be expressed as a single-valued function of the *integrity basis*, the elements of which are the three invariants (2.5a,b,c) or, alternatively, the three irreducible invariants

$$S_\lambda \equiv \mathrm{tr}\,\mathbf{V}^\lambda, \qquad \lambda = 1, 2, 3 \qquad (2.8)$$

of the argument tensor.

By applying the partial derivative $\partial W(S_k)/\partial V_{ii}$ we immediately obtain the constitutive equation:

$$\sigma_{ij} = \frac{\partial W}{\partial S_1}\delta_{ij} + 2\frac{\partial W}{\partial S_2}V_{ij} + 3\frac{\partial W}{\partial S_3}V_{ij}^{(2)}. \qquad (2.9)$$

The more general case in which W is allowed to depend also on *temperature* or *entropy*, and in which heat flux is permitted, leads to the theory of *thermoelasticity* (Betten, 1993). We do not discuss this theory in this article.

If we compare the *minimum polynomial representation* (2.3) with the result (2.9) based on the *elastic potential theory*, we arrive at the following identities:

$$\varphi_0 \equiv \partial W/\partial S_1, \qquad \varphi_1 \equiv 2\partial W/\partial S_2, \qquad \varphi_2 \equiv 3\partial W/\partial S_3. \qquad (2.10)$$

By eliminating the elastic potential W in (2.10) we achieve the following sufficient and necessary conditions

$$2\frac{\partial \varphi_0}{\partial S_2} \equiv \frac{\partial \varphi_1}{\partial S_1}, \quad 3\frac{\partial \varphi_1}{\partial S_3} = 2\frac{\partial \varphi_2}{\partial S_2}, \quad 3\frac{\partial \varphi_0}{\partial S_3} = \frac{\partial \varphi_2}{\partial S_1}, \tag{2.11}$$

as has already been pointed out in more detail by Betten (1985a).

The results presented above illustrate that the elastic potential hypothesis (2.9) is *compatible* with the tensor function theory (2.3) provided additional conditions (2.11) of integrability have been fulfilled. This is *not* true for *anisotropic materials* (Betten, 1985a).

B. PLASTIC BEHAVIOR

For the analysis of strain and stress distributions in a material loaded beyond the elastic limit, a constitutive theory of plasticity must specify the *yield condition* under multiaxial states of stress, since the uniaxial condition $\sigma = \sigma_F$, where σ_F is the yield stress, is inadequate if there is more than one stress component σ. Furthermore, the theory of plasticity must specify the *postyield* behavior. Thus, the following questions arise:

1. *Yield criterion.* What stress combinations cause plastic deformations?

2. *Postyield Behavior.* How are the plastic deformation increments related to the stress components? How does the yield condition change with workhardening?

A yield condition for a material under multiaxial states of stress can be seen as a law defining the limit of elasticity under any possible combination of stresses. A yield condition is a special form of a failure criterion, which indicates the onset of plastic deformation.

A failure criterion for a material under multiaxial states of stress can be geometrically interpreted as a limiting envelope in the stress space, and failure occurs when the given "stress vector" (not to be confused with the term "traction") penetrates the failure surface (Figure 1). The principal values of Cauchy's stress tensor are indicated by $\sigma_I, \sigma_{II}, \sigma_{III}$.

The general form of a failure criterion is given by a *scalar-valued tensor function*

$$f(\sigma_{ij}; A_{ij}, A_{ijkl}, \ldots) = 1 \tag{2.12}$$

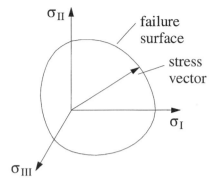

FIG. 1. Failure surface in the stress space.

of the Cauchy stress tensor (σ_{ij}) and several material tensors (A_{ij}), (A_{ijkl}), etc., of rank two, four, etc., characterizing the *anisotropy* of the material.

Due to the requirement of invariance under any orthogonal transformation, the function f in (2.12) can be expressed as a single-valued function of the integrity basis, the elements of which are the irreducible invariants. Together with the invariants of the single argument tensors, the set of simultaneous or joint invariants forms the integrity basis. The theory of invariants for several argument tensors of rank 2 has been developed, for instance, by Rivlin (1970) and Spencer (1971, 1987). Integrity bases for tensors of an order higher than two are discussed by Betten (1982a, 1987c, 1993, 1998).

To formulate a failure criterion (2.12), it is not necessary to consider the complete set of irreducible invariants. In finding the "essential invariants" we first formulate the constitutive equation

$$d\varepsilon_{ij}^{p} = f_{ij}(\sigma_{pq}; A_{pq}, A_{pqrs}, \ldots) \tag{2.13}$$

based on the representation theory of tensor-valued functions, where ε_{ij}^{p} are the Cartesian components of the plastic part of the strain tensor ε. Since the plastic work $dW = \sigma_{ji} \, d\varepsilon_{ij}^{p}$ characterizes the yield process, we read from its scalar expression the "relevant invariants" to formulate a yield criterion, as has been pointed out by Betten (1988a) in detail.

Some experimental data of several isotropic and anisotropic metals and polymers were compared with numerical results based on the theoretical formulation by Betten (1982c). For instance, the phenomenon of variation in yield strength under compression and tension (strength differential or SD effect) accompanied by the pressure-dependent yield was investigated by

Betten (1982c, 1993) and by Betten and Borrmann (1984) or by Betten *et al.* (1999).

Extended surveys of experimental results concerning yield surfaces were given by Ikegami (1975) and Phillips and Das (1985). Additional investigations should be mentioned, for instance, the papers by Litewka and Morzyńska (1985), Mazilu and Meyers (1985), Allirot and Boehler (1982), and many other scientists.

Constitutive equations describing plastic deformation can be formulated, for instance, in two different ways as pointed out in the following.

In the *classical theory of plasticity* the flow rule

$$d\varepsilon_{ij}^p = (\partial F/\partial \sigma_{ij})\, d\lambda \tag{2.14}$$

is used, where the factor $d\lambda$ can be interpreted as a Lagrange multiplier, as illustrated by Betten (1985b, 1993), and the scalar function F represents the *plastic potential*. From the theory of isotropic tensor functions discussed by Spencer (1971), it is evident that for an isotropic medium the scalar function F in (2.14) might be expressed as a single-valued function of the irreducible basic invariants

$$S_\nu \equiv \mathrm{tr}\boldsymbol{\sigma}^\nu \equiv \sigma_{kk}^{(\nu)}; \qquad \nu = 1, 2, 3 \tag{2.15}$$

of the stress tensor. Thus, considering the plastic potential

$$F = F(\sigma_{ij}) = F(S_\nu); \qquad \nu = 1, 2, 3 \tag{2.16}$$

and using the classical flow rule (2.14), one immediately obtains the following constitutive equation:

$$d\varepsilon_{ij}^p = \left(\frac{\partial F}{\partial S_1} \delta_{ij} + 2 \frac{\partial F}{\partial S_2} \sigma_{ij} + 3 \frac{\partial F}{\partial S_3} \sigma_{ij}^{(2)} \right) d\lambda. \tag{2.17}$$

Instead of the *plastic potential theory*, one can represent the tensor $d\varepsilon_{ij}^p$ as a *tensor-valued function* of one second rank argument tensor:

$$d\varepsilon_{ij}^p = f_{ij}(\sigma_{pq}) = \varphi_1 \delta_{ij} + \varphi_1 \sigma_{ij} + \varphi_2 \sigma_{ij}^{(2)}, \tag{2.18}$$

where φ_0, φ_1, φ_2 are scalar-valued functions of the integrity basis (2.15).

Comparing eqs. (2.17) and (2.18), we find the following identities:

$$\varphi_0 \equiv \frac{\partial F}{\partial S_1} d\lambda, \quad \varphi_1 \equiv 2 \frac{\partial F}{\partial S_2} d\lambda, \quad \varphi_2 \equiv 3 \frac{\partial F}{\partial S_3} d\lambda. \tag{2.19}$$

By eliminating the plastic potential F in (2.19), one can impose additional restrictions on the scalar functions φ_0, φ_1, φ_2, if the existence of a plastic potential is assumed. This problem was solved by Betten (1985a). Thus, in the isotropic special case (2.16), the plastic potential theory is compatible with the tensor function theory if some conditions of "integrability" have been fulfilled. However, for anisotropic materials the plastic potential theory provides only restricted forms of constitutive equations, even if a general plastic potential has been assumed. Consequently, the classical flow rule must be modified for anisotropic solids. Appropriate modifications have been discussed and the resulting conditions of "compatibility" were derived by Betten (1985a).

C. Creep Behavior

In creep mechanics one can differentiate between three stages: the *primary*, *secondary*, and *tertiary* creep stage. These terms correspond to a decreasing, constant, and increasing creep strain-rate, respectively, and were introduced by Andrade (1910).

To describe the creep behavior of metals in the primary stage, *tensorial nonlinear constitutive equations* involving the *strain-hardening hypothesis* have been proposed by Betten *et al.* (1989). Based on these general relations, the primary creep behavior of a thin-walled circular cylindrical shell subjected to internal pressure has also been analyzed by Betten *et al.* (1989). The creep buckling of cylindrical shells subjected to internal pressure and axial compression was investigated by Betten and Butters (1990) by considering *tensorial nonlinearities* and *anisotropic primary creep*.

In the following, the secondary creep stage of isotropic and anisotropic solids in a state of multiaxial stress will first be discussed. Creep deformations of metals usually remain unaffected if hydrostatic pressure is superimposed. To describe the secondary creep behavior of isotropic materials some authors use a *creep potential* (Rabotnov, 1969; Betten, 1975a, 1981a), which is a scalar-valued function of *Cauchy's stress tensor*. One can show that the *creep potential theory* is compatible with the *tensor function theory* provided the material is *isotropic* and additional conditions are fulfilled (Betten, 1985a). However, the creep potential hypothesis only furnishes restricted forms of constitutive equations and, therefore, has only limited justification if the material is *anisotropic*, as discussed in the following.

When investigating secondary creep behavior one can presume the existence of a creep potential. The *creep potential hypothesis* is based on the *principle of maximum dissipation rate*, from which one obtains [similar to (2.14)] the *flow rule*

$$d_{ij} = \dot{\Lambda} \partial F / \partial \sigma_{ij} \tag{2.20}$$

by applying *Lagrange's method* in connection with a *creep condition* F = const. as a subsidiary condition, where d_{ij} are the Cartesian components of the so-called *rate-of-deformation tensor* **d** and σ_{ij} are the Cartesian components of *Cauchy's stress tensor*. The tensor **d** is not to be confused with the *material time derivative* $\dot{\varepsilon}$ of the *infinitesimal strain tensor* ε as described in detail by Betten (1986b). Only in cases of small displacement gradients and small velocity gradient tensors do we find $d_{ij} \approx \dot{\varepsilon}_{ij}$. Based on the *hypotheses of the equivalent dissipation rate*, the *Lagrange multiplier* $\dot{\Lambda}$ has been determined for isotropic materials by Betten (1975a) and for anisotropic materials also by Betten (1981b) by taking the *Norton–Bailey creep law* into account. For isotropic materials the creep potential $F = F(\sigma_{ij})$ is a scalar-valued tensor function only of the *Cauchy stress tensor*. It is evident from the theory of isotropic tensor functions that the creep potential can be expressed as a single-valued function of the *integrity basis*, the elements of which are the *irreducible basic invariants* $S_\nu \equiv \text{tr}\,\mathbf{\sigma}^\nu$, $\nu = 1, 2, 3$, of the stress tensor. Thus, by considering the form $F = F(S_1, S_2, S_3)$ and using the *classical flow rule* (2.20), one immediately obtains the following constitutive equation:

$$d_{ij} = \dot{\Lambda} \left[\left(\frac{\partial F}{\partial S_1} \right) \delta_{ij} + 2 \left(\frac{\partial F}{\partial S_2} \right) \sigma_{ij} + 3 \left(\frac{\partial F}{\partial S_3} \right) \sigma_{ij}^{(2)} \right], \tag{2.21}$$

where the abbreviation $\sigma_{ij}^{(2)} \equiv \sigma_{ik}\sigma_{kj}$ has been introduced and Einstein's summation convention is used.

Instead of the *creep potential hypothesis*, one can represent the rate-of-deformation tensor as a symmetric *tensor-valued function* of one second-order argument tensor:

$$d_{ij} = f_{ij}(\mathbf{\sigma}) = \varphi_0 \delta_{ij} + \varphi_1 \sigma_{ij} + \varphi_2 \sigma_{ij}^{(2)}, \tag{2.22}$$

where φ_0, φ_1, φ_2 are scalar-valued functions of the integrity basis S_1, S_2, S_3.

If one compares the *minimum polynomial representation* (2.22) with the result (2.21) based on the *creep potential hypothesis*, one arrives at the following identities:

$$\varphi_0 \equiv \dot{\Lambda}\frac{\partial F}{\partial S_1}, \quad \varphi_1 \equiv 2\dot{\Lambda}\frac{\partial F}{\partial S_2}, \quad \varphi_2 \equiv 3\dot{\Lambda}\frac{\partial F}{\partial S_3}. \tag{2.23}$$

By eliminating the creep potential F in (2.23) one achieves the following sufficient and necessary condition:

$$\left(3\frac{\partial\varphi_1}{\partial S_3} - 2\frac{\partial\varphi_2}{\partial S_2}\right)\varphi_0 + \left(\frac{\partial\varphi_2}{\partial S_1} - 3\frac{\partial\varphi_0}{\partial S_3}\right)\varphi_1 + \left(2\frac{\partial\varphi_0}{\partial S_2} - 2\frac{\partial\varphi_1}{\partial S_1}\right)\varphi_2 = 0, \tag{2.24}$$

as has already been pointed out in more detail by Betten (1985a).

As one can see from the results given above, the theory of the creep potential is compatible with the tensor function theory provided condition (2.24) has been fulfilled. That is why this condition has been called the *condition of compatibility* by Betten (1985a).

The creep potential hypothesis is *not* compatible with the tensor function theory for more complicated examples than those mentioned above. It furnishes only restricted forms of constitutive equations for *anisotropic materials* in particular, even if a general creep potential has been assumed. Consequently, the classical flow rule (2.20) must be modified, and for each such modification *conditions of compatibility* must be found. Some examples are discussed by Betten (1985a), for instance a case in which oriented solids can be described by a *material tensor of rank two*. In such cases the creep potential is a scalar-valued function of two symmetric second-order tensors, $F = F(\sigma_{ij}, A_{ij})$, the integrity basis of which consists of 10 irreducible elements. A more general case is also described by Betten (1985a) by using a *material tensor of rank four*.

It must be noted that, from the physical point of view (Rice, 1970), the assumption of a creep potential has only limited justification, especially in the anisotropic case and in the tertiary creep stage. Fortunately, constitutive equations can be represented in full as tensor-valued functions.

For example, the tensor function theory has been applied by Betten and Waniewski (1986, 1989, 1998), Sawczuk and Anisimowicz (1981), Sawczuk and Trampczyński (1982), and Waniewski (1985) in order to study the effect of *plastic prestrain* on the secondary creep behavior. Plastic prestrain can produce *anisotropy* in materials that are initially isotropic. This anisotropy greatly influences the creep behavior of such materials. *Tensorial nonlinear constitutive equations* for this secondary creep behavior are proposed by Betten and Waniewski (1986, 1995), based on the representation theory of tensor functions. The induced anisotropy is characterized by introducing a second rank *prestrain tensor* as an additional argument tensor of the constitutive equations. Numerical results are compared with own creep tests on plastic prestrained specimens of INCONEL 617 at 950°C.

A *simplified theory* has been developed by Betten (1981a,b) in order to study anisotropic effects. This theory is based on the isotropic concept (2.21) and (2.22), where the *Cauchy stress tensor* σ_{ij} is replaced by the *mapped stress tensor*

$$\tau_{ij} = \beta_{ijkl}\sigma_{kl}. \tag{2.25}$$

According to this assumption the anisotropy of the material is entirely involved in the fourth-rank tensor β_{ijkl}. For example, the orthotropic case is considered, and the *Poynting effect* is taken into account by Betten (1981a). Furthermore, the influence of the *second-order effect* and of the *anisotropy* on the creep behavior of a thin-walled tube subjected to internal pressure has been investigated by Betten (1981a) and Betten and Waniewski (1989).

In the following, uniaxial and multiaxial *tertiary creep behavior* are discussed in detail.

The *tertiary creep phase* is accompanied by the formation of microscopic cracks on the grain boundaries, so that damage accumulation occurs. In some cases voids are caused by a given stress history and, therefore, they are distributed anisotropically among the grain boundaries. Thus, the mechanical behavior will be *anisotropic* and it is therefore necessary to investigate this kind of anisotropy by introducing appropriately defined *anisotropic damage tensors* into constitutive equations (Betten, 1982b, 1983a).

Problems of *creep damage* have been investigated by many authors. Very extensive surveys into recent advances in *damage mechanics* are given by Bodner and Hashin (1986), Krajcinovic (1996), and Krajcinovic and Lemaitre (1987), for instance. Further contributions to the theory of *continuum damage mechanics* should be mentioned in the literature, for example, Betten (1983b, 1986b), Chaboche (1984), Chrzanowski (1976), Krajcinovic (1983), Litewka and Hult (1989), Litewka and Morzýnska (1989), Murakami (1983, 1987), Murakami and Ohno (1981), Murakami *et al.* (1986), Murakami and Sawczuk (1981), and Onat (1986).

In the past two decades, considerable progress and significant advances have been made in the development of fundamental concepts of damage mechanics and their application to solving practical engineering problems. For instance, new concepts have been effectively applied to characterize creep damage, low and high cycle fatigue damage, creep–fatigue interaction, brittle/elastic damage, ductile/plastic damage, strain softening, strain-rate-sensitivity damage, impact damage, and other physical phenomena. The materials include rubbers, concretes, rocks, polymers, composites, ceramics, and metals. This area has attracted the interest of a broad spectrum of international research scientists in micromechanics, continuum mechanics, mathematics, materials science, physics, chemistry, and numerical analysis. However, sustained rapid growth in the development of damage mechanics requires the prompt dissemination of original research results, not only for the benefit of the researchers themselves, but also for the practicing engineers who are under continued pressure to incorporate the latest research results in their design procedures and processing techniques with newly developed materials.

Because of the broad applicability and versatility of the concept of damage mechanics, the research results have been published in more than 30 English and non-English technical journals. This multiplicity has imposed an unnecessary burden on scientists and engineers alike to keep abreast with the latest development in the subject area. The new *International Journal of Damage Mechanics* has been inaugurated to provide an effective mechanism hitherto unavailable to them, which will accelerate the dissemination of information on damage mechanics not only within the research community but also between the research laboratory and industrial design department, and it should promote and contribute to future development of the concept of damage mechanics.

Furthermore, one should emphasize that special Conferences on Damage Mechanics has contributed significantly to the development of theories and experiments in damage mechanics, for instance, the Conference on Damage Mechanics held in Cachan (1981) or the IUTAM Symposium on Mechanics of Damage and Fatigue held in Haifa and Tel Aviv (1985), to name just a few, gave many impulses. Certainly, there will be organized important conferences on damage mechanics in the future, too.

In the following the uniaxial and multiaxial *tertiary creep behavior* is discussed in detail.

In a uniaxial tension specimen, material deterioration can be described by introducing an additional variable ω or, alternatively, $\psi \equiv 1 - \omega$ into constitutive equations, i.e., the strain rate d can be expressed as $d = f(\sigma, \omega)$, or $d = f(\sigma, \psi)$, where σ is the uniaxial stress (Kachanov, 1958, 1986; Rabotnov, 1969). The material parameters ω and ψ describe the current *damage* state and the *continuity* of the material, respectively. The parameter of *continuity*, ψ, represents that fraction of the cross-sectional area that is not occupied by either voids or internal fissures. The *net stress* acting over the cross section of the uniaxial specimen is then $\hat{\sigma} = \sigma/\psi$. When $\psi = 1$ the material is in its virgin undamaged state, and when $\psi = 0$, the material can no longer sustain any load. In the latter case the constitutive equation would be required in order to approach an infinite strain rate. Furthermore, it is assumed that the damage rate $\dot{\omega}$, or alternatively the rate of continuity change $\dot{\psi}$, is also governed by the uniaxial stress and by the current state of continuity, i.e., $\dot{\omega} = g(\sigma, \omega)$ or $\dot{\psi} = -g(\sigma, \psi)$.

The forms of the functions f and g have been discussed in detail by many scientists, for instance by Rabotnov (1969), Chrzanowski (1973), Leckie and Ponter (1974), Leckie and Hayhurst (1977), Goel (1975), and Hayhurst *et al.* (1980). Often the forms

$$\frac{d}{d_0} = \frac{(\sigma/\sigma_0)^n}{(1 - \omega)^m}, \qquad \frac{\dot{\omega}}{\dot{\omega}_0} = \frac{(\sigma/\sigma_0)^v}{(1 - \omega)^\mu} \qquad (2.26a,b)$$

are used, where $n \geqslant v$, m, μ, d_0, $\dot{\omega}_0$, and σ_0 are constants. The undamaged case ($\omega = 0$) hereby leads to *Norton's power law*, which is assumed to be valid for the secondary creep stage, while the creep rate d approaches infinity as ω approaches 1.

Integrating the *kinetic equation* (2.26b) under the initial condition $\omega(t = 0) = 0$ and inserting the result into (2.26a), one arrives at the following relation (Betten, 1992):

$$\frac{d}{d_0} = \left(\frac{\sigma}{\sigma_0}\right)^n \left[1 - k\left(\frac{\sigma}{\sigma_0}\right)^v \dot{\omega}_0 t\right]^{-m/k} \qquad \text{with } k = 1 + \mu. \qquad (2.27)$$

A further integration leads to the tertiary creep strain

$$\varepsilon_t = \frac{a}{b(1 - c)} [1 - (1 - bt)^{1-c}], \qquad (2.28)$$

if we take the initial condition $\varepsilon_t(0) = 0$ into account. The abbreviations

$$a \equiv d_0 \left(\frac{\sigma}{\sigma_0}\right)^n, \qquad b \equiv k\left(\frac{\sigma}{\sigma_0}\right)^v \dot{\omega}_0, \qquad c \equiv \frac{m}{k} \qquad (2.29a,b,c)$$

have been introduced in (2.28).

Because creep rupture is characterized by $\omega = 1$ or $d \to \infty$ we can immediately find the time to rupture from (2.27):

$$t_r = [k(\sigma/\sigma_0)^v \dot{\omega}_0]^{-1}. \qquad (2.30)$$

If, for the sake of convenience, the constants m and μ are taken to be equal to the parameters n and v, respectively, the relations (2.26a,b) can then be simplified to

$$d = K\hat{\sigma}^n, \qquad \dot{\omega} = L\hat{\sigma}^v, \qquad (2.31a,b)$$

where $\hat{\sigma} = \sigma/(1 - \omega)$ is interpreted as the *net-stress* acting over the current cross-sectional area of a uniaxial specimen. Thus, the simplification (2.31a,b) can be called the *net stress concept*. From (2.26a,b) and (2.31a,b) we read:

$$K \equiv d_0/\sigma_0^n \quad \text{and} \quad L \equiv \dot{\omega}_0/\sigma_0^v.$$

One immediately arrives at the first relation (2.31a) from *Norton's law* $d = K\sigma^n$ if we replace the nominal stress σ by the net stress $\hat{\sigma}$. Furthermore, a *tensorial generalization* of (2.31a,b) can be achieved in a very similar manner to that described in Section IV, where the *Norton creep law* is generalized. This generalization has been illustrated by Betten (1991b).

Because of the simplifications $m = n$ and $\mu = v$, which lead to (2.31a,b), the creep rupture time (2.30) takes the form

$$t_r = [(1 + v)L\sigma^v]^{-1}, \qquad (2.32)$$

where the nominal stress σ can be interpreted as the actual stress at the beginning of the tertiary creep stage ($\omega = 0$), i.e., considering *Norton's law* $d = K\sigma^n$ and starting from (2.32) we arrive at the formula

$$d_{\min}^{v/n} t_r = K^{v/n}/L(1 + v). \qquad (2.33a)$$

The quantity $d_{\min}$ is the steady-state or minimum creep rate. Assuming $v = n$, one arrives at the Monkman–Grant (1956) relationship

$$d_{\min} t_r = K/L(1 + n) = \text{const.} \qquad (2.33b)$$

Thus, the *net stress concept* (2.31a,b) with identical exponents, $v \equiv n$, is compatible with the model of Monkman–Grant. The justification of this model has, for example, been analyzed by Ilschner (1973), Edward and Ashby (1979), Evans (1984) and Riedel (1987).

Under certain conditions, the grain boundaries in polycrystals slide during creep deformation. Edward and Ashby (1979) illustrate that this sliding can be accommodated in various ways: elastically, by diffusion, or by nonuniform creep or plastic flow of the grains themselves. In other cases, holes or cracks appear at the grain boundaries and grow until they link, leading to an *intergranular creep fracture*. When fracture is of this sort, the *Monkman–Grant rule* can be approximatively confirmed. Often, however, the *Monkman–Grant product*, $d_{\min} t_r$, is proportional to the *strain-to-rupture*, ε_r, as has been observed by Ilschner (1973) and Riedel (1987) or in our own experiments.

Because of its microscopic nature, *damage* generally has an *anisotropic* character even if the material was originally isotropic. The fissure orientation and length cause anisotropic macroscopic behavior. Therefore, damage in an isotropic or anisotropic material that is in a state of multiaxial stress can only be described in a tensorial form.

When generalizing the uniaxial concept (2.26a,b) or (2.31a,b), constitutive equations and anisotropic growth equations are expressed as the tensor-valued functions

$$d_{ij} = f_{ij}(\boldsymbol{\sigma}, \boldsymbol{\omega}), \qquad \overset{\circ}{\omega}_{ij} = g_{ij}(\boldsymbol{\sigma}, \boldsymbol{\omega}), \qquad (2.34a,b)$$

respectively, where $\circ$ denotes the *Jaumann derivative*, $\boldsymbol{\sigma}$ is the *Cauchy stress tensor*, and $\boldsymbol{\omega}$ represents an appropriately defined *damage tensor*.

Damage tensors are constructed, for instance, by Betten (1981b, 1983a,b). Furthermore, we also refer to the work of Murakami and Ohno (1981). They assumed that damage accumulating in the process of creep can be expressed through a symmetric tensor of rank two.

Some details about *damage tensors* and *tensors of continuity* are discussed in Section VI. Furthermore, the influence of material deterioration on the stresses in a continuum is studied in Section VII.

Rabotnov (1968) has also introduced a symmetric second-order tensor of damage and defined a symmetric *net stress tensor* $\hat{\boldsymbol{\sigma}}$ by way of a linear transformation

$$\sigma_{ij} = \Omega_{ijkl} \hat{\sigma}_{kl}, \qquad (2.35)$$

where the fourth-order tensor $\boldsymbol{\Omega}$ is assumed to be symmetric.

However, it has been pointed out in more detail by Betten (1982b) that the fourth-order tensor in (2.35) is only symmetric with reference to the first index pair ij, but not to the second, kl. Thus, the net stress tensor is *not* symmetric in a case of anisotropic damage. The net stress tensor can be decomposed into a symmetric part and into an antisymmetric one, where only the symmetric part is equal to the net stress tensor introduced by Rabotnov (1968), as shown by Betten (1982b).

Starting from a third-order skew-symmetric tensor of continuity to represent area vectors (*bivectors*) of *Cauchy's tetrahedron* in a damaged state, one finally arrives at a second-order damage tensor, which has the *diagonal form* with respect to the rectangular *Cartesian* coordinate system under consideration, as has been pointed out in detail by Betten (1983a).

The symmetric tensor-valued functions (2.34a,b) are valid for an *isotropic material* in an *anisotropic damage state*. Furthermore, one must differentiate between *anisotropic damage growth* and the *initial anisotropy* resulting from a forming process, for instance, rolling. Constitutive equations and anisotropic damage growth equations are then represented by expressions such as

$$d_{ij} = f_{ij}(\sigma, \omega; A) \quad \text{and} \quad \dot{\omega}_{ij} = g_{ij}(\sigma, \omega; A), \qquad (2.36a,b)$$

respectively, where **A** is a fourth-order constitutive tensor with components A_{pqrs} characterizing the anisotropy from, for example, rolling, i.e., the anisotropy of the material in its undamaged state.

A general representation of (2.36a) and similarly (2.36b) is given through a linear combination

$$\mathbf{d} = \sum_{\alpha} \varphi_{\alpha} {}^{\alpha}\mathbf{G} \quad \text{or} \quad d_{ij} = \sum_{\alpha} \varphi_{\alpha} {}^{\alpha}G_{ij}, \qquad (2.37)$$

where the **G**'s are symmetric *tensor generators* of rank two involving the argument tensors σ, ω, **A**. Some possible methods in arriving at such tensor generators have been discussed by Betten (1982b, 1983a, 1987c, 1998), for instance. The coefficients φ_{α} in (2.37) are scalar-valued functions of the integrity basis associated with the representation of (2.37). They must also contain experimental data measured in uniaxial creep tests. The main problems are to construct an irreducible set of tensor generators and to determine the scalar coefficients involving the integrity basis and experimental data.

A further aim is to represent the constitutive equation (2.37) in the *canonical form*

$$d_{ij} = {}^{0}H_{ijkl}\delta_{kl} + {}^{1}H_{ijkl}\sigma_{kl} + {}^{2}H_{ijkl}\sigma_{kl}^{(2)}, \qquad (2.38)$$

where ${}^{0}H_{ijkl}, \ldots, {}^{2}H_{ijkl}$ are the Cartesian components of fourth-order tensor-valued functions ${}^{0}\mathbf{H}, \ldots, {}^{2}\mathbf{H}$ depending on the *damage tensor* ω and the *anisotropy tensor* **A**. The canonical form of (2.38) is a representation in three terms, which are the contributions of zero, first, and second orders in the stress tensor σ influenced by the functions ${}^{0}\mathbf{H}$, ${}^{1}\mathbf{H}$, and ${}^{2}\mathbf{H}$, respectively.

In the isotropic special case the coefficient tensors ${}^{0}\mathbf{H}, \ldots, {}^{2}\mathbf{H}$ can be expressed as fourth-order spherical tensors:

$$H_{ijkl} = \tfrac{1}{2}(\delta_{ik}\delta_{jl} + \delta_{il}\delta_{jk})\varphi_{\lambda}, \qquad \lambda = 0, 1, 2. \qquad (2.39)$$

In that case, the *canonical form* (2.38) simplifies to the *standard form* (2.22).

It may be impossible to find a canonical form (2.38) for all types of anisotropy. However, for the most important kinds of anisotropy, namely, *transversely isotropic* and *orthotropic behavior*, the constitutive equation (2.37) can be expressed in the *canonical form* (2.38) as has been illustrated by Betten (1982b, 1983a, 1987c, 1998), for instance.

When formulating constitutive equations such as (2.36a) one has to take the following into account: The undamaged case ($\omega \to 0$) immediately leads to the secondary creep stage, while the *rate-of-deformation tensor* **d** approaches infinity as ω approaches the unit tensor δ. In view of polynomial representations of constitutive equations it is convenient to use the tensor

$$D_{ij} := (\delta_{ij} - \omega_{ij})^{(-1)} \equiv \psi_{ij}^{(-1)} \tag{2.40}$$

as an argument tensor instead of the tensorial damage variable ω. Thus, expressions such as $d_{ij} = f_{ij}(\sigma, \mathbf{D}, \mathbf{A})$ must be taken into consideration. Some possible representations of such functions have been discussed in detail by Betten (1982b, 1983a, 1987c, 1998).

III. Determination of Scalar Coefficients in Constitutive and Evolutional Equations

In the following the scalar coefficients in constitutive and evolutional equations are determined. This can be achieved by applying *interpolation methods for tensor functions*. It is illustrated in detail that the scalar coefficients can be expressed as functions of the irreducible invariants of the argument tensors and of the empirical constitutive relations found in uniaxial tests.

A. Polynomial Representation of Tensor Functions

Let

$$Y_{ij} = f_{ij}(\mathbf{X}) = \varphi_0 \delta_{ij} + \varphi_1 X_{ij} + \varphi_2 X_{ij}^{(2)} \tag{3.1}$$

be an isotropic tensor function where φ_0, φ_1, φ_2 are scalar-valued functions of the integrity basis, the elements of which are the irreducible invariants of the argument tensor **X**. Furthermore, they depend on experimental data.

First, it is possible to express the scalar functions through the principal values $X_I, \ldots, X_{III}$ and $Y_I, \ldots, Y_{III}$ if we solve the system of linear equations

$$\left.\begin{aligned}
Y_I &= \varphi_0 + \varphi_1 X_I + \varphi_2 X_I^2, \\
Y_{II} &= \varphi_0 + \varphi_1 X_{II} + \varphi_2 X_{II}^2, \\
Y_{III} &= \varphi_0 + \varphi_1 X_{III} + \varphi_2 X_{III}^2.
\end{aligned}\right\} \tag{3.2}$$

The solution can be written in the form

$$\varphi_0 = \sum_{\alpha=I}^{III} P_\alpha X_{(\alpha+I)} X_{(\alpha+II)} Y_{(\alpha)}, \tag{3.3a}$$

$$\varphi_1 = \sum_{\alpha=I}^{III} P_\alpha (X_{(\alpha+I)} X_{(\alpha+II)}) Y_{(\alpha)}, \tag{3.3b}$$

$$\varphi_2 = \sum_{\alpha=I}^{III} P_\alpha Y_{(\alpha)}, \tag{3.3c}$$

where the abbreviation

$$P_\alpha := \prod_{\substack{\beta=I \\ \beta \neq \alpha}}^{III} 1/(X_\alpha - X_\beta) \tag{3.4}$$

is introduced. A similar representation was used by Sobotka (1984) based on the Sylvester theorem (Sedov, 1966).

Because of the products P_α, expressions (3.3a)–(3.3c) can only be used if all principal values are different. Therefore, in the following an *interpolation method* is used in order to determine the scalar coefficients, even if two principal values coincide.

B. Interpolation Methods for Tensor Functions

In extending the Lagrange interpolation method to a tensor-valued function, we consider the *principal values* of the argument tensor as *interpolating points* and find the tensorial representation

$$Y_{ij} = f_{ij}(\mathbf{X}) = \sum_{\alpha=1}^{III} {}^\alpha L_{ij} Y_\alpha + R_{ij}(\mathbf{X}) \tag{3.5}$$

with the tensor polynomials

$${}^\alpha L_{ij} := P_\alpha (X_{ik} - X_{(\alpha+I)} \delta_{ik})(X_{kj} - X_{(\alpha+III)} \delta_{kj}). \tag{3.6}$$

Due to the *Hamilton–Cayley theorem*, the tensor-valued remainder term R_{ij} in (3.5) is always equal to the zero tensor (Betten, 1984b, 1987c). As an alternate approach, we find, by extending the Newton formula, the tensorial representation

$$Y_{ij} = \alpha_0 \delta_{ij} + \alpha_1 (X_{ij} - X_I \delta_{ij}) + a_2 (X_{ik} - X_I \delta_{ik})(X_{kj} - X_{II} \delta_{kj}), \quad (3.7)$$

Further terms in (3.7) are not possible because of the *Hamilton–Cayley theorem*. The coefficients in (3.7) can be found by inserting the principal values:

$$a_0 = Y_I, \qquad a_1 = (Y_I - Y_{II})/(X_I - X_{II}), \qquad (3.8a,b)$$

$$a_2 = [a_1 - (Y_{III} - Y_I)/(X_{III} - X_I)]/(X_{II} - X_{III}). \qquad (3.8c)$$

The interpolation formula (3.7) can be written as an isotropic tensor function (3.1) if we define

$$\varphi_0 \equiv a_0 - a_1 X_I + a_2 X_I X_{II}, \qquad (3.9a)$$

$$\varphi_I \equiv a_1 - a_2 (X_I + X_{II}), \qquad \varphi_2 \equiv a_2. \qquad (3.9b,c)$$

In the case of *coincident points*, we need the derivatives of the tensor function (3.1):

$$f'_{ij} := \partial Y_{ip}/\partial X_{pj} = \varphi_1 \delta_{ij} + 2\varphi_2 X_{ij}, \qquad (3.10a)$$

$$f''_{ij} := \partial f'_{iq}/\partial X_{qj} = 2\varphi_2 \delta_{ij}. \qquad (3.10b)$$

For example, in the case of $X_I \neq X_{II} = X_{III}$, we find from (3.7) and (3.10a) the coefficients

$$a_0 = Y_I, \qquad a_1 = (Y_I - Y_{II})/(X_I - X_{II}), \qquad (3.11a,b)$$

$$a_2 = (a_1 - f'_{II})/(X_I - X_{II}), \qquad (3.11c)$$

if we substitute $Y_I = f_{II}(X_{11} \equiv X_I)$, $Y_{II} = f_{22}(X_{22} \equiv X_{II})$, and $f'_{22}(X_{22} \equiv X_{II}) \equiv f'_{II}$.

Finally, if all principal values coincide, we calculate

$$a_0 = f_I, \qquad a_1 = f'_I, \qquad a_2 = f''_I/2. \qquad (3.12a,b,c)$$

However, in this special case the argument tensor is a spherical one $X_{ij} = X_I \delta_{ij}$, and therefore formula (3.7) reduces to the trivial result: $Y_{ij} = f_{ij} = f_I \delta_{ij}$. Note that the interpolation formula for a scalar function $y = f(x)$ approaches the Taylor expansion for $f(x)$ at x_0 if we make x_α,

$\alpha = 1, 2, \ldots, n$, coincide at x_0. An interpolation method for tensor functions with two argument tensors can be developed in a similar way (Betten, 1987b,c).

C. SIMPLE APPLICATIONS

The interpolation method for tensor functions is a very useful and powerful tool. In the following some applications and results should be discussed. Besides many applications in tensor algebra or tensor analysis discussed by Betten (1987b,c), engineering applications are also very important.

In the theory of finite deformation the tensorial Hencky measure of strain and strain rate plays a central role [see Fitzgerald (1980) and Betten (1985b, 1993)] because it can be decomposed into a sum of an isochoric distortion and a volume change. The problem to represent the logarithmic function

$$\mathbf{Y} = \ln \mathbf{X} \quad \text{or} \quad Y_{ij} = \{\ln \mathbf{X}\}_{ij} \tag{3.13}$$

as an isotropic tensor function (3.1) is solved by determining the scalar functions φ_0, φ_1, φ_2. This can be done by using the interpolation method described before by Betten (1985b, 1993).

Other examples are $\mathbf{Y} = \exp \mathbf{X}$ or $\mathbf{Y} = \sin \mathbf{X}$, etc., which can be treated in the same way. These functions play a central role, for instance, in problems concerning *vibro creep* (Jakowluk, 1993).

IV. Tensorial Generalization of Uniaxial Relations to Multiaxial States of Stress

In the following paragraphs some uniaxial relations that are important for engineering applications are generalized to multiaxial states of stress.

A. TENSORIAL GENERALIZATION OF NORTON'S CREEP LAW

The following example is concerned with the generalization of Norton's power law

$$d/d_0 = (\sigma/\sigma_0)^n \quad \text{or} \quad d = K\sigma^n \tag{4.1a,b}$$

to multiaxial states of stress where d is the strain rate, σ the uniaxial true stress, and d_0, σ_0, n, K are constants. To solve this problem, we use an isotropic tensor function:

$$d_{ij} = f_{ij}(\boldsymbol{\sigma}) = \varphi_0^* \delta_{ij} + \varphi_1^* \sigma_{ij} + \varphi_2^* \sigma_{ij}^{(2)} \qquad (4.2)$$

and determine the scalar coefficients $\varphi_0^*, \ldots, \varphi_2^*$ as functions of experimental data (K, n) in (4.1b) and of the integrity basis, the elements of which are the irreducible invariants of the Cauchy stress tensor $\boldsymbol{\sigma}$.

Alternatively, we can represent the constitutive equation in the form

$$d_{ij} = f_{ij}(\boldsymbol{\sigma}') = \varphi_0 \delta_{ij} + \varphi_1 \sigma'_{ij} + \varphi_2 \sigma'^{(2)}_{ij}, \qquad (4.3)$$

where $\sigma'_{ij} := \sigma_{ij} - \sigma_{kk} \delta_{ij}/3$ are the Cartesian components of the *stress deviator* $\boldsymbol{\sigma}'$. For the special case of incompressible behavior $(d_{kk} \equiv 0)$, we find from (4.3) the condition

$$3\varphi_0 + \varphi_2 \sigma'^{(2)}_{kk} = 0 \quad \Rightarrow \quad \varphi_0 = -2\varphi_2 J'_2/3 \qquad (4.4)$$

with the quadratic invariant $J'_2 \equiv \sigma'_{ik} \sigma'_{ki}/2$ of the stress deviator, so that the constitutive equation (4.3) is reduced to the simple form

$$d_{ij} = \varphi_1 \sigma'_{ij} + \varphi_2 \sigma''_{ij} \qquad (4.5)$$

containing the traceless tensors

$$\sigma'_{ij} \equiv \partial J'_2/\partial \sigma_{ij} \quad \text{and} \quad \sigma''_{ij} \equiv \partial J'_3/\partial \sigma_{ij} \qquad (4.6a,b)$$

with the cubic invariant $J'_3 \equiv \sigma'_{ij} \sigma'_{jk} \sigma'_{ki}/3$ of the stress deviator. The uniaxial equivalent state of stress (index V) is characterized through the tensor variables

$$(\sigma_{ij})_V = \text{diag}\{\sigma, 0, 0\}, \qquad (4.7a)$$

$$(\sigma'_{ij})_V = \text{diag}\{2\sigma/3, -\sigma/3, -\sigma/3\}, \qquad (4.7b)$$

$$(d_{ij})_V = \text{diag}\{d, -vd, vd\}, \qquad (4.7c)$$

where v is the transverse contraction ratio.

In the following the diagonal elements in eqs. (4.7) are considered as interpolating points where two points coincide. Since the two coincident points in (4.7a) are zero, it may be more convenient to determine the coefficients $\varphi_0, \ldots, \varphi_2$ in the constitutive equation (4.3) instead of (4.2). Thus, we use the uniaxial creep law

$$d = (3/2)^n K(\sigma')^n \qquad (4.8)$$

instead of (4.1b). Because of (4.7b), i.e., $X_{II} = X_{III} \equiv -\sigma/3$, and (4.8), we find from (3.11a) the coefficient

$$a_0 = Y_1 \equiv (3/2)^n K(\sigma')^n = K\sigma^n. \tag{4.9a}$$

Furthermore, because of (4.7b), (4.9a), and $Y_{II} = -vd = -vK\sigma^n$, we find from (3.11b) the coefficient

$$a_1 = (1 + v)K\sigma^{n-1}. \tag{4.9b}$$

The derivative f'_{II} at the coincident points $X_{II} = X_{III}$ can be determined in the following way. From (4.8) we derive

$$f' \equiv \partial d/\partial \sigma' = n(3/2)^n(\sigma')^{n-1} = nd/\sigma', \tag{4.10a}$$

$$f'_{II} = nd_{II}/\sigma'_{II}. \tag{4.10b}$$

From (4.7a) and (4.7b) we read $\sigma' = -\sigma/3$ and $d_{II} = -vd_1 \equiv -vd = -vK\sigma^n$, so that (4.10b) can be written as

$$f'_{II} = 3vnK\sigma^{n-1}. \tag{4.10c}$$

Considering (4.7b) and (4.9b), we calculate from Eq. (3.11c) the coefficient

$$a_2 = (1 + v - 3vn)K\sigma^{n-2}. \tag{4.9c}$$

Inserting (4.9a)–(4.9c) in (3.9a)–(3.9c), we finally determine the scalar functions:

$$\varphi_0 = \tfrac{1}{9}(1 - 8v + 6vn)K\sigma^n, \tag{4.11a}$$

$$\varphi_1 = \tfrac{2}{3}(1 + v + \tfrac{3}{2}vn)K\sigma^{n-1}, \tag{4.11b}$$

$$\varphi_2 = (1 + v - 3vn)K\sigma^{n-2}. \tag{4.11c}$$

Assuming the incompressibility (4.5) and neglecting tensorial nonlinearity ($\varphi_2 = 0 \Rightarrow a_2 = 0$, $\varphi_0 = 0$, and $\varphi_1 = a_1$) we find from (4.3) the simplified constitutive equation

$$d_{ij} = a_1\sigma'_{ij} \quad \text{or} \quad d_{ij} = \tfrac{3}{2}K\sigma^{n-1}\sigma'_{ij}, \tag{4.12a,b}$$

if we use (4.9b) with $v = 1/2$. The result (4.12b) is identical to a constitutive equation proposed by Leckie and Hayhurst (1977). If we insert the von Mises equivalent stress $\sigma = \sqrt{3J'_2}$ into (4.12b), we can find the constitutive equation

$$d_{ij} = \tfrac{3}{2}K(3J'_2)^{(n-1)/2}\sigma'_{ij} \tag{4.12c}$$

used by Odquist and Hult (1962).

The equivalent stress σ in (4.11a)–(4.11c) can be determined as a function f the stress invariants if we use the *hypothesis of the equivalent dissipation rate*:

$$\dot{D} := \sigma_{ij} d_{ji} \overset{!}{=} \sigma d, \qquad (4.13)$$

where $\dot{D}$ is called the rate of dissipation of creep energy. The result is

$$\sigma^3 + A\sigma^2 + B\sigma + C = 0, \qquad (4.14)$$

where the abbreviations

$$A \equiv -(1 - 8v + 6vn)J_1/9, \qquad (4.15a)$$

$$B \equiv -4(1 + v + 3vn/2)J'_2/3, \qquad (4.15b)$$

$$C \equiv -(1 + v - 3vn)(3J'_3 + 2J_1 J'_2/3) \qquad (4.15c)$$

have been used. Thus, the scalar coefficients (4.11a)–(4.11c) are functions of the *irreducible invariants*

$$J_1 \equiv \sigma_{kk}, \qquad J'_2 \equiv \sigma'_{ik}\sigma'_{ki}/2, \qquad J'_3 \equiv \sigma'_{ij}\sigma'_{jk}\sigma'_{ki}/3 \qquad (4.16a,b,c)$$

and of *experimental data* (K, n, v):

$$\varphi_\alpha = \varphi_\alpha(J_1, J'_2, J'_3; K, n, v), \qquad \alpha = 0, 1, 2. \qquad (4.17)$$

This statement is compatible with the representation theory of tensor-valued functions (2.37) in which the coefficients φ_α are scalar-valued functions of the *integrity basis* (4.16a,b,c).

In the case of incompressible behavior ($v = 1/2$), the first invariant J_1 has no influence. The cubic equation (4.14) then takes the reduced form

$$\sigma^3 + B^*\sigma^2 + C^* = 0 \qquad (4.14^*)$$

with the abbreviations

$$B^* \equiv -(2 + n)J'_2 \quad \text{and} \quad C^* \equiv \tfrac{9}{2}(n - 1)J'_3 \qquad (4.15^*b,c)$$

depending on the irreducible invariants (4.16b,c) of the stress deviator.

Some authors (Brown *et al.*, 1986) are losing faith in *Norton's law* since they feel that their new θ *projection concept* provides a far more comprehensive description of creep behavior for design. In this new approach, normal creep curves are envisaged as the sum of a decaying primary and an ascending tertiary stage, i.e., the *secondary stage* is merely the period of ostensibly constant rate observed when the decay in the creep rate during the primary stage is offset by the gradual acceleration caused by tertiary processes. This concept neglects the secondary component and may be valid for some special materials, e.g., $\frac{1}{2}Cr\frac{1}{2}Mo\frac{1}{4}V$, as has been discussed in detail by Brown *et al.* (1986). However, an extended secondary creep stage can be observed for many materials. Thus, in spite of the discussion by Brown *et al.* (1986), it is very important that *Norton's law* be generalized to multiaxial states of stress. This can be achieved by applying a *tensorial interpolation method* as illustrated above.

B. Tensorial Generalization of a Creep Law including Damage

Involving the damage state in the tertiary creep stage the uniaxial relation

$$d/d_0 = (\sigma/\sigma_0)^n D^m \quad \text{with} \quad D := 1/(1 - \omega) \tag{4.18}$$

should be generalized to multiaxial states of stress where ω is the damage parameter (material deterioration) introduced by Kachanov (1958) and also used by Rabotnov (1969).

To generalize (4.18), we consider the tensor-valued function

$$d_{ij} = \begin{cases} f_{ij}(\boldsymbol{\sigma}, \mathbf{D}) \\ \dfrac{1}{2} \sum_{v,\mu=0}^{2} \psi_{[v,\mu]}(\sigma_{ik}^{(v)} D_{kj}^{(\mu)} + D_{ik}^{(\mu)} \sigma_{kj}^{(v)}), \end{cases} \tag{4.19}$$

where v and μ are exponents of the Cauchy stress tensor $\boldsymbol{\sigma}$ and the second-rank tensor $\mathbf{D}$ with the components $D_{ij} = (\delta_{ij} - \omega_{ij})^{(-1)}$ given by the damage tensor $\boldsymbol{\omega}$.

Now, the main problem is to determine the scalar coefficients $\psi_{[v,\mu]}$ as functions of the integrity basis containing 10 *irreducible invariants* (Betten, 1987b,c) and experimental data. To solve this problem, we suggest the following method, which may be useful for practical applications as has been discussed by Betten (1988b).

A representation with the same tensor generators as contained in the function (4.19) can be found by seperating the two variables $\boldsymbol{\sigma}$ and $\mathbf{D}$ in the following way:

$$d_{ij} = f_{ij}(\boldsymbol{\sigma}, \mathbf{D}) = \frac{1}{2}(X_{ik} Y_{kj} + Y_{ik} X_{kj}), \tag{4.20}$$

where the isotropic tensor functions

$$\left. \begin{array}{l} X_{ij} = X_{ij}(\boldsymbol{\sigma}) = \varphi_0^* \delta_{ij} + \varphi_1^* \sigma_{ij} + \varphi_2^* \sigma_{ij}^{(2)} \\ \varphi_v^* = \varphi_v^*(\text{tr }\boldsymbol{\sigma}^\lambda) = \varphi_v^*(\sigma_I, \sigma_{II}, \sigma_{III}) \end{array} \right\}, \tag{4.21}$$

$$\left. \begin{array}{l} Y_{ij} = Y_{ij}(\mathbf{D}) = \Phi_0 \delta_{ij} + \Phi_1 D_{ij} + \varphi_2 D_{ij}^{(2)} \\ \Phi_\mu = \Phi_\mu(\text{tr }\mathbf{D}^\lambda) = \Phi_\mu(D_I, D_{II}, D_{III}) \end{array} \right\} \tag{4.22}$$

($\mu, v = 0, 1, 2$ and $\lambda = 1, 2, 3$) are used.

Thus, we find the representation (4.19) with the scalar coefficients

$$\psi_{[v,\mu]} = \varphi_v^* \Phi_\mu, \qquad \mu, v = 0, 1, 2, \tag{4.23}$$

where the scalars φ_v^* are determined by Betten (1986a,b):

$$\varphi_0^* = \varphi_0 - J_1 \varphi_1/3 + J_1^2 \varphi_2/9, \tag{4.24a}$$

$$\varphi_1^* = \varphi_1 - 2J_1 \varphi_2/3, \qquad \varphi_2^* \equiv \varphi_2. \tag{4.24b,c}$$

The coefficients Φ_μ can be found by solving the following system of linear equations:

$$\left. \begin{array}{l} \Phi_0 + D_I \varphi_1 + D_I^2 \Phi_2 = (D_I)^{m_I}, \\ \Phi_0 + D_{II} \Phi_1 + D_{II}^2 \Phi_2 = (D_{II})^{m_{II}}, \\ \Phi_0 + D_{III} \Phi_1 + D_{III}^2 \Phi_2 = (D_{III})^{m_{III}}. \end{array} \right\} \tag{4.25}$$

The exponents $m_I, \dots, m_{III}$ in (4.25) are determined by using the creep law (4.18) in tests on specimens cut along the mutually perpendicular directions x_1, x_2, x_3.

Because of $D_{ij} := (\delta_{ij} - \omega_{ij})^{(-1)} \equiv \psi_{ij}^{(-1)}$ and $\psi_{ij} = \text{diag}\{\alpha, \beta, \gamma\}$ according to (2.40), the principal values in (4.25) can be expressed through

$$D_I = 1/\alpha, \qquad D_{II} \equiv 1/\beta, \qquad D_{III} \equiv 1/\gamma, \tag{4.26}$$

where the essential components α, β, γ are fractions that represent the net cross-sectional elements of Cauchy's tetrahedron perpendicular to the coordinate axes (Betten, 1983a). In the case of two equal parameters, for instance $\alpha \neq \beta = \gamma$, the scalars Φ_μ, $\mu = 0, 1, 2$, in (4.25) can be determined by using the interpolation method described above in (3.1)–(3.12).

Instead of (4.21), we can use the isotropic tensor function

$$X_{ij} = X_{ij}(\boldsymbol{\sigma}') = \varphi_0 \delta_{ij} + \varphi_1 \sigma'_{ij} + \varphi_2 \sigma'^{(2)}_{ij} \tag{4.27}$$

and find the representation

$$d_{ij} = \frac{1}{2} \sum_{\nu,\mu=0}^{2} \varphi_\nu \Phi_\mu (\sigma_{ik}^{\prime(\nu)} D_{kj}^{(\mu)} + D_{ik}^{(\mu)} \sigma_{kj}^{\prime(\nu)}), \tag{4.28}$$

where the scalar coefficients φ_ν are determined in the functions (4.11a)–(4.11c) and the Φ_μ are taken from (4.25).

The scalar coefficients $\psi_{[\nu,\mu]} \equiv \varphi_\nu \Phi_\mu$ in the representation (4.28) must be functions of the integrity basis

$$\left. \begin{array}{l} J_1 \equiv \sigma_{kk}, \quad J_2' \equiv \sigma_{ij}' \sigma_{ji}'/2, \quad J_3' \equiv \sigma_{ij}' \sigma_{jk}' \sigma_{ki}'/3, \\ L_1 \equiv D_{kk}, \quad L_2 \equiv D_{kk}^{(2)}, \quad L_3 \equiv D_{kk}^{(3)}, \quad \Omega_1' \equiv \sigma_{ij}' D_{ji}, \\ \Omega_2' \equiv \sigma_{ij}'^{(2)} D_{ji}, \quad \Omega_3' \equiv \sigma_{ij}' D_{ji}^{(2)}, \quad \Omega_4' \equiv \sigma_{ij}'^{(2)} D_{ji}^{(2)} \end{array} \right\} \tag{4.29}$$

and experimental data. To show this we can start from hypothesis (4.13) and find similarly to (4.14) the cubic equation

$$\sigma^3 + A^* \sigma^2 + B^* \sigma + C^* = 0, \tag{4.30}$$

if we insert (4.20), (4.22), and (4.11a)–(4.11c) into hypothesis (4.13). In (4.30) the following abbreviations are used:

$$A^* \equiv -\tfrac{1}{9}(1 - 8v + 6vn)(\Phi_0 J_1 + \Phi_1(\Omega_1' + \tfrac{1}{3}J_1 L_1) + \Phi_2(\Omega' + \tfrac{1}{3}J_1 L_2))/D^m, \tag{4.31a}$$

$$B^* \equiv -\tfrac{2}{3}(1 + v + \tfrac{3}{2}vn)(2\Phi_0 J_2' + \Phi_1(\Omega_2' + \tfrac{1}{3}J_1 \Omega_1') + \Phi_2(\Omega_4' + \tfrac{1}{3}J_1 \Omega_3'))/D^m, \tag{4.31b}$$

$$C^* \equiv -(1 + v - 3vn)(3\Phi_0(J_3' + \tfrac{2}{9}J_1 J_2') + \Phi_1(J_2'\Omega_1' + J_3' L_1 + \tfrac{1}{3}J_1 \Omega_2')$$
$$+ \Phi_2(J_2'\Omega_3' + J_3' L_2 + \tfrac{1}{3}J_1 \Omega_4'))/D^m, \tag{4.31c}$$

$$D \equiv (D_I D_{II} D_{III})^{1/3}, \quad m \equiv (m_I + m_{II} + m_{III})/3. \tag{4.31d,e}$$

We see that the elements of the integrity basis (4.29) and experimental data are contained in (4.31a)–(4.31e). Thus the coefficients $\psi_{[\nu,\mu]} \equiv \varphi_\nu \Phi_\mu$ in (4.28) are scalar functions of the integrity basis (4.29) and experimental data $K, n, v; m_I, m_{II}, m_{III}; D_I, D_{II}, D_{III}$ found in creep tests on specimens cut along three mutually perpendicular directions.

In the case (2.36) of damage and initial anisotropy we can use for simplification the constitutive equation

$$d_{ij} = f_{ij}(\mathbf{t}, \tau) = \frac{1}{2} \sum_{\nu,\mu=0}^{2} \psi_{[\nu,\mu]}^{*}(t_{ik}^{(\nu)} \tau_{kj}^{(\mu)} + \tau_{ik}^{(\mu)} t_{kj}^{(\nu)}), \tag{4.32}$$

where the linear transformations

$$t_{ij} = D_{ijpq}\sigma_{pq} = t_{ji} \quad \text{with} \quad D_{ijpq} := (D_{ip}D_{jq} + D_{iq}D_{jp})/2, \quad (4.33)$$

$$\tau_{ij} = A_{ijpq}\sigma_{pq} = \tau_{ji} \quad\quad\quad\quad\quad (4.34)$$

have been introduced. Then the scalar functions in (4.32) can be determined in a very similar way as described above.

C. Tensorial Generalization of the Elastic-Plastic Transition

The elastic-plastic behavior of solids loaded under uniaxial stress σ can be expressed by the stress-strain relations

$$\sigma/\sigma_F = [\tanh(E\varepsilon/\sigma_F)^n]^{1/n}, \quad\quad\quad (4.35a)$$

$$\sigma/\sigma_F = (E\varepsilon/\sigma_F)/[1 + (E\varepsilon/\sigma_F)^n]^{1/n}, \quad\quad (4.35b)$$

proposed by Betten (1975b), where σ_F is the yield stress determined in a uniaxial tension test, and E represents the modulus of elasticity—often called "Young's modulus" (1807), although this modulus had already been used by Euler in 1760. The exponent n regulates the elastic-plastic transition. For instance, an elastic-perfectly plastic behavior is characterized by $n \to \infty$ (Figure 2).

It has been shown by Betten (1975c) that independently of the parameter n the "limit carrying capacity" coincides with that for a perfectly plastic body ($n \to \infty$). Hence, a new aspect of the uniqueness of the "limit load" can be formulated as we can read in the book of Życzkowski (1981, p. 210):

Uniqueness understood as the independence of that load of the assumed stress-strain diagram belonging to the class of asymptotically perfect plasticity. Such independence may be observed in many cases.

For engineering applications, it is very important to generalize the relations (4.35a) and (4.35b) to multiaxial states of stress. This can be achieved by using an isotropic tensor function (3.1) in the form

$$\sigma_{ij} = f_{ij}(\boldsymbol{\varepsilon}) = \sigma(\psi_0^*\delta_{ij} + \psi_1^*\varepsilon_{ij} + \psi_2^*\varepsilon_{ij}^{(2)}), \quad\quad (4.36)$$

or, alternatively,

$$\sigma_{ij} = f_{ij}(\boldsymbol{\varepsilon}') = \sigma(\psi_0\delta_{ij} + \psi_1\varepsilon'_{ij} + \psi_2\varepsilon_{ij}^{'(2)}), \quad\quad (4.37)$$

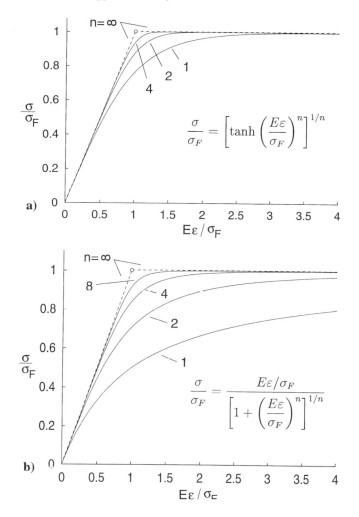

FIG. 2. Elastic-plastic transition: (a) eq. (4.35a) and (b) eq. (4.35b).

where σ is given by (4.35a) and (4.35b) and ε' is the deviator of the strain tensor ε. The relations between the scalar coefficients in (4.36) and (4.37) are given by

$$\psi_0^* = \psi_0 - I_1\psi_1/3 + I_1^2\psi_2/9, \tag{4.38a}$$

$$\psi_1^* = \psi_1 - 2I_1\psi_2/3, \qquad \psi_2^* = \psi_2, \tag{4.38b,c}$$

where the scalars ψ_0, ψ_1, ψ_2 can be determined in a similar way as the coefficients of (4.11a)–(4.11c). Thus, we find the results

$$\psi_0 = (1 - 2I'_2\psi_2)/3, \qquad \psi_1 = [c + 2/(1 + v)]/2\varepsilon, \qquad \text{(4.39a,b)}$$

$$\psi_2 = [1/(1 + v) - c]/(1 + v)^2\varepsilon^2, \qquad \text{(4.39c)}$$

if we generalized the uniaxial relation (4.35a), for instance. In (4.39b) and (4.39c) the constant c is defined as

$$c \equiv \frac{1}{2}\frac{9 + 4v}{2(1 + v)^2}\left(1 \pm \sqrt{1 - \frac{8(1 + v)}{(9 + 4v)^2}}\right) \qquad \text{(4.40)}$$

and the equivalent strain ε can be calculated by assuming the *hypothesis of the equivalent strain energy*

$$\sigma_{ij}\varepsilon_{ji} \overset{!}{=} \sigma\varepsilon. \qquad \text{(4.41)}$$

Thus, we find similarly to (4.14) the cubic equation

$$\varepsilon^2 + P\varepsilon^2 + Q\varepsilon + R = 0, \qquad \text{(4.42)}$$

where the abbreviations

$$P \equiv -I_1/3, \quad Q \equiv -\frac{2}{3}cI'_2 - \frac{4}{3(1 + v)}I'_2, \qquad \text{(4.43a,b)}$$

$$R \equiv \frac{3}{1 + v}\left(c - \frac{1}{1 + v}\right)I'_3 \qquad \text{(4.43c)}$$

have been introduced. In (4.39a,b) and (4.39c) and (4.43a,b) and (4.43c) the irreducible invariants

$$I_1 \equiv \varepsilon_{kk}, \qquad I_2 = \varepsilon'_{ij}\varepsilon'_{ji}/2, \qquad I_3 = \varepsilon'_{ij}\varepsilon'_{jk}\varepsilon'_{ki}/3 \qquad \text{(4.44a,b,c)}$$

are the elements of the *integrity basis*. Thus, similarly to (4.11a,b,c), the scalar coefficients of (4.39a,b) and (4.39c) are functions of the integrity basis (4.44a,b,c) and experimental data (σ_F, E, n, v).

D. Tensorial Generalization of the Ramberg–Osgood Relation

Another example is the tensorial generalization of the Ramberg–Osgood (1943) stress-strain relation

$$\varepsilon = \sigma/E + k(\sigma/E)^n \qquad \text{(4.45a)}$$

or, alternatively, the modified relations (Betten, 1989a, 1993)

$$\frac{E\varepsilon}{\sigma_F} = \frac{\sigma}{\sigma_F} + k\left(\frac{\sigma}{\sigma_F}\right)^n \quad \text{and} \quad \varepsilon = \frac{\sigma}{E} + k\left(\frac{\sigma}{\sigma_F}\right)^n, \qquad (4.45\text{b,c})$$

where k and n are parameters found in uniaxial tests (Figure 3).

Now, we consider the isotropic tensor function

$$\varepsilon_{ij} = f_{ij}(\boldsymbol{\sigma}) = \varphi_0^* \delta_{ij} + \varphi_1^* \sigma_{ij} + \varphi_2^* \sigma_{ij}^{(2)}, \qquad (4.46)$$

or, alternatively,

$$\varepsilon_{ij} = f_{ij}(\boldsymbol{\sigma}') = \varphi_0 \delta_{ij} + \varphi_1 \sigma'_{ij} + \varphi_2 \sigma_{ij}^{'(2)} \qquad (4.47)$$

and find, in a similar way as pointed out above, the following results:

$$\varphi_0^* = \varphi_0 - J_1 \varphi_1/3 + J_1^2 \varphi_2/9, \qquad (4.48\text{a})$$

$$\varphi_1^* = \varphi_1 - 2J_1\varphi_2/3, \qquad \varphi_2^* \equiv \varphi_2 \qquad (4.48\text{b,c})$$

and

$$\varphi_0 = \frac{1}{3}(1-n)\frac{\sigma}{E} + \frac{1}{9}(1 - 8v + 6vn)\left(\frac{\sigma}{E} + k\left(\frac{\sigma}{E}\right)^n\right), \qquad (4.49\text{a})$$

$$\varphi_1 = \frac{1-n}{2E} + \frac{2(1+v)/3 + vn}{E}\left(1 + k\left(\frac{\sigma}{E}\right)^{n-1}\right), \qquad (4.49\text{b})$$

$$\varphi_2 = \frac{1}{E\sigma}\left\{(1 + v - 3vn)\left(1 + k\left(\frac{\sigma}{E}\right)^{n-1}\right) - \frac{3}{2}(1-n)\right\}. \qquad (4.49\text{c})$$

The equivalent stress in (4.49a)–(4.49c) can be determined again by using the hypothesis (4.41). Then we find again a cubic equation (4.14) with corresponding abbreviations A, B, C, which are, similar to (4.15a)–(4.15c), functions of the *integrity basis* (4.16a)–(4.16c) and *experimental data* E, k, n, found in a uniaxial test.

In Section VIII, Figure 12a, it has been illustrated that the *modified* stress-strain relations (4.45b,c) proposed by Betten (1989a, 1993) agree very well with his own experiments on *aluminum alloy* AA7075 T 7351, whereas the Ramberg–Osgood stress-strain relation (4.45a) is less suitable.

V. Material Tensors of Rank Four

In the theories of elasticity and plasticity [(2.13) for instance] or in creep mechanics of anisotropic solids (2.36a,b), material *tensors of rank four* with

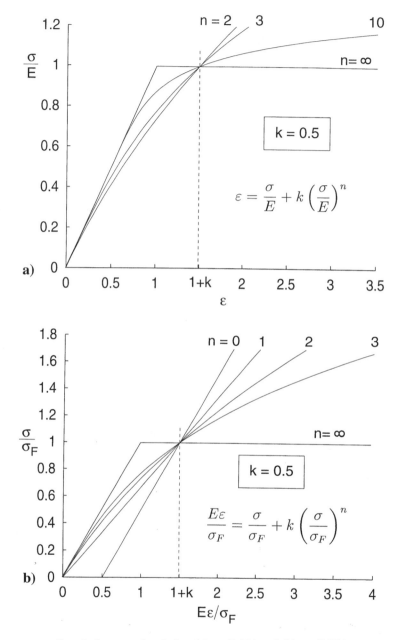

FIG. 3. Stress-strain relation: (a) eq. (4.45a) and (b) eq. (4.45b).

the usual symmetry conditions

$$A_{ijkl} = A_{jikl} = A_{ijlk} = A_{klij} \tag{5.1}$$

play a central role.

To take into account a material tensor of rank four, realistic examples were discussed in detail elsewhere by Betten (1976, 1982a, 1986b, 1987a,b,c). In those cases, the yield condition (2.12) has the form

$$f(\sigma_{ij}, A_{ijkl}) = 1, \tag{5.2}$$

and the constitutive equation (2.13)

$$d\varepsilon_{ij}^p = f_{ij}(\sigma_{pq}, A_{pqrs}) \tag{5.3}$$

is a symmetric ($f_{ij} = f_{ji}$) tensor-valued tensor function that has to fulfill the condition of form invariance

$$a_{ik}a_{jl}f_{kl}(\sigma_{pq}, A_{pqrs}) \equiv f_{ij}(a_{pt}a_{qu}\sigma_{tu}, a_{pt}a_{qu}a_{rv}a_{sw}A_{tuvw}) \tag{5.4}$$

under the transformations of the orthogonal group ($a_{ik}a_{jk} = \delta_{ij}$). A suitable form of the constitutive equation (5.3) is given by a linear combination

$$d\varepsilon_{ij}^p = \sum_{\lambda,\mu,v} \Phi_{[\lambda,\mu,v]} G_{ij}^{[\lambda,\mu,v]} \tag{5.5}$$

of several symmetric *tensor generators*

$$G_{ij}^{[\lambda,\mu,v]} = (M_{ij}^{[\lambda,\mu,v]} + M_{ji}^{[\lambda,\mu,v]})/2 \tag{5.6}$$

formed by matrix products of the forms

$$M_{ij}^{[\lambda,\mu,v]} \equiv \sigma_{ik}^{(\lambda)} A_{kjpq}^{(\mu)} \sigma_{pq}^{(v)}. \tag{5.7}$$

which fulfill the condition of form invariance (5.4). Note that in (5.7) those values of λ, μ, v that are enclosed by parentheses are exponents, whereas a square bracket gives any set λ, μ, v the character of a label to indicate several matrix products.

In finding an irreducible set of tensor generators, we note that, according to the *Hamilton–Cayley theorem*, none of the exponents λ, μ of the second-order tensor $\boldsymbol{\sigma}$ needs to be larger than 2. In generalizing the *Hamilton–Cayley theorem*, we find (Betten, 1982a, 1987a) that a symmetric fourth-order tensor of power six and all higher powers can be expressed in terms of lower powers, that is, $\mu \leqslant 5$. For $\mu = 0$, we must notice that the combinations $[\lambda, 0, v]$ are symmetric with respect to λ, v, and that the restriction $\lambda + v \leqslant 2$ follows. Thus we have 48 irreducible matrix products of the form (5.7), which are listed in Table 1 (Betten, 1982a, 1987a).

TABLE 1
MATRIX PRODUCTS OF (5.7)

$\mu = 0;$ $M_{ij}^{[\lambda,0,\nu]}$ with $\lambda + \nu \leqslant 2$		
λ ⟍ ν 0	1	2
0 δ_{ij}	σ_{ij}	$\sigma_{ij}^{(2)}$
1 σ_{ij}	$\sigma_{ij}^{(2)}$	
2 $\sigma_{ij}^{(2)}$		

$\mu = 1, 2, \ldots, 5;$ $M_{ij}^{[\lambda,\mu,\nu]}$		
λ ⟍ ν 0	1	2
0 $A_{ijrr}^{(\mu)}$	$A_{ijpq}^{(\mu)}\sigma_{pq}$	$A_{ijpq}^{(\mu)}\sigma_{pq}^{(2)}$
1 $\sigma_{ik}A_{kjrr}^{(\mu)}$	$\sigma_{ik}A_{kjpq}^{(\mu)}\sigma_{pq}^{(2)}$	$\sigma_{ik}A_{kjpq}^{(\mu)}\sigma_{pq}^{(2)}$
2 $\sigma_{ik}^{(2)}A_{kjrr}^{(\mu)}$	$\sigma_{ik}^{(2)}A_{kjpq}^{(\mu)}\sigma_{pq}$	$\sigma_{ik}^{(2)}A_{kjpq}^{(\mu)}\sigma_{pq}^{(2)}$

The coefficients $\Phi_{[\lambda,\mu,\nu]}$ in (5.5) are scalar-valued functions of the *integrity basis* associated with the representation given in (5.5). In finding this integrity basis, we notice the theorem which says that the trace of any product formed from matrices and their transposes is a polynomial invariant under the full or proper orthogonal group. Thus we form the following traces:

$$\sigma_{ji}\, d\varepsilon_{ij}^{p} = \sum_{\lambda,\mu,\nu} \Phi_{[\lambda,\mu,\nu]}\sigma_{ji}G_{ij}^{[\lambda,\mu,\nu]} \tag{5.8}$$

and

$$d\varepsilon_{ij}^{p}A_{jirs}\sigma_{rs} = \sum_{\lambda,\mu,\nu} \Phi_{[\lambda,\mu,\nu]}G_{ij}^{[\lambda,\mu,\nu]}A_{jirs}\sigma_{rs}, \tag{5.9}$$

from which we read the following systems of invariants:

$$\mathscr{A}_{[\lambda,\mu,\nu]} \equiv \sigma_{ij}^{(\lambda+1)}A_{jikl}^{(\mu)}\sigma_{lk}^{(\nu)} \tag{5.10}$$

and

$$\mathscr{B}_{[\lambda,\mu,\nu]} \equiv \sigma_{ij}^{(\lambda)}A_{jrkl}^{(\mu)}A_{rimn}\sigma_{lk}^{(\nu)}\sigma_{nm}, \tag{5.11}$$

respectively, if we consider the abbreviations (5.6) and (5.7). The 28

TABLE 2

IRREDUCIBLE INVARIANTS OF THE SYSTEM OF (5.10)

$\sigma_{ij}M_{ij}^{[\lambda,0,\nu]} = \sigma_{kk}^{(\lambda+\nu+1)}$ where $\lambda + \nu \leqslant 2$		
λ \ ν	0	1	2
0	$S_1 \equiv \operatorname{tr}\sigma$	$S_2 \equiv \operatorname{tr}\sigma^2$	$S_3 \equiv \operatorname{tr}\sigma^3$
1	S_2	S_3	
2	S_3		

$\sigma_{ij}M_{ij}^{[\lambda,\mu,\nu]} = \sigma_{ij}^{(\lambda+1)}A_{ijkl}^{(\mu)}\sigma_{lk}^{(\nu)};\quad \mu = 1,2,\ldots,5$		
λ \ ν	0	1	2
0	$\sigma_{ij}A_{ijkk}^{(\mu)}$	$\sigma_{ij}A_{ijkl}^{(\mu)}\sigma_{kl}$	$\sigma_{ij}A_{ijkl}^{(\mu)}\sigma_{kl}^{(2)}$
1	$\sigma_{ij}^{(2)}A_{ijkl}^{(\mu)}$	$\longrightarrow$	$\sigma_{ij}^{(2)}A_{ijkl}^{(\mu)}\sigma_{kl}^{(2)}$

irreducible elements of the system of (5.10) are listed in Table 2 and discussed by Betten (1987a).

Since (5.8) represents the *plastic work*, the invariants of Table 2 are most important. Thus we use them to formulate the yield conditions

$$f(\sigma_{ij}^{(\lambda+1)}A_{jikl}^{(\mu)}\sigma_{lk}^{(\nu)}) = 1, \tag{5.12}$$

where $\lambda = 0,\ 1$ and $\nu = 0,\ 1,\ 2$ for $\mu = 1, 2,\ldots, 5$. For $\mu = 0$, we have the isotropic special case with

$$A_{ijkl}^{(0)} = (\delta_{ik}\delta_{jl} + \delta_{il}\delta_{jk})/2. \tag{5.13}$$

Then the restriction $\lambda + \nu \leqslant 2$ must be noted (Table 2).

A. SPECIAL FORMULATIONS OF CONSTITUTIVE EQUATIONS AND YIELD CRITERIA

The special case of *transverse isotropy* is contained in (5.12). This case can be specified by a symmetric tensor of rank two as pointed out in detail by Boehler and Sawczuk (1977) or by Betten (1988a). We can also describe the yield behavior of such a material by using a fourth-rank tensor specified by

$$A_{ijkl} \equiv (A_{ik}A_{jl} + A_{il}A_{jk})/2, \tag{5.14}$$

which has the symmetry properties (5.1). Then we find the tensor power

$$A_{ijkl}^{(\mu)} = (A_{ik}^{(\mu)}A_{jl}^{(\mu)} + A_{il}^{(\mu)}A_{jk}^{(\mu)})/2. \tag{5.15}$$

Inserting this relation in (5.7), we immediately find the constitutive equation

$$d\varepsilon_{ij}^{P} = \frac{1}{2} \sum_{\lambda,v=0}^{2} \Phi_{[\lambda,v]}(\sigma_{ik}^{(\lambda)}A_{kj}^{(v)} + A_{ik}^{(v)}\sigma_{kj}^{(\lambda)}) \tag{5.16}$$

from (5.5) and (5.6), if we select reducible elements like $\sigma A \sigma$ or $A \sigma A$ or $A \sigma^2 A$. Furthermore, substituting the special tensor (5.14) in (5.10) and (5.12), we find the *integrity basis*

$$\left. \begin{array}{ll} S_v \equiv \operatorname{tr} \sigma^v, & T_v \equiv \operatorname{tr} A^v; \quad v = 1, 2, 3 \\ \Omega_1 \equiv \operatorname{tr} \sigma A, & \Omega_2 \equiv \operatorname{tr} \sigma A^2, \\ \Omega_3 \equiv \operatorname{tr} A\sigma^2, & \Omega_4 \equiv \operatorname{tr} \sigma^2 A^2 \end{array} \right\}, \tag{5.17}$$

after omitting redundant elements. The tensor A in (5.17) is a second-order tensor.

To describe the *strength-differential effect* and (or) the *Bauschinger phenomenon*, it is convenient to substitute the difference tensor

$$\bar{\sigma}_{ij} = \sigma_{ij} - B_{ij} \tag{5.18}$$

for the stress tensor σ in all the above equations. Then, for instance, the simultaneous invariant $\mathscr{A}_{[0,1,1]}$ from Table 2 takes the form

$$\mathscr{A}_{[0,1,1]} = A_{ijkl}(\sigma_{ij} - B_{ij})(\sigma_{kl} - B_{kl}), \tag{5.19}$$

and a yield criterion containing this invariant only can be expressed in the simplest relation

$$A_{ijkl}(\sigma_{ij} - B_{ij})(\sigma_{kl} - B_{kl}) = 1. \tag{5.20}$$

In the isotropic special case characterized by

$$A_{ijkl} = a_1 \delta_{ij}\delta_{kl} + a_2(\delta_{ik}\delta_{jl} + \delta_{il}\delta_{jk}),$$

and $B_{ij} = b\delta_{ij}$, the criterion in (5.20) reduces to the form

$$J_2' + c_1 J_1^2 + c_2 J_1 = c_3, \tag{5.21}$$

where the constants c_1, c_2, and c_3 can be determined by experimental data (Betten, 1982c).

Furthermore, substituting $c_1 = -\alpha^2$, $c_2 = 2\alpha\beta$, and $c_3 = \beta$ from (5.21) we recover the well-known Drucker–Prager criterion

$$\sqrt{J'_2} + \alpha J_1 = \beta, \tag{5.22}$$

where the parameters α and β are determined by experimental data:

$$\alpha = (1/\sqrt{3})(\sigma_{FD} - \sigma_{FZ})/(\sigma_{FD} + \sigma_{FZ}), \tag{5.23}$$

$$\beta = (2/\sqrt{3})\sigma_{FD}\sigma_{FZ}/(\sigma_{FD} + \sigma_{FZ}). \tag{5.24}$$

The symbols σ_{FZ} and σ_{FD} characterize the *yield stresses* in *tension* and *compression*, respectively. Thus, the parameter in (5.23) expresses the strength differential or SD effect.

B. Incompressibility and Volume Change

The constitutive equation (5.5) can be expressed in the form

$$d\varepsilon^p_{ij} = \sum_{\lambda,\mu,\nu} \Phi_{[\lambda,\mu,\nu]} A^{[\mu]}_{ijpqrs} \sigma^{(\nu)}_{pq} \sigma^{(\lambda)}_{rs} \tag{5.25}$$

by introducing several tensors $A^{[\mu]}$ of rank six, the components of which can be determined from the given fourth-order tensor (5.1):

$$A^{[\mu]}_{ijpqrs} \equiv \tfrac{1}{4}(A^{(\mu)}_{irpq}\delta_{js} + A^{(\mu)}_{ispq}\delta_{jr} + \delta_{is}A^{(\mu)}_{jspq} + \delta_{is}A^{(\mu)}_{jrpq}). \tag{5.26}$$

These tensors have the symmetry conditions

$$A^{[\mu]}_{ijpqrs} = A^{[\mu]}_{jipqrs} = A^{[\mu]}_{ijqprs} = A^{[\mu]}_{ijpqsr} \tag{5.27a}$$

and

$$A^{[\mu]}_{ijpqrs} = A^{[\mu]}_{rspqij}. \tag{5.27b}$$

Further symmetry conditions are not valid because the form of (5.25) is not symmetric with respect to λ and ν, except for $\mu = 0$, i.e., for the isotropic special case given by (5.13).

To study the magnitude of *volume change* or an *incompressible* deformation, it is convenient to decompose (5.25) into two functions, a *scalar* one and a *deviatoric* one:

$$d\varepsilon^p_{rr} = \sum_{\lambda,\mu,\nu} \Phi_{[\lambda,\mu,\nu]} \sigma^{(\lambda)}_{ij} A^{(\mu)}_{ijkl} \sigma^{(\nu)}_{kl}, \tag{5.28a}$$

$$d\varepsilon^{p'}_{ij} = \sum_{\lambda,\mu,\nu} \Phi_{[\lambda,\mu,\nu]} A^{[\mu]}_{\{ij\}pqrs} \sigma^{(\nu)}_{pq} \sigma^{(\lambda)}_{rs}, \tag{5.28b}$$

where the tensors

$$A^{[\mu]}_{\{ij\}pqrs} \equiv A^{[\mu]}_{ijpqrs} - \tfrac{1}{3}\delta_{ij}A^{(\mu)}_{pqrs} \tag{5.29}$$

are deviatoric with respect to the braced index pair $\{ij\}$, and $\varepsilon^{p'}_{ij} = \varepsilon^{p}_{ij} - \varepsilon^{p}_{rr}\delta_{ij}/3$ is the plastic strain deviator.

Another decomposition of (5.25) is given by

$$d\varepsilon^{p}_{rr} = \sum_{\mu=0}^{5} A^{(\mu)}_{ijkl}P^{(\mu)}_{ijkl}, \tag{5.30a}$$

$$d\varepsilon^{p'}_{ij} = \sum_{\mu=0}^{5} A^{(\mu)}_{\{ij\}pqrs}P^{(\mu)}_{pqrs}, \tag{5.30b}$$

where the fourth-rank tensors $\mathbf{P}^{[\mu]}$ can be presented in the following matrix product form:

$$P^{[\mu]}_{ijkl} = (\delta_{kl}, \sigma'_{kl}, \sigma''_{kl})\begin{bmatrix} \Psi_{(0,\mu,0)} & \Psi_{(0,\mu,1)} & \Psi_{(0,\mu,2)} \\ \Psi_{(1,\mu,0)} & \Psi_{(1,\mu,1)} & \Psi_{(1,\mu,2)} \\ \Psi_{(2,\mu,0)} & \Psi_{(2,\mu,1)} & \Psi_{(2,\mu,2)} \end{bmatrix}\begin{bmatrix} \delta_{ij} \\ \sigma'_{ij} \\ \sigma''_{ij} \end{bmatrix}, \tag{5.31}$$

i.e., we have a representation in the *traceless tensors*

$$\sigma'_{ij} \equiv \partial J'_2/\partial\sigma_{ij} = \sigma_{ij} - \sigma_{kk}\delta_{ij}/3, \tag{5.32}$$

$$\sigma''_{ij} \equiv \partial J'_3/\partial\sigma_{ij} = \sigma'^{(2)}_{ij} - 2J'_2\delta_{ij}/3. \tag{5.33}$$

The elements $\Psi_{[.,\mu,.]}$ in the matrix of (5.31) can be expressed by the scalar-valued functions $\Phi_{[\lambda,\mu,\nu]}$ for example,

$$\Psi_{[1,\mu,1]} \equiv \Phi_{[1,\mu,1]} + \tfrac{2}{3}J_1(\Phi_{[1,\mu,1]} + \Phi_{[2,\mu,1]} + \tfrac{2}{3}J_1\Phi_{[2,\mu,2]}),$$

$$\equiv \Phi_{[1,\mu,2]} + 2J_1\Phi_{[2,\mu,2]}/3,$$

$$\Psi_{[2,\mu,1]} \equiv \Phi_{[2,\mu,1]} + 2J_1\Phi_{[2,\mu,2]}/3,$$

etc. The coefficients $\Phi_{[\lambda,\mu,\nu]}$ are themselves scalar-valued functions of the integrity basis (5.10) and (5.11) associated with the representation (5.5).

Using representation (5.30), we read from the plastic work $\sigma_{ij}d\varepsilon^{p}_{ij}$ the following 34 irreducible invariants (Betten, 1982c):

$$\left.\begin{array}{l} J_1, J'_2, J'_3, J_1A^{(\mu)}_{iijj}, J_1A^{(6)}_{iijj}, \sigma'_{ij}A^{(\mu)}_{ijkk}, \sigma'_{ij}A^{(\mu)}_{ijkl}\sigma'_{kl}, \\ \sigma'_{ij}A^{(\mu)}_{ijkl}\sigma''_{kl}, \sigma''_{ij}A^{(\mu)}_{ijkk}, \sigma''_{ij}A^{(\mu)}_{ijkl}\sigma''_{kl}; \quad \mu = 1, 2, \ldots, 5. \end{array}\right\} \tag{5.34}$$

For practical use, it may be sufficient to consider the anisotropy by $\mu = 1$ only. Then, we have the *yield condition*

$$f(J_1, J'_2, J'_3, \Pi_1, \Pi'_1, \Pi'_2, \Pi'_3, \Pi''_1, \Pi''_2) = 1, \tag{5.35}$$

where the $J_1 \equiv S_1$, J'_2, J'_3 are defined by (4.16a,b,c). The invariant $\Pi_1 = J_1 A_{iijj}$ expresses the hydrostatic pressure accompanied by anisotropy. The remaining simultaneous invariants are defined by

$$\left.\begin{array}{ll} \Pi'_1 \equiv \sigma'_{ij} A_{ljkk}, & \Pi'_2 = \sigma'_{ij} A_{ijkl} \sigma'_{kl}, \\[2mm] \Pi'_3 \equiv \sigma'_{ij} A_{ijkl} \sigma_{kl}, & \\[2mm] \Pi''_1 \equiv \sigma''_{ij} A_{ijkk}, & \Pi''_2 \equiv \sigma''_{ij} A_{ijkl} \sigma''_{kl}. \end{array}\right\} \tag{5.36}$$

In the isotropic special case characterized by (5.13), the invariants $\Pi_1, \dots, \Pi''_3$ in the condition (5.35) reduce to

$$\Pi_1 = 3J_1, \qquad \Pi'_1 = 0, \qquad \Pi'_2 = 2J'_2,$$

$$\Pi'_3 = 3J'_3, \qquad \Pi''_1 = 0, \qquad \Pi''_2 = 2J'^2_2/3,$$

i.e., they are expressible by the invariants (4.16a,b,c) and are therefore redundant elements in this case.

For oriented solids we can specify the fourth-rank tensor A_{ijkl} in (5.36) by

$$A_{ijkl} \equiv \tfrac{1}{4}(A_{ik}\delta_{jl} + A_{il}\delta_{jk} + \delta_{ik}A_{jl} + \delta_{il}A_{jk}), \tag{5.37}$$

and the yield condition (5.35) reduces to the form

$$f(J_1, J'_2, J'_3, \sigma'_{ij}A_{ji}, \sigma'^{(2)}_{ij}A_{ji}) = 1, \tag{5.38}$$

which is an alternative representation of the yield criterion formulated by Boehler and Sawczuk (1977). The results given by (5.16), (5.6), (5.12), and (5.25) are summarized in Table 3.

Finally, we remark that in addition to an initial anisotropy as given in (5.1) corresponding to a forming process, for instance, rolling, a deformation-induced anisotropy can appear, for instance, by the formation of microscopic cracks. Because of its microscopic nature, damage has, in general, an anisotropic character even if the material is originally isotropic. The orientation of fissures and their length result in anisotropic macroscopic behavior. Thus it is necessary to consider this kind of anisotropy by introducing appropriately defined *anisotropic damage tensors* into the constitutive equations. In view of polynomial representations of constitutive equations, it is convenient to use the tensor

$$D_{ij} \equiv (\delta_{ij} - \omega_{ij})^{(-1)} \equiv \psi^{(-1)}_{ij} \tag{5.39}$$

TABLE 3
TENSOR GENERATORS AND YIELD CRITERIA

	Tensor generators	Yield criteria
General anisotropy	$G_{ij}^{[\lambda,\mu,\nu]} = A_{ijpqrs}^{[\mu]}\sigma_{pq}^{(\nu)}\sigma_{rs}^{(\lambda)}$ $\lambda,\mu = 0,1,2$ $\mu = 1,2,\ldots,5$ for	$f(d_{ji}G_{ij}^{[\lambda,\mu,\nu]}) = 1$ $\lambda = 0,1;\quad \nu = 0,1,2$ $\mu = 0 \Rightarrow \lambda + \nu \leqslant 2$
$\nu = 0$ $\Downarrow$ Oriented solids	$G^{[\lambda,\mu]} = A_{ijrs}^{[\mu]}\sigma_{rs}^{(\lambda)}$ $\lambda,\mu = 0,1,2$	$f(\sigma_{ji}G_{ij}^{[\lambda,\mu]}) = 1$ $\lambda,\mu = 0,1,2$ but for $\mu \neq 0 \Rightarrow \lambda < 2$
$\mu = 0$ $\Downarrow$ Isotropy	$G_{ij}^{[\lambda]} = \sigma_{ij}^{(\lambda)};\quad \lambda = 0,1,2$	$f(\sigma_{ji}G_{ij}^{[\lambda]}) = 1$

$$A_{ijpqrs}^{(\mu)} \equiv \tfrac{1}{4}(A_{irpq}^{(\mu)}\sigma_{js} + A_{ispq}^{(\mu)}\sigma_{jr} + \sigma_{ir}A_{jspq}^{(\mu)} + \sigma_{is}A_{jrpq}^{(\mu)})$$
$$A_{ijrs}^{(\mu)} \equiv \tfrac{1}{4}(A_{ir}^{(\mu)}\sigma_{js} + A_{is}^{(\mu)}\sigma_{jr} + \sigma_{ir}A_{js}^{(\mu)} + \sigma_{is}A_{jr}^{(\mu)})$$

as an additional argument tensor where ω and ψ are the second-rank *damage tensor*, and the *tensor of continuity*, respectively. The undamaged state is characterized by $\omega \to 0$ or $\psi \to \delta$. Considering (5.39), the constitutive equation takes the form

$$d\varepsilon_{ij}^{p} = f_{ij}(\sigma_{pq}, D_{pq}, A_{pqrs}) \tag{5.40}$$

instead of (5.3). The representation of (5.40), that is, of a symmetric second-order tensor-valued function of two symmetric tensors of rank two and one symmetric tensor of rank four is pointed out in detail by Betten (1983a), and, again, the essential invariants to formulate the yield criterion can be read from the *plastic work* $\sigma_{ji}\,d\varepsilon_{ij}^{p}$.

Based on the plastic work (5.8) one can arrive at a system of invariants (5.10), which is most important to formulate yield conditions of type (5.12). However, the system (5.10) contains irreducible invariants, but this system cannot be complete, because a lot of irreducible invariants cannot be expressed through (5.10). Thus, we have to look for other methods in order to find complete sets of invariants of a single fourth-order tensor and mixed invariants. These methods are discussed in the following three sections.

C. CHARACTERISTIC POLYNOMIAL FOR A FOURTH-ORDER TENSOR

To construct an *irreducible* set of principal invariants of a fourth-order tensor **A**, the *characteristic polynomial*

$$P_n(\lambda) \equiv \det(A_{ijkl} - \lambda A_{ijkl}^{(0)}) = \sum_{\nu=0}^{n} J_\nu(A)\lambda^{n-\nu} \tag{5.41}$$

is considered by Betten (1982a).

The *principal invariants* J_ν in (5.41) can be determined by performing the operation of alternation

$$(-1)^{n-\nu}J_\nu \equiv A_{\alpha_1[\alpha_1]}A_{\alpha_2[\alpha_2]}\cdots A_{\alpha_\nu[\alpha_\nu]} \tag{5.42}$$

where $(-1)^n J_0 \equiv 1$. The right-hand side in (5.42) is equal to the sum of all principal minors of order $\nu \leq n$, where $\nu = 1$, and $\nu = n$ lead to $\operatorname{tr}\mathbf{A}$ and $\det \mathbf{A}$, respectively.

Assuming the usual symmetry conditions (5.1), the number n in (5.41) and (5.42) is equal to 6. Thus we have found six irreducible invariants defined in (5.42). However, this system cannot be complete because some irreducible invariants like A_{iijj}, $A_{iipq}A_{pqjj}$, $A_{ijip}A_{pjqr}A_{rsqs}$, etc., cannot be expressed through (5.42). Consequently the characteristic polynomial (5.41) must be generalized. This can be achieved by using the fourth-order tensor

$$I_{ijkl} = \lambda\delta_{ij}\delta_{kl} + \mu(\delta_{ik}\delta_{jl} + \delta_{il}\delta_{jk}) \tag{5.43}$$

instead of the spherical tensor $\lambda A_{ijkl}^{(0)}$ where $A_{ijkl}^{(0)}$ is the zero-power tensor defined in (5.13). Thus, we take into consideration the *characteristic polynomial*

$$P(\lambda, \mu) = \det(A_{ijkl} - I_{ijkl}) = 0, \tag{5.44}$$

that is, we formulate the eigenvalue problem

$$A_{ijkl}X_{kl} = I_{ijkl}X_{kl}. \tag{5.45}$$

The isotropic tensor (5.43) yields an image dyad

$$Y_{ij} = I_{ijkl}X_{kl} = \lambda X_{rr}\delta_{ij} + 2\mu X_{ij}, \tag{5.46}$$

which is coaxial with the dyad **X**.

In continuum mechanics, relation (5.46) is known as the constitutive equation for an isotropic linear elastic solid. Such a material is characterized by the two elastic constants λ and μ, which are called the Lamé constants. Using the notation $Y_\alpha = I_{\alpha\beta}X_\beta$, $\alpha, \beta = 1, 2, \ldots, 6$, we see that the isotropic tensor **I** in (5.46) can be represented in matrix form:

$$
I_{ijkl} \equiv
\begin{bmatrix}
2\mu + \lambda & \lambda & \lambda & 0 & 0 & 0 \\
\lambda & 2\mu + \lambda & \lambda & 0 & 0 & 0 \\
\lambda & \lambda & 2\mu + \lambda & 0 & 0 & 0 \\
0 & 0 & 0 & 2\mu & 0 & 0 \\
0 & 0 & 0 & 0 & 2\mu & 0 \\
0 & 0 & 0 & 0 & 0 & 2\mu
\end{bmatrix},
\tag{5.47}
$$

which is quasi-diagonal of the structure $\{3, 1, 1, 1\}$. The inverse tensor of (5.43) can be found from $X_{ij} = I_{ijkl}^{(-1)} Y_{kl}$ in connection with (5.46):

$$
I_{ijkl}^{(-1)} = \frac{1}{4\mu}(\delta_{ik}\delta_{jl} + \delta_{il}\delta_{jk}) - \frac{\lambda}{2\mu(3\lambda + 2\mu)}\delta_{ij}\delta_{kl}.
\tag{5.48}
$$

To control this result, we prove

$$
I_{ijpq}^{(-1)}I_{pqkl} = I_{ijpq}I_{pqkl}^{(-1)} \equiv A_{ijkl}^{(0)},
\tag{5.49}
$$

where the zero-power tensor has the diagonal form

$$
A_{ijkl}^{(0)} = \{1, 1, 1, 1, 1, 1\}.
\tag{5.50}
$$

We see that in the special case ($\lambda = 0$, $\mu = 1/2$) the isotropic tensor (5.47) tends to the unit tensor as given in (5.50). Then we find from (5.44) the simplified characteristic polynomial given in (5.41). The determinant of the form (5.47) is

$$
\det(I_{ijkl}) = 32(2\mu + 3\lambda)\mu^5,
$$

which tends to $\det(A_{ijkl}^0) = 1$ in the special case ($\lambda = 0$, $\mu = 1/2$).

In the following we shall calculate irreducible invariants as coefficients in the characteristic polynomial (5.44). Due to (5.1) and (5.47), the characteristic polynomial (5.44) can be written as

$$
\begin{vmatrix}
A_{1111}-\Omega & A_{1122}-\lambda & A_{1133}-\lambda & 2A_{1112} & 2A_{1123} & 2A_{1131} \\
A_{2211}-\lambda & A_{2222}-\Omega & A_{2233}-\lambda & 2A_{2212} & 2A_{2223} & 2A_{2231} \\
A_{3311}-\lambda & A_{3322}-\lambda & A_{3333}-\Omega & 2A_{3312} & 2A_{3323} & 2A_{3331} \\
A_{1211} & A_{1222} & A_{1233} & 2A_{1212}-2\mu & 2A_{1223} & 2A_{1231} \\
A_{2311} & A_{2322} & A_{2333} & 2A_{2312} & 2A_{2323}-2\mu & 2A_{2331} \\
A_{3111} & A_{3122} & A_{3133} & 2A_{3112} & 2A_{3123} & 2A_{3131}-2\mu
\end{vmatrix} = 0
$$

$$
\tag{5.51a}
$$

or as

$$
\begin{bmatrix}
A_{1111}-\Omega & A_{1122}-\lambda & A_{1133}-\lambda & 2A_{1112} & 2A_{1123} & 2A_{1131} \\
& A_{2222}-\Omega & A_{2233}-\lambda & 2A_{2212} & 2A_{2223} & 2A_{2231} \\
& & A_{3333}-\Omega & 2A_{3312} & 2A_{3323} & 2A_{3331} \\
& & & 2A_{1212}-2\mu & 2A_{1223} & 2A_{1231} \\
& & & & 2A_{2323}-2\mu & 2A_{2331} \\
\text{symm.} & & & & & 2A_{3131}-2\mu
\end{bmatrix} = 0,
$$

(5.51b)

where the abbreviation $\Omega \equiv 2\mu + \lambda$ is used. The determinant (5.51) is a polynomial of degree 3 in λ and degree 6 in μ:

$$
\left.
\begin{aligned}
P(\lambda, \mu) &= \sum_{p,q} C_{(p,q)}\lambda^p\mu^q, \\
p &= 0, 1, 2, 3; \quad q = 0, 1, \ldots, 6, \quad p+q \leqslant 6
\end{aligned}
\right\}
$$

(5.52)

with 22 coefficients $C_{(p,q)}$. However, the characteristic equation $P(\lambda, \mu) = 0$ can be divided by the coefficient $C_{(0,6)}$, so that we can find no more than 21 invariants. Now the main problem is to expand the determinant (5.51) and to find out if all 21 coefficients are irreducible invariants.

To avoid the lengthy expansion of the determinant (5.51), we propose the following method. If we consider a second-order tensor, we can start from the diagonal form

$$
\sigma_{ij} = \text{diag}\{\sigma_I, \sigma_{II}, \sigma_{III}\}.
$$

(5.53)

Then the determinant, $\det(\sigma_{ij} - \lambda\delta_{ij})$, is identical to the characteristic polynomial

$$
P_3(\lambda) = (\lambda - \sigma_I)(\lambda - \sigma_{II})(\lambda - \sigma_{III}) - \lambda^3 - J_1\lambda^2 - J_2\lambda - J_3.
$$
(5.54)

Similarly, we could consider the orthotropic case with

$$
A_{1112} = A_{1123} = A_{1131} = A_{2212} = A_{2223} = A_{2231} = A_{3312}
$$

$$
= A_{3323} = A_{3331} = 0.
$$

Thus the dyads $\mathbf{X}$ and $\mathbf{Y}$ in $Y_{ij} = A_{ijkl}X_{kl}$ are coaxial and, instead of the polynomial of (5.51), the following determinant is used:

$$
\begin{vmatrix}
A_I - (2\mu + \lambda) & B_I - \lambda & B_{II} - \lambda & 0 & 0 & 0 \\
B_I - \lambda & A_{II} - (2\mu + \lambda) & B_{III} - \lambda & 0 & 0 & 0 \\
B_{II} - \lambda & B_{III} - \lambda & A_{III} - (2\mu + \lambda) & 0 & 0 & 0 \\
0 & 0 & 0 & A_{IV} - 2\mu & 0 & 0 \\
0 & 0 & 0 & 0 & A_V - 2\mu & 0 \\
0 & 0 & 0 & 0 & 0 & A_{VI} - 2\mu
\end{vmatrix} = 0.
$$

$$(5.55)$$

From the determinant of (5.55) we find the characteristic polynomial

$$
P(\lambda, \mu) = 64\mu^6 - 32I_1\mu^5 + 16I_2\mu^4 - 8I_3\mu^3 + 4I_4\mu^2 - 2I_5\mu + I_6\lambda + I_7\lambda\mu
$$

$$
+ I_8\lambda\mu^2 + I_9\lambda\mu^3 + I_{10}\lambda\mu^4 + 96\lambda\mu^5 + I_{11}, \qquad (5.56)
$$

where the 11 invariants $I_1, \ldots, I_{11}$ are as follows:

$$
I_1 \equiv K_1 + M_1, \quad I_2 \equiv K_2 + M_2 + K_1 M_1 - L_1^2 + 2L_2,
$$

$$
I_3 \equiv D + M_3 + K_1 M_2 + M_1(K_2 - L_1^2 + 2L_2),
$$

$$
I_4 \equiv M_1 D + K_1 M_3 + M_2(K_2 - L_1^2 + 2L_2),
$$

$$
I_5 = M_2 D + M_3(K_2 - L_1^2 + 2L_2), \quad I_6 \equiv M_3(L_1^2 - 4L_2 - K_2 + 2N),
$$

$$
I_7 \equiv 4M_3(K_1 - L_1) - 2M_2(L_1^2 - 4L_2 - K_2 + 2N), \qquad (5.57)
$$

$$
I_8 \equiv 4M_1(L_1^2 - 4L_2 - K_2 - 2N) - 8M_2(K_1 - L_1) - 12M_3,
$$

$$
I_9 \equiv 8\{2M_1(K_1 - L_1) - L^2 + 4L_2 + K_2 - 2N + 3M_2\},
$$

$$
I_{10} \equiv -16\{2(K_1 - L_1) + 3M_1\}, \quad I_{11} = M_3 D.
$$

In the system (5.57) the following abbreviations are used:

$$
K_1 \equiv A_I + A_{II} + A_{III}, \quad M_1 \equiv A_{IV} + A_V + A_{VI},
$$

$$
K_2 \equiv A_I A_{II} + A_{II} A_{III} + A_{III} A_I,
$$

$$
M_2 \equiv A_{IV} A_V + A_V A_{VI} + A_{VI} A_{IV},
$$

$$
K_3 \equiv A_I A_{II} A_{III}, \quad M_3 \equiv A_{IV} A_V A_{VI},
$$

$$
L_1 \equiv B_I + B_{II} + B_{III}, \qquad (5.58)
$$

$$L_2 \equiv B_I B_{II} + B_{II} B_{III} + B_{III} B_I, \quad L_3 \equiv B_I B_{II} B_{III},$$

$$N \equiv A_I B_{III} + A_{II} B_{II} + A_{III} B_I,$$

$$D \equiv K_3 + 2L_3 - (A_I B_{III}^2 + A_{II} B_{II}^2 + A_{III} B_I^2).$$

We see that in system (5.57) the invariants I_1 and I_{11} are the trace and determinant of the fourth-order tensor, respectively.

In the special case $(\lambda = 0)$, only the six invariants $I_1, \ldots, I_5, I_{11}$ are relevant, and they are identical to the six invariants of (5.42), if $B_I = B_{II} = B_{III} = 0$.

In the two-dimensional case $(i, j, k, l = 1, 2)$, the eigenvalue problem (5.45) leads to a characteristic equation $P(\lambda, \mu) = 0$ from which we read five irreducible invariants as illustrated by Betten (1987c).

D. The Lagrange Multiplier Method

To find the characteristic equation of a fourth-order tensor, one can utilize the *Lagrange multiplier method*. To do this, we start from the scalar-valued function

$$F = A_{ijkl} X_{ij} X_{kl}. \tag{5.59}$$

Since the second-order tensor **X** in (5.59) has three irreducible invariants as given in (2.4),

$$S_1 \equiv \text{tr}\, \mathbf{X}, \qquad S_2 \equiv \text{tr}\, \mathbf{X}^2, \qquad S_3 \equiv \text{tr}\, \mathbf{X}^3 \tag{5.60}$$

being the elements of the integrity basis, we take into consideration the following three "auxiliary" conditions:

$$\left. \begin{aligned} L &= \delta_{ij}\delta_{kl} X_{ij} X_{kl} - S_1^2 = 0, \\ M &= (\delta_{ik}\delta_{jl} + \delta_{il}\delta_{jk}) X_{ij} X_{kl} - 2S_2 = 0, \\ N &= (\delta_{jk}\delta_{pl} + \delta_{jl}\delta_{pk}) X_{ip} X_{ij} X_{kl} - 2S_3 = 0. \end{aligned} \right\} \tag{5.61}$$

Thus the quadratic form (5.59) of the dyad **X** is modified to the form

$$\Phi = F - \lambda L - \mu M - \nu N, \tag{5.62}$$

which is to be made stationary, that is, we require

$$\partial \Phi / \partial X_{rs} = 0_{rs}. \tag{5.63}$$

From (5.63), we find the system of nonlinear equations in the dyad **X**,

$$(A_{ijkl} - I_{ijkl})X_{kl} = 3vX_{ij}^{(2)}, \tag{5.64}$$

where I_{ijkl} are the components of the fourth-order isotropic tensor defined in (5.43). In the special case when $v = 0$, we find from (5.64) the eigenvalue problem (5.45), which yields the characteristic equation (5.44). Some more details concerning the system (5.64) will be investigated by Betten (2001). Furthermore, applications of the *Lagrange multiplier method* to tensors of order six will also be discussed by Betten (2001).

E. COMBINATORIAL METHOD

To find *irreducible* sets of invariants for a fourth-order material tensor, two methods have been discussed:

1. By way of an extended characteristic polynomial, and

2. By application of a modified Lagrange multiplier method.

In the following a third one,

3. A combinatorial method,

is proposed, which is most effective and able to produce both *irreducible* and *complete* sets of invariants. As an example, let us form irreducible invariants of a fourth-order symmetric tensor

$$A_{ijkl} = A_{jikl} = A_{ijlk} = A_{klij}. \tag{5.1}$$

Because

$$\delta_{ij}\delta_k\delta_{pq}\delta_{rs} = \delta_{ji}\delta_{kl}\delta_{qp}\delta_{rs} = \cdots = \delta_{sr}\delta_{kl}\delta_{qp}\delta_{ij} \tag{5.65}$$

represent $P_8 = \dfrac{8!}{2^4 \cdot 4!} = 105$ independent combinations, one can form by *transvections* like

$$\delta_{ip}\partial_{jq}\delta_{kr}\delta_{ls}A_{ijkl}A_{pqrs} \equiv A_{ijkl}A_{ijkl} \equiv I_{2,4},$$

$$\delta_{is}\delta_{jr}\delta_{kq}\delta_{lp}A_{ijkl}A_{pqrs} \equiv A_{ijkl}A_{lkji} \equiv I_{2,4}, \tag{5.66a}$$

$$\delta_{ip}\delta_{lq}\delta_{jr}\delta_{ks}A_{ijkl}A_{pqrs} \equiv A_{ijkl}A_{iljk} \equiv I_{2,5},$$

altogether 105 invariants. However, considering the index symmetries (5.1), this number is reduced to the following 5 different invariants of degree 2:

$$I_{2,1}, I_{2,2}, \ldots, I_{2,5},$$

where

$$I_{2,1} \equiv A_{iijl} A_{jkkl},$$

$$I_{2,2} \equiv A_{iijk} A_{jkll},$$

$$I_{2,3} \equiv A_{ilij} A_{jkkl}, \qquad (5.66b)$$

$$I_{2,4} \equiv A_{ijkl} A_{ijkl},$$

$$I_{2,5} \equiv A_{ijkl} A_{iljk}.$$

In 3D Euclidian space the fourth-order permutation tensor ε_{ijkl} is a zero tensor so that the determinant

$$\varepsilon_{ijkl} \varepsilon_{pqrs} \equiv \begin{vmatrix} \delta_{ip} & \delta_{iq} & \delta_{ir} & \delta_{is} \\ \delta_{jp} & \delta_{jq} & \delta_{jr} & \delta_{js} \\ \delta_{kp} & \delta_{kq} & \delta_{kr} & \delta_{ks} \\ \delta_{lp} & \delta_{lq} & \delta_{lr} & \delta_{ls} \end{vmatrix} = 0 \qquad (5.67)$$

vanishes. Thus, from

$$\varepsilon_{ijkl} \varepsilon_{pqrs} A_{ijkl} A_{pqrs} = 0 \qquad (5.68)$$

we find the relation

$$8I_{2,1} = 4I_{2,2} + 4I_{2,3} + 2I_{2,4} + 2I_{2,5}. \qquad (5.69)$$

Consequently, the invariant $I_{2,1}$ is redundant, and there are only 4 irreducible invariants of degree 2 for a fourth-order symmetric tensor (5.1). Following this way by using a specially developed computer program, we find more than 65 irreducible invariants of a fourth-order tensor listed in Table 4. The highest degree is 6, which has to be taken into account (Betten, 1982a). However, the complete set of invariants of degree 6 for a fourth-order symmetric tensor could not be calculated up to now because of missing computer capacities.

The combinatorial method is not restricted to the formation of invariants, but also leads to results on tensor-valued terms called *tensor generators*. For

TABLE 4

NUMBERS OF IRREDUCIBLE INVARIANTS OF A FOURTH-ORDER SYMMETRIC TENSOR

Degree	1	2	3	4	5	6	
Invariants	2	4	10	16	33	?	$\Sigma > 65$

instance, the expressions

$$\delta_{ij}\delta_{kl}A_{kl} \equiv A_{kk}\delta_{ij}, \qquad \delta_{ik}\delta_{jl}A_{kl} \equiv A_{ij},$$
$$\delta_{ip}\delta_{qr}\delta_{sj}A_{pq}A_{rs} \equiv A_{ir}A_{rj} \equiv A_{ij}^{(2)} \tag{5.70}$$

are the three irreducible tensor generators of a second-order tensor. Furthermore, the *transvections*

$$\delta_{ip}\delta_{qr}\delta_{sj}A_{pq}B_{rs} \equiv A_{ir}B_{rj},$$
$$\delta_{ip}\delta_{qr}\delta_{st}\delta_{uv}\delta_{wj}A_{pqrs}A_{tuvw} \equiv A_{irrs}A_{suuj} \tag{5.71}$$

are index combinations with two free indices ij, i.e., second-order tensor-valued terms. Our specially developed computer program forms all possible index combinations, such as (5.71), and selects all redundant elements by considering index symmetries. Thus, we find sets of irreducible tensor generators, which are complete, too. Some results are listed in Table 5.

Further mathematical considerations and possible engineering applications are discussed and illustrated, for instance, by Betten (1998) and Betten and Helisch (1995a,b, 1996), or by Zheng and Betten (1995a,b), or Zheng *et al.* (1992). An extended survey of the theory of representations for tensor functions was provided by Zheng (1994).

TABLE 5

NUMBERS OF IRREDUCIBLE INVARIANTS AND TENSOR GENERATORS

Symmetric argument tensor	Irreducible invariants	Tensor functions $f_{ij} = \Sigma_\alpha \varphi_\alpha {}^\alpha G_{ij}$	Tensor generators ${}^\alpha G_{ij} = {}^\alpha G_{ji}$
X_{pq}	3	$Y_{ij} = f_{ij}(X_{pq})$	3
X_{pq}, A_{pq}	10	$f_{ij}(X_{pq}, A_{pq})$	9
X_{pq}, A_{pq}, B_{pq}	28	$f_{ij}(X_{pq}, A_{pq}, B_{pq})$	46
A_{pqrs}	>65	$f_{ij}(A_{pqrs})$	>108
X_{pq}, A_{pqrs}	>156	$f_{ij}(X_{pq}, A_{pqrs})$	>314
X_{pq}, D_{pq}, A_{pqrs}	>512	$f_{ij}(X_{pq}, D_{pq}, A_{pqrs})$	>884

As one can see from Table 5, constitutive equations and anisotropic damage growth equations of the forms (2.36a,b) containing complete sets of irreducible invariants and tensor generators may be too complicated for practical use. Therefore, we have to look for simplified representations as proposed in the following.

F. SIMPLIFIED REPRESENTATIONS

As an example, we suggest the following *simplified* constitutive equation:

$$d_{ij} = f_{ij}(\mathbf{t}, \tau) = \frac{1}{2} \sum_{\nu,\mu=0}^{2} \Phi_{[\nu,\mu]}[t_{ik}^{(\nu)}\tau_{kj}^{(\mu)} + \tau_{ik}^{(\mu)}t_{kj}^{(\nu)}], \qquad (5.72)$$

where the linear transformations

$$t_{ij} = D_{ijpq}\sigma_{pq} = t_{ji}, \qquad D_{ijpq} := \tfrac{1}{2}(D_{ip}D_{jq} + D_{iq}D_{jp}) \qquad (5.73)$$

and

$$\tau_{ij} = A_{ijpq}\sigma_{pq} = \tau_{ji} \qquad (5.74)$$

have been introduced by Betten (1983a). We can see that the first linear transformation (5.73) considers the *anisotropic damage state*, while the second one (5.74) expresses the *initial anisotropy* of the material. The main problem now is to determine the scalar coefficients $\Phi_{[\nu,\mu]}$ in (5.72) as functions of the integrity basis and experimental data. This can be achieved by using the *interpolation method* developed by Betten (1984b, 1987c) and applied also by Betten (1986b, 1989a).

The tensor **t** defined in (5.73) is called the *pseudo-net stress tensor*. This tensor is symmetric even in the anisotropic damage case. The nonsymmetric property of the *actual net stress tensor* $\hat{\sigma}$ in (2.35) is a disadvantage, and this tensor is awkward to use in constitutive equations.

In recent years much effort has been devoted to the elaboration of *evolutional equations* such as (2.36b). The results are discussed by Betten and Meydanli (1990). For the sake of simplicity, we can once again use the linear transformations (5.73) and (5.74); as in (5.72), we have to represent the tensor-valued function $\overset{\circ}{D}_{ij} = g_{ij}(\mathbf{t}, \tau)$ where $\overset{\circ}{D}_{ij}$ is the *Jaumann derivative* of the tensor (2.40).

The representation (5.72) involves nine tensor generators, where $\Phi_{[\nu,\mu]}$ are scalar-valued functions of the integrity basis associated with the representation (5.72), which can be found by forming $\operatorname{tr}\mathbf{td}$ and $\operatorname{tr}\boldsymbol{\tau}\mathbf{d}$. In this way we find the following invariants:

$$t_{ij}^{(\nu+1)}\tau_{ji}^{(\mu)} \equiv \operatorname{tr}\mathbf{t}^{\nu+1}\boldsymbol{\tau}^{\mu} \tag{5.75a}$$

and

$$t_{ij}^{(\nu)}\tau_{ji}^{(\mu+1)} \equiv \operatorname{tr}\mathbf{t}^{\nu}\boldsymbol{\tau}^{\mu+1}, \tag{5.75b}$$

some of them are reducible or equivalent, which can be found by using the following lemmas:

If $\mathbf{G}$ is a reducible generator, then $\operatorname{tr}\mathbf{XG}$ and $\operatorname{tr}\mathbf{GX}$ are reducible, where $\mathbf{X}$ is an arbitrary tensor.

The trace of a matrix product is unaltered by cyclic permutations of its factors.

The trace of a matrix product is equal to the trace of the transpose of the product.

The transpose of a matrix product π is the matrix product formed by writing down the transposes of the factors of π in reverse order, for instance: $\operatorname{tr}\mathbf{abc} = \operatorname{tr}\mathbf{c'b'a'}$.

Further deliberations concerning reducibility and equivalence and consequences from the Hamilton–Cayley theorem are given, for instance, by Betten (1987c) or by Zheng (1994).

In selecting the redundant elements, we find from (5.75a,b) the *integrity basis*

$$\left.\begin{matrix} \operatorname{tr}\mathbf{t}, & \operatorname{tr}\mathbf{t}^2, & \operatorname{tr}\mathbf{t}^3, & \operatorname{tr}\boldsymbol{\tau}, & \operatorname{tr}\boldsymbol{\tau}^2, & \operatorname{tr}\boldsymbol{\tau}^3, \\ \operatorname{tr}\mathbf{t}\boldsymbol{\tau}, & \operatorname{tr}\mathbf{t}\boldsymbol{\tau}^2, & \operatorname{tr}\mathbf{t}^2\boldsymbol{\tau}, & \operatorname{tr}\mathbf{t}^2\boldsymbol{\tau}^2, \end{matrix}\right\} \tag{5.76}$$

associated with the representation (5.72).

We can write the constitutive equation (5.72) in a *canonical form*, like (2.38), if we introduce the following identities for tensor generators: Let $\mathbf{X}$ and $\mathbf{Y}$ be two symmetric second-order tensors, i.e., $X_{ij} = X_{ji}$ and $Y_{ij} = Y_{ji}$, respectively. Then, the identities

$$\{\tfrac{1}{2}(\mathbf{X}^{\mu}\mathbf{Y}^{\nu} + \mathbf{Y}^{\nu}\mathbf{X}^{\mu})\}_{ij} \equiv \eta_{ijkl}^{[\nu]}X_{kl}^{(\mu)} \equiv \xi_{ijkl}^{[\mu]}Y_{kl}^{(\nu)} \tag{5.77}$$

are valid for arbitrary values μ and ν, where the symbols $\xi_{ijkl}^{[\mu]}$ and $\eta_{ijkl}^{[\nu]}$ are μ and ν several fourth-order tensors defined by

$$\xi_{ijkl}^{[\mu]} \equiv \tfrac{1}{4}(X_{ik}^{(\mu)}\delta_{jl} + X_{il}^{(\mu)}\delta_{jk} + \delta_{ik}X_{jl}^{(\mu)} + \delta_{il}X_{jk}^{(\mu)}) \tag{5.78a}$$

and

$$\eta_{ijkl}^{[\nu]} \equiv \tfrac{1}{4}(Y_{ik}^{(\nu)}\delta_{jl} + Y_{il}^{(\nu)}\delta_{jk} + \delta_{ik}Y_{jl}^{(\nu)} + \delta_{il}Y_{jk}^{(\nu)}), \tag{5.78b}$$

respectively. Note, that in (5.77), (5.78a), and (5.78b) those values of μ and ν which are bracketed by parentheses are exponents, while a square bracket gives any μ or ν the character of a label.

Furthermore, from (5.77), (5.78a), and (5.78b) we read the simple identities

$$\delta_{ij} \equiv \tfrac{1}{2}(\delta_{ik}\delta_{jl} + \delta_{il}\delta_{jk})\delta_{kl} \equiv E_{ijkl}\delta_{kl} \tag{5.79}$$

and

$$X_{ij}^{(\mu)} \equiv E_{ijkl}X_{kl}^{(\mu)} \equiv \zeta_{ijkl}^{[\mu]}\delta_{kl}, \tag{5.80}$$

where E_{ijkl} is the zero power tensor of fourth order defined by

$$E_{ijkl} = \tfrac{1}{2}(\delta_{ik}\delta_{jl} + \delta_{il}\delta_{jk}). \tag{5.81}$$

In using the introduced algebraic relations (5.77) to (5.80), we finally find a *canonical* form

$$d_{ij} = {}^{0}h_{ijkl}\delta_{kl} + {}^{1}h_{ijkl}t_{kl} + {}^{2}h_{ijkl}t_{kl}^{(2)} \tag{5.82}$$

of the constitutive equation (5.72) in the *pseudo-net stress tensor* **t**, where the fourth-order tensor-valued functions ${}^{0}\mathbf{h}$, ${}^{1}\mathbf{h}$, ${}^{2}\mathbf{h}$ are defined by

$$ {}^{0}h_{ijkl} \equiv \Phi_{[0,0]}T_{ijkl}^{(0)} + \Phi_{[0,1]}T_{ijkl} + \Phi_{[0,2]}T_{ijkl}^{[2]}, \tag{5.83a}$$

$$ {}^{1}h_{ijkl} \equiv \Phi_{[1,0]}T_{ijkl}^{(0)} + \Phi_{[1,1]}T_{ijkl} + \Phi_{[1,2]}T_{ijkl}^{[2]}, \tag{5.83b}$$

$$ {}^{2}h_{ijkl} \equiv \Phi_{[2,0]}T_{ijkl}^{(0)} + \Phi_{[2,1]}T_{ijkl} + \Phi_{[2,2]}T_{ijkl}^{[2]}, \tag{5.83c}$$

which can be expressed in the compact form

$$ {}^{\lambda}h_{ijkl} \equiv \sum_{\nu=0}^{2} \Phi_{[\lambda,\nu]}T_{ijkl}^{[\nu]}, \qquad \lambda = 0, 1, 2, \tag{5.84}$$

where, by analogy of (5.78b), the symmetric tensors $\mathbf{T}^{[\nu]}$ of rank four are defined by

$$T_{ijkl}^{[\nu]} \equiv \tfrac{1}{4}(\tau_{ik}^{(\nu)}\delta_{jl} + \tau_{il}^{(\nu)}\delta_{jk} + \delta_{ik}\tau_{jl}^{(\nu)} + \delta_{il}\tau_{jk}^{(\nu)}). \tag{5.85}$$

For example, for $\nu = 0$ we read from (5.85) the symmetric zero power tensor of rank four $T_{ijkl}^{[0]} \equiv T_{ijkl}^{(0)} = E_{ijkl}$ defined by (5.81).

The result of (5.84) elucidates that the anisotropy expressed by the tensors $\mathbf{T}^{[v]}$ according to (5.85) has equal influence on the contributions of orders zero, first, and second in the *pseudo-net stresses* t_{kl} contained in (5.82). The functions (5.83a)–(5.83c) or (5.84) express the anisotropy of the undamaged state (5.74), while the effect of anisotropic damage is considered in (5.82) by the *pseudo-net stress tensor* $\mathbf{t}$ according to (5.73) with (2.40). In the total damaged state, i.e., when $\boldsymbol{\omega} \to \boldsymbol{\delta}$ or $\boldsymbol{\psi} \to \mathbf{O}$, the tensors $\mathbf{D}$ and $\mathbf{t}$ are *singular*, and the strain rates (5.82) approach infinity.

In the undamaged state ($\boldsymbol{\omega} \to \mathbf{0}$ or $\mathbf{D} \to \boldsymbol{\delta}$) the pseudo-net stress tensor (5.73) is identical to Cauchy's stress tensor $\boldsymbol{\sigma}$. If, furthermore, the initial state is isotropic [$\mathbf{A} \to \mathbf{E}$ according to (5.81) or $\boldsymbol{\tau} \to \boldsymbol{\sigma}$ in (5.74)], then the constitutive equation yields the simple form of (2.22), valid for isotropic materials in the secondary creep stage.

Finally, because of (5.84), one can write the constitutive equation (5.82) in the very short form

$$d_{ij} \equiv \sum_{\lambda,v=0}^{2} \Phi_{[\lambda,v]} T_{ijkl}^{[v]} t_{kl}^{(\lambda)}. \tag{5.86}$$

Replacing the pseudo-net stress tensor $\mathbf{t}$ according to (5.73) in (5.82) we find the representation

$$d_{ij} = {}^{0}K_{ij} + {}^{1}K_{ijkl}\sigma_{kl} + {}^{2}K_{ijklmn}\sigma_{kl}\sigma_{mn} \tag{5.87}$$

of the constitutive equation (5.72), if we define:

$${}^{0}K_{ij} \equiv {}^{0}h_{ijpq}\delta_{pq}, \tag{5.88a}$$

$${}^{1}K_{ijkl} \equiv {}^{1}h_{ijpq}D_{pqkl}, \tag{5.88b}$$

$${}^{2}K_{ijklmn} \equiv {}^{2}h_{ijpq}D_{prkl}D_{rqmn}. \tag{5.88c}$$

The coefficient tensors of (5.88b) and (5.88c) are decomposed into anisotropy characterized by the fourth-rank tensors of (5.83b) and (5.83c) and into anisotropic damage determined by the fourth-rank tensor (D_{ijkl}) defined in (5.73) in a multiplicative way. In the undamaged case [$D_{ij} \to \delta_{ij}$ according to (2.40)] the tensor (D_{ijkl}) in (5.73) is identical to the symmetric zero power tensor of fourth order defined by (5.81), and in the total damage state ($\boldsymbol{\omega} \to \boldsymbol{\delta}$) the fourth-order tensor $\mathbf{D}$ in (5.73) is singular, i.e., the strain rates according to (5.72) or (5.87) approach infinity as the damage tensor $\boldsymbol{\omega}$ approaches the unit tensor $\boldsymbol{\delta}$ or as the continuity tensor $\boldsymbol{\psi}$ in (2.40) approaches the zero tensor $\mathbf{0}$.

VI. Damage Tensors and Tensors of Continuity

In three-dimensional space a parallelogram formed by the vectors A_i and B_i and can be represented by

$$S_i = \varepsilon_{ijk} A_j B_k \qquad (6.1a)$$

or in the dual form

$$S_{ij} = \varepsilon_{ijk} S_k \Leftrightarrow S_i = \tfrac{1}{2}\varepsilon_{ijk} S_{jk}, \qquad (6.1b)$$

where ε_{ijk} is the third-order alternating tensor ($\varepsilon_{ijk} = 1$, or -1 if i, j, k are even or odd permutations of 1, 2, 3, respectively, otherwise the components ε_{ijk} are equal to zero). From (6.1a) and (6.1b) we immediately find

$$S_{ij} = 2! A_{[i} B_{j]} = \begin{vmatrix} A_i & A_j \\ B_i & B_j \end{vmatrix}. \qquad (6.2)$$

Because of the decomposition (6.2) as an alternating product of two vectors the bivector **S** is called *simple* and has the following three nonvanishing essential components

$$S_{12} = A_1 B_2 - A_2 B_1, \qquad S_{23} = A_2 B_3 - A_3 B_2, \qquad (6.3a,b)$$

$$S_{31} = A_3 B_1 - A_1 B_3. \qquad (6.3c)$$

In rectilinear components in three-dimensional space, we see that the absolute values of the components (6.3) are the projections of the area of the parallelogram, considered above, on the coordinate planes. Thus S_{ij}, according to (6.2), represents an area vector in three-dimensional space and has an orientation fixed by (6.1a).

According to (6.1b) a surface element dS with an unit normal n_i, i.e., $dS_i = n_i \, dS$, is expressed by

$$dS_{ij} = \varepsilon_{ijk} \, dS_k \Leftrightarrow dS_i = \tfrac{1}{2}\varepsilon_{ijk} \, dS_{jk} \qquad (6.4a)$$

and

$$n_{ij} = \varepsilon_{ijk} n_k \Leftrightarrow n_i = \tfrac{1}{2}\varepsilon_{ijk} n_{jk}. \qquad (6.4b)$$

The components of the bivector **n** are the direction cosines n_1, n_2, n_3:

$$n_{ij} = \begin{pmatrix} 0 & n_3 & -n_2 \\ -n_3 & 0 & n_1 \\ n_2 & -n_1 & 0 \end{pmatrix}. \qquad (6.5)$$

The principal invariants of (6.5), defined as

$$J_1 \equiv n_{ii}, \qquad -J_2 \equiv n_{i[i]}n_{j[j]}, \qquad (6.6a,b)$$

$$J_3 \equiv n_{i[i]}n_{j[j]}n_{k[k]}, \qquad (6.6c)$$

take the following values:

$$J_1(\mathbf{n}) = 0, \qquad -J_2(\mathbf{n}) = n_1^2 + n_2^2 + n_3^2 = 1, \qquad (6.7a,b)$$

$$J_3(\mathbf{n}) = 0, \qquad (6.7c)$$

i.e., the only nonvanishing invariant is determined by the length of the vector n_i.

Imagine that at a point o in a continuous medium a set of rectangular coordinate axes is drawn and a differential tetrahedron is bounded by parts of the three coordinate planes through o and a fourth plane not passing through o, as shown in Figure 4. Such a tetrahedron can be characterized by a system of bivectors,

$$d^1S_i = -\tfrac{1}{2}\varepsilon_{ijk}(dx_2)_j(dx_3)_k,$$

$$d^2S_i = -\tfrac{1}{2}\varepsilon_{ijk}(dx_3)_j(dx_1)_k,$$

$$\qquad\qquad (6.8)$$

$$d^3S_i = -\tfrac{1}{2}\varepsilon_{ijk}(dx_1)_j(dx_2)_k,$$

$$d^4S_i = -\tfrac{1}{2}\varepsilon_{ijk}[(dx_1)_j - (dx_3)_j][(dx_2)_k - (dx_3)_k],$$

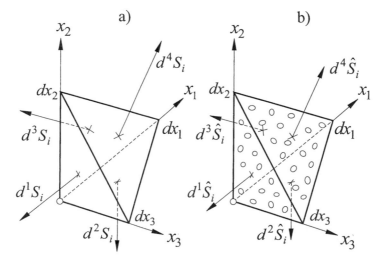

FIG. 4. Cauchy's tetrahedron (a) in an undamaged state and (b) in a damaged state.

where the sum is the zero vector:

$$d^1S_i + d^2S_i + d^3S_i + d^4S_i = 0_i. \tag{6.9}$$

In a damaged continuum we define a *net cross section* $\hat{S} \equiv \psi S$ where $\psi \leqslant 1$ describes the "continuity" of the material, as mentioned in Section II. Then, by analogy of (6.8), a tetrahedron in a damaged continuum (Figure 4b) can be characterized by the following system of bivectors:

$$d^1\hat{S}_i = -\tfrac{1}{2}\alpha_{ijk}(dx_2)_j(dx_3)_k \equiv \alpha\, d^1S_i,$$

$$d^2\hat{S}_i = -\tfrac{1}{2}\beta_{ijk}(dx_3)_j(dx_1)_k \equiv \beta\, d^2S_i,$$

$$\tag{6.10}$$

$$d^3\hat{S}_i = -\tfrac{1}{2}\gamma_{ijk}(dx_1)_j(dx_2)_k \equiv \gamma\, d^3S_i,$$

$$d^4\hat{S}_i = -\tfrac{1}{2}\kappa_{ijk}[(dx_1)_j - (dx_3)_j][(dx_2)_k - (dx_3)_k] \equiv \kappa\, d^4S_i,$$

where $\alpha_{ijk} \equiv \alpha\varepsilon_{ijk}$, $\beta_{ijk} \equiv \beta\varepsilon_{ijk}$, etc., are total skew-symmetric tensors of order 3, which have the essential components $\alpha_{123} \equiv \alpha$, $\beta_{123} \equiv \beta$, etc., respectively.

From Figures 4a and 4b we find that only $dS_1 = -n_1\, dS$, $d^1\hat{S}_1 = \alpha\, d^1S_1$, $d^2S_2 = -n_2\, dS$, etc., are nonvanishing components of the bivector systems (6.8) and (6.10). Then the sum of (6.10) yields the vector

$$\Sigma_i \equiv d^1\hat{S}_1 + \cdots + d^4\hat{S}_i = \begin{pmatrix} (\kappa - \alpha)n_1 \\ (\kappa - \beta)n_2 \\ (\kappa - \gamma)n_3 \end{pmatrix} dS, \tag{6.11}$$

which is not the zero vector, unless in the isotropic damage case ($\alpha = \beta = \gamma = \kappa$) or in the undamaged case ($\alpha = \beta = \gamma = \kappa = 1$) according to (6.9).

Furthermore, because of $d^1\hat{S}_1 \neq 0$, $d^1\hat{S}_2 = d^1\hat{S}_3 = 0$, etc., the damage state of the continuum at a point is characterized by the bivectors

$$\alpha_{1ij} = \begin{pmatrix} 0 & 0 & 0 \\ 0 & 0 & \alpha \\ 0 & -\alpha & 0 \end{pmatrix}, \qquad \beta_{2ij} = \begin{pmatrix} 0 & 0 & -\beta \\ 0 & 0 & 0 \\ \beta & 0 & 0 \end{pmatrix}, \tag{6.12a,b}$$

$$\gamma_{3ij} = \begin{pmatrix} 0 & \gamma & 0 \\ -\gamma & 0 & 0 \\ 0 & 0 & 0 \end{pmatrix}. \tag{6.12c}$$

In the following we will examine whether the bivector

$$\psi_{ij} = \alpha_{1ij} + \beta_{2ij} + \gamma_{3ij} = \begin{pmatrix} 0 & \gamma & -\beta \\ -\gamma & 0 & \alpha \\ \beta & -\alpha & 0 \end{pmatrix}, \tag{6.13a}$$

$$= \alpha\varepsilon_{1ij} + \beta\varepsilon_{2ij} + \gamma\varepsilon_{3ij} \tag{6.13b}$$

could be a suitable *tensor of continuity*. Then the damage tensor $\boldsymbol{\omega}$ would be of the form

$$\omega_{ij} = \delta_{(k)k}\varepsilon_{kij} - \psi_{ij} = \begin{pmatrix} 0 & 1-\gamma & -(1-\beta) \\ -(1-\gamma) & 0 & 1-\alpha \\ 1-\beta & -(1-\alpha) & 0 \end{pmatrix} \tag{6.14a}$$

(no sum on the bracketed index k) or

$$\omega_{ij} = (1-\alpha)\varepsilon_{1ij} + (1-\beta)\varepsilon_{2ij} + (1-\gamma)\varepsilon_{3ij}. \tag{6.14b}$$

If a tensor is symmetric or antisymmetric, respectively, in one Cartesian coordinate system, it is symmetric or antisymmetric in all such systems; thus symmetry and antisymmetry are really tensor properties. Therefore, the skew-symmetric tensor (6.13a) has only three essential components in any Cartesian system, for instance, α, β, γ in relation to the system x_i or α^*, β^*, γ^* with respect to the system x_i^*.

The only nonvanishing invariants of bivectors (6.13) and (6.14) are determined by their lengths:

$$-J_2(\boldsymbol{\psi}) \equiv -\tfrac{1}{2}\operatorname{tr}\boldsymbol{\psi}^2 \equiv -\tfrac{1}{2}\psi_{ij}\psi_{ji} = \alpha^2 + \beta^2 + \gamma^2, \tag{6.15}$$

$$-J_2(\boldsymbol{\omega}) = (1-\alpha)^2 + (1-\beta)^2 + (1-\gamma)^2. \tag{6.16}$$

In the undamaged state ($\alpha = \beta = \gamma = 1$), we have $-J_2(\boldsymbol{\psi}) = 3$, $-J_2(\boldsymbol{\omega}) = 0$, and

$$\psi_{ij} \to \eta_{ij} \equiv \varepsilon_{1ij} + \varepsilon_{2ij} + \varepsilon_{3ij} = \begin{pmatrix} 0 & 1 & -1 \\ -1 & 0 & 1 \\ 1 & -1 & 0 \end{pmatrix}. \tag{6.17}$$

We see that the undamaged state does not yield an isotropic tensor, because the components of $\boldsymbol{\eta}$ in (6.17) transform under the change of the coordinate system.

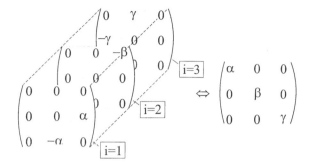

FIG. 5. Third-order tensor of continuity and its dual form.

Thus the bivector ψ defined by (6.13a) and (6.13b) is not suitable to describe the state of continuity of a damaged continuum, and we have to find another tensor composed by the bivectors of eqs. (6.12). As shown below, a suitable tensor of continuity can be defined by

$$\boxed{\psi_{ijk} \equiv \psi_{i[jk]}} \quad \text{with} \quad \begin{array}{l} \psi_{1jk} \equiv \alpha_{1jk} = \alpha\varepsilon_{1jk} \\ \psi_{2jk} \equiv \beta_{2jk} = \beta\varepsilon_{2jk}. \\ \psi_{3jk} \equiv \gamma_{3jk} = \gamma\varepsilon_{3jk} \end{array} \tag{6.18}$$

This tensor is skew-symmetric only with respect to the two bracketed indices $[jk]$ and possesses three essential components (α, β, γ), as illustrated in Figure 5.

In the isotropic damage state $(\alpha = \beta = \gamma = \kappa)$ the tensor (6.18) is total skew-symmetric, and the undamaged continuum $(\alpha = \beta = \gamma = 1)$ is characterized by the third-order alternating tensor ε_{ijk}. Supplementary to (6.18) we introduce the "damage tensor"

$$\boxed{\omega_{ijk} \equiv \varepsilon_{ijk} - \psi_{ijk}} \quad \text{where} \quad \begin{array}{l} \omega_{1jk} = (1 - \alpha)\varepsilon_{1jk} \\ \omega_{2jk} \equiv (1 - \beta)\varepsilon_{2jk}. \\ \omega_{3jk} = (1 - \gamma)\varepsilon_{3jk} \end{array} \tag{6.19}$$

By analogy with (6.1b) or (6.4a,b) the dual relations

$$\psi_{ijk} \equiv \psi_{i[jk]} = \varepsilon_{jkr}\psi_{ir} \Leftrightarrow \psi_{ir} = \tfrac{1}{2}\varepsilon_{rjk}\psi_{ijk}, \tag{6.20}$$

$$\omega_{ijk} \equiv \omega_{i[jk]} = \varepsilon_{jkr}\omega_{ir} \Leftrightarrow \omega_{ir} = \tfrac{1}{2}\varepsilon_{rjk}\omega_{ijk} \tag{6.21}$$

are valid.

Contrary to (6.13) and (6.14) the *dual tensor of continuity* ψ_{ij} according to (6.20) and the *dual damage tensor* ω_{ij} according to (6.21) have the diagonal forms

$$\psi_{ij} = \text{diag}\{\alpha, \beta, \gamma\} \tag{6.22}$$

and

$$\omega_{ij} = \text{diag}\{(1 - \alpha), (1 - \beta), (1 - \gamma)\}, \tag{6.23}$$

respectively. For the undamaged continuum ($\psi_{ijk} \rightarrow \varepsilon_{ijk}$) the dual tensor of continuity ψ_{ij} is equal to Kronecker's tensor δ_{ij}, as we can see from (6.20) or immediately from (6.22). The relations (6.20) and (6.22) are illustrated in Figure 5.

In particular, from Figure 5 we can see the skew-symmetric character of the third-order tensor of continuity indicated in (6.20) and its three essential components α, β, γ. These values are fractions that represent the net cross-sectional elements perpendicular to the coordinate axes x_1, x_2, x_3 (Figure 4b) and which can be measured in tests on specimens cut along three mutually perpendicular directions x_1, x_2, x_3.

According to (6.4a) a damaged surface element $d\hat{S}$ can be expressed in the dual form

$$d\hat{S}_{ij} = \varepsilon_{ijk} d\hat{S}_k \Leftrightarrow d\hat{S}_i = \tfrac{1}{2}\varepsilon_{ijk} d\hat{S}_{jk}, \tag{6.24}$$

and using the tensor of continuity (6.18) we find

$$d\hat{S}_{ij} = \psi_{ijk} dS_k \Leftrightarrow d\hat{S}_i = \tfrac{1}{2}\psi_{ijk} dS_{jk}. \tag{6.25}$$

Note, that the bivector $d\hat{S}_{ij}$ or dS_{jk} in (6.25) must have the same indices with respect to which the tensor (6.18) is skew-symmetric. Combining (6.4a) and (6.25) we have the linear transformations

$$d\hat{S}_{ij} = \tfrac{1}{2}\psi_{ijpq} dS_{pq}, \qquad d\hat{S}_i = \psi_{ir} dS_r, \tag{6.26a,b}$$

where ψ_{ir} is the tensor (6.20), (6.22), while ψ_{ijpq} is a fourth-order non-symmetric tensor defined as

$$\psi_{ijpq} \equiv \psi_{kij}\varepsilon_{kpq}, \tag{6.27a}$$

which, by using (6.20), can be expressed through

$$\psi_{ijpq} = (\delta_{ip}\delta_{jq} - \delta_{iq}\delta_{jp})\psi_{rr} - (\psi_{ip}\delta_{jq} - \psi_{iq}\delta_{jp}) - (\delta_{ip}\psi_{jq} - \delta_{iq}\psi_{jp}). \tag{6.27b}$$

This tensor has the antisymmetric properties

$$\psi_{ijpq} = -\psi_{jipq} = -\psi_{ijqp} = \psi_{jiqp}, \tag{6.28}$$

and is symmetric only with respect to the index pairs, i.e.,

$$\psi_{ijpq} = \psi_{pqij}. \tag{6.29}$$

More briefly, the properties of (6.28) and (6.29) can be indicated by

$$\psi_{ijpq} = \psi_{([ij][pq])}. \tag{6.30}$$

The essential components of the tensor (6.27) are given by

$$\psi_{ijpq} = \begin{cases} \alpha, \beta, \gamma, & \text{if } ij \text{ is an even permutation of } pq \\ -\alpha, -\beta, -\gamma, & \text{if } ij \text{ is an odd permutation of } pq \\ 0, & \text{otherwise} \end{cases} \tag{6.31a}$$

which means

$$\left. \begin{array}{l} \psi_{2323} = \psi_{3232} \equiv \alpha, \quad \psi_{3131} = \psi_{1313} \equiv \beta, \quad \psi_{1212} = \psi_{2121} \equiv \gamma, \\ \psi_{3223} = \psi_{2332} = -\alpha, \quad \psi_{1331} = \psi_{3113} \equiv -\beta, \quad \psi_{2112} - \psi_{1221} \equiv -\gamma. \end{array} \right\} \tag{6.31b}$$

In the isotropic damage state ($\alpha = \beta = \gamma = \kappa$) the tensor (6.27) is proportional to Kronecker's generalized delta

$$\delta_{ijpq} = \varepsilon_{kij}\varepsilon_{kpq} = \begin{vmatrix} \delta_{kk} & \delta_{kp} & \delta_{kq} \\ \delta_{ik} & \delta_{ip} & \delta_{iq} \\ \delta_{jk} & \delta_{jp} & \delta_{jq} \end{vmatrix},$$

$$= \delta_{ip}\delta_{jq} - \delta_{iq}\delta_{jp},$$

and is identical to that one in the undamaged continuum characterized by ($\alpha = \beta = \gamma = 1$).

To construct the tensor of continuity (6.22) we can use the following way. In addition to Figure 4 let us consider a fictitious undamaged configuration as illustrated in Figure 6c, which is, similar to (6.8) and (6.10), characterized by the following system of bivectors:

$$d^1\tilde{S}_i = -\tfrac{1}{2}\varepsilon_{ijk}(d\tilde{x}_2)_j(d\tilde{x}_3)_k \equiv d^1\hat{S}_i,$$

$$d^2\tilde{S}_i = -\tfrac{1}{2}\varepsilon_{ijk}(d\tilde{x}_3)_j(d\tilde{x}_1)_k \equiv d^2\hat{S}_i,$$

$$d^3\tilde{S}_i = -\tfrac{1}{2}\varepsilon_{ijk}(d\tilde{x}_1)_j(d\tilde{x}_2)_k \equiv d^3\hat{S}_i, \tag{6.32}$$

$$d^4\tilde{S}_i = -\tfrac{1}{2}\varepsilon_{ijk}[(d\tilde{x}_1)_j - (d\tilde{x}_3)_j][(d\tilde{x}_2)_k - (d\tilde{x}_3)_k],$$

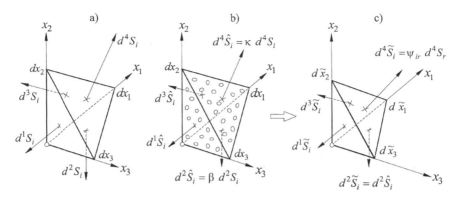

FIG. 6. Cauchy's tetrahedron: (a) undamaged configuration, (b) anisotropic damaged configuration, and (c) fictitious undamaged configuration.

where, by analogy of (6.9), the vector sum is equal to the zero vector:

$$d^1\tilde{S}_i + d^2\tilde{S}_i + d^3\tilde{S}_i + d^4\tilde{S}_i = 0_i. \tag{6.33}$$

The three area vectors $d^1\tilde{S}_i, \ldots, d^3\tilde{S}_i$ in (6.32) are identical to the corresponding vectors in (6.10) of the damaged configuration. The fourth vector $d^4\tilde{S}_i$ in (6.32), having the same magnitude as $d^4\hat{S}_i$ in (6.10), differs from the vector d^4S_i in (6.8) not only in length, but also in its direction. Therefore, the vectors $d^4\tilde{S}_i$ and d^4S_i are connected by a linear operator ψ of rank two (second-order tensor):

$$d^4\tilde{S}_i = \psi_{ir}\,d^4S_r. \tag{6.34}$$

Comparing the three systems of bivectors (6.8), (6.10), (6.32) and using eqs. (6.9) and (6.33) in connection with transformation (6.34), we find the relation:

$$\psi_{ir}\varepsilon_{rjk}[(dx_2)_j(dx_3)_k + (dx_3)_j(dx_1)_k + (dx_1)_j(dx_2)_k]$$

$$= \alpha_{ijk}(dx_2)_j(dx_3)_k + \beta_{ijk}(dx_3)_j(dx_1)_k + \gamma_{ijk}(dx_1)_j(dx_2)_k, \tag{6.35}$$

where the transvection $\psi_{ir}\varepsilon_{rjk}$ leads to the third-order tensor of continuity:

$$\psi_{ir}\varepsilon_{rjk} \equiv \psi_{ijk} = \psi_{i[jk]}, \tag{6.36}$$

which is skew-symmetric with respect to the bracketed index pair $[jk]$. Result (6.36) is contained in (6.18) and (6.20).

Because of $\alpha_{ijk} \equiv \alpha\varepsilon_{ijk}$, etc., the terms on the right-hand side of (6.35) are vectors with magnitudes

$$|d^1\tilde{S}_i| = \tfrac{1}{2}\alpha_{1jk}(dx_2)_j(dx_3)_k, \text{ etc.}$$

and with the directions of the basis vectors ^1e_i, ^2e_i, ^3e_i of the Cartesian coordinate system. Therefore, in connection with (6.36), relation (6.35) can be written in the following form:

$$\psi_{ijk}[(dx_2)_j(dx_3)_k + \cdots] = {}^1e_i\alpha_{1jk}(dx_2)_j(dx_3)_k + \cdots, \tag{6.37}$$

from which we immediately read the decomposition:

$$\psi_{ijk} = {}^1e_i\alpha_{1jk} + {}^2e_i\beta_{2jk} + {}^3e_i\gamma_{3jk}, \tag{6.38a}$$

or because of $\alpha_{1jk} \equiv \alpha\varepsilon_{1jk}$, etc.:

$$\psi_{ijk} = \alpha^1 e_i\varepsilon_{1jk} + \beta^2 e_i\varepsilon_{2jk} + \gamma^3 e_i\varepsilon_{3jk}. \tag{6.38b}$$

By analogy of (6.1b) we find the *dual relation* from (6.36):

$$\psi_{ijk} = \psi_{i[jk]} = \varepsilon_{jkr}\psi_{ir} \Leftrightarrow \psi_{ir} = \tfrac{1}{2}\varepsilon_{rjk}\psi_{ijk}, \tag{6.39}$$

and finally the diagonal form:

$$\psi_{ir} = \tfrac{1}{2}\psi_{ipq}\varepsilon_{jpq} = \operatorname{diag}\{\alpha, \beta, \gamma\} \tag{6.40a}$$

in accordance with (6.22). Inserting (6.38b) into (6.40a) and replacing $\delta_{1j} \equiv {}^1e_j$, etc., we see that the second rank tensor of continuity can be decomposed in terms of dyadics formed from the basis vectors:

$$\psi_{ij} = \alpha({}^1\mathbf{e} \otimes {}^1\mathbf{e})_{ij} + \beta({}^2\mathbf{e} \otimes {}^2\mathbf{e})_{ij} + \gamma({}^3\mathbf{e} \otimes {}^3\mathbf{e})_{ij}. \tag{6.40b}$$

The relations (6.39) and (6.40a) are illustrated in Figure 5. In particular, from Figure 5 we can see the skew-symmetric character of a third-order tensor of continuity indicated in (6.36) and its three essential components α, β, γ. These values are fractions that represent the net cross-sectional elements perpendicular to the coordinate axes x_1, x_2, x_3 (Figure 6b) and which can be measured in tests on specimens cut along three mutually perpendicular directions x_1, x_2, x_3. Such experiments are carried out by Betten and his coworkers, as discussed in Section VIII.

The damage may sometimes develop *isotropically*, as observed by Johnson (1960) for R.R. 59 aluminum alloy. In this special case

($\alpha = \beta = \gamma \equiv \psi$), the second rank tensor of continuity (6.39) is a *spherical tensor*:

$$\psi_{ijk} = \psi \varepsilon_{jkr} \delta_{ir} = \psi \varepsilon_{ijk} \Leftrightarrow \psi_{ir} = \frac{\psi}{2} \varepsilon_{rjk} \varepsilon_{ijk} = \psi \delta_{ir} \qquad (6.41)$$

and, contrary to (6.39), the third-order tensor of continuity is now totally skew-symmetric ($\psi_{ijk} \equiv \psi_{[ijk]}$).

Instead of the continuity tensor $\boldsymbol{\psi}$ according to (6.39) we can use the damage tensor $\boldsymbol{\omega}$ defined by (6.19) and (6.23) and characterized by the dual relation (6.21). In view of polynominal representations of constitutive equations it is convenient to use the tensor (2.40), as discussed in Section II in more detail.

VII. Stresses in a Damaged Continuum

In the undamaged continuum (Figure 7) Cauchy's formula

$$p_i = \sigma_{ji} n_j \qquad (7.1)$$

is derived from equilibrium, where p_i and n_i are the components of the stress vector $\mathbf{p}$ and the unit vector normal $\mathbf{n}$, respectively. In the same way we get to the corresponding relation for a damaged continuum,

$$\hat{p}_i \psi(n) = \psi_{jk} \hat{\sigma}_{ki} n_j, \qquad (7.2)$$

where ψ_{jk} are the components of the continuity tensor $\boldsymbol{\psi}$ according to (7.20).

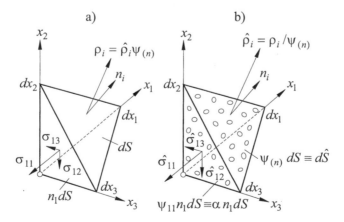

FIG. 7. Stress tensor regarding (a) an undamaged and (b) a damaged continuum.

The surface elements dS and $d\hat{S}$ in Figure 7 are subjected to the *same* force vector:

$$dP_i = p_i \, dS \equiv \hat{p}_i \, d\hat{S} = d\hat{P}_i. \tag{7.3}$$

Thus, considering (7.1) and (7.2), we finally find the actual net stress tensor $\hat{\boldsymbol{\sigma}}$ as a transformation from Cauchy's tensor:

$$\sigma_{ij} = \psi_{ir}\hat{\sigma}_{rj} = \sigma_{ji} \Leftrightarrow \hat{\sigma}_{ij} = \psi_{ir}^{(-1)}\sigma_{rj} \neq \hat{\sigma}_{ji}. \tag{7.4}$$

By suitable transvections we find $\sigma_{ij}\sigma_{jk}^{(-1)} = \psi_{ik}$ and $\hat{\sigma}_{ij}\sigma_{jk}^{(-1)} = \psi_{ik}^{(-1)}$.

As indicated in (7.4), the actual net stress tensor $\hat{\boldsymbol{\sigma}}$ is *nonsymmetric*, unless we have isotropic damage expressed by $\psi_{ir} = \psi\delta_{ir}$.

Because of the symmetry $\sigma_{ij} = (\sigma_{ij} + \sigma_{ji})/2$ of Cauchy's stress tensor $\boldsymbol{\sigma}$ we find the representations

$$\sigma_{ij} = \tfrac{1}{2}(\psi_{ip}\delta_{jq} + \delta_{iq}\psi_{jp})\hat{\sigma}_{pq} \equiv \varphi_{ijpq}\hat{\sigma}_{pq}, \tag{7.5a}$$

$$\hat{\sigma}_{ij} = \tfrac{1}{2}(\psi_{ip}^{(-1)}\delta_{jq} + \psi_{iq}^{(-1)}\delta_{jp})\sigma_{pq} \equiv \Phi_{ijpq}\sigma_{pq} \tag{7.5b}$$

from (7.4). We see that the fourth-order tensors $\boldsymbol{\varphi}$ and $\boldsymbol{\Phi}$ defined as (7.5a) and (7.5b) are only symmetric with respect to two indices:

$$\varphi_{ijpq} = \varphi_{jipq}, \qquad \Phi_{ijpq} = \Phi_{ijqp}, \tag{7.6a,b}$$

that is, the actual net stress tensor $\hat{\boldsymbol{\sigma}}$ is nonsymmetric in the anisotropic damage case:

$$\frac{\hat{\sigma}_{12}}{\hat{\sigma}_{21}} = \frac{\beta}{\alpha}, \qquad \frac{\hat{\sigma}_{23}}{\hat{\sigma}_{32}} = \frac{\gamma}{\beta}, \qquad \frac{\hat{\sigma}_{31}}{\hat{\sigma}_{13}} = \frac{\alpha}{\gamma}. \tag{7.7}$$

This fact is a disadvantage, and it is awkward to use the actual net stress tensor $\hat{\boldsymbol{\sigma}}$ in constitutive equations with a symmetric strain rate tensor $\mathbf{d}$. Therefore, we introduce a *transformed* net stress tensor $\mathbf{t}$ defined by the operation

$$t_{ij} = \tfrac{1}{2}(\hat{\sigma}_{ik}\psi_{kj}^{(-1)} + \psi_{ki}^{(-1)}\hat{\sigma}_{jk}), \tag{7.8}$$

which is *symmetric*. Inserting (7.5b) into (7.8) we have

$$t_{ij} = C_{ijpq}^{(-1)}\sigma_{pq}, \tag{7.9}$$

where

$$C_{ijpq}^{(-1)} = \tfrac{1}{2}(\psi_{ip}^{(-1)}\psi_{jq}^{(-1)} + \psi_{iq}^{(-1)}\psi_{jp}^{(-1)}) \tag{7.10}$$

is a symmetric fourth-order tensor

$$C_{ijpq}^{(-1)} = C_{jipq}^{(-1)} = C_{ijqp}^{(-1)} = C_{pqij}^{(-1)}, \qquad (7.11)$$

which is identical to the tensor D_{ijpq} in (5.73). In the undamaged ($\psi \to \delta$) and total damaged state ($\psi \to 0$) we have

$$C_{ijpq}^{(-1)} \to E_{ijpq} \Rightarrow t_{ij} \to \sigma_{ij} \qquad (7.12)$$

and

$$C_{ijpq}^{(-1)} \to \infty_{ijpq} \Rightarrow t_{ij} \to \infty_{ij} \text{ (singular)}, \qquad (7.13)$$

respectively, where

$$E_{ijpq} = \tfrac{1}{2}(\delta_{ip}\delta_{jq} + \delta_{iq}\delta_{jp}) \qquad (7.14)$$

is the zero power tensor of rank four.

The inverse form of (7.9) is given by

$$\sigma_{ij} = C_{ijpq}t_{pq}, \qquad (7.15)$$

where

$$C_{ijpq} = \tfrac{1}{2}(\psi_{ip}\psi_{jq} + \psi_{iq}\psi_{jp}) \qquad (7.16)$$

is a symmetric fourth-order tensor of continuity,

$$C_{ijpq} = C_{jipq} = C_{ijqp} = C_{pqij}, \qquad (7.17)$$

which is connected with the tensor (7.10) by the relation

$$C_{ijpq}C_{pqkl}^{(-1)} = C_{ijpq}^{(-1)}C_{pqkl} = C_{ijkl}^{(0)} \equiv E_{ijkl}. \qquad (7.18)$$

Because of the symmetry properties (7.11) and (7.17) the fourth-order tensor of continuity (7.16) and its inversion (7.10) can be represented by 6×6 square matrices, which, because of (7.22), have the diagonal forms:

$$C_{ijkl} = \text{diag}\{C_{1111}, C_{2222}, C_{3333}, C_{1212}, C_{2323}, C_{3131}\}, \qquad (7.19\text{a})$$

$$C_{ijkl} = \text{diag}\{\alpha^2, \beta^2, \gamma^2, \tfrac{1}{2}\alpha\beta, \tfrac{1}{2}\beta\gamma, \tfrac{1}{2}\gamma\alpha\}, \qquad (7.19\text{b})$$

and

$$C_{ijkl}^{(-1)} = \text{diag}\{C_{1111}^{(-1)}, C_{2222}^{(-1)}, C_{3333}^{(-1)}, C_{1212}^{(-1)}, C_{2323}^{(-1)}, C_{3131}^{(-1)}\}, \qquad (7.20\text{a})$$

$$C_{ijkl}^{(-1)} = \text{diag}\left\{\frac{1}{\alpha^2}, \frac{1}{\beta^2}, \frac{1}{\gamma^2}, \frac{1}{2\alpha\beta}, \frac{1}{2\beta\gamma}, \frac{1}{2\gamma\alpha}\right\}; \qquad (7.20\text{b})$$

that is, the components of the *pseudo-net stress tensor* **t**, according to (7.9), are given in the following manner:

$$t_{11} = \frac{1}{\alpha^2}\sigma_{11}, \qquad t_{12} = \frac{1}{\alpha\beta}\sigma_{12}, \qquad t_{13} = \frac{1}{\alpha\gamma}\sigma_{13},$$

$$t_{21} = t_{12}, \qquad t_{22} = \frac{1}{\beta^2}\sigma_{22}, \qquad t_{23} = \frac{1}{\beta\gamma}\sigma_{23}, \qquad (7.21)$$

$$t_{31} = t_{13}, \qquad t_{32} = t_{23}, \qquad t_{33} = \frac{1}{\gamma^2}\sigma_{33}.$$

The results of (7.9) and (7.15) can also be found in the following way. Using the linear transformations

$$t_{ij} = \tfrac{1}{2}(\delta_{ir}\psi_{js}^{(-1)} + \psi_{is}^{(-1)}\delta_{jr})\hat{\sigma}_{rs} \qquad (7.22)$$

and

$$\hat{\sigma}_{pq} = \tfrac{1}{2}(\delta_{pr}\psi_{qs} + \delta_{ps}\psi_{qr})t_{rs}, \qquad (7.23)$$

which connect a fictitious symmetric tensor **t** with the actual nonsymmetric net stress tensor $\hat{\boldsymbol{\sigma}}$, we immediately find (7.9) by inserting (7.5b) into (7.22) and (7.15) by inserting (7.23) into (7.5a), respectively.

Because of the nonsymmetric property of the actual net stress tensor we find from (7.23) the decomposition

$$\hat{\sigma}_{pq} = \hat{\sigma}_{(pq)} + \hat{\sigma}_{[pq]}, \qquad (7.24)$$

where the symmetric and antisymmetric parts are given by

$$\hat{\sigma}_{(pq)} = (t_{pr}\psi_{rq} + \psi_{pr}t_{rq})/2 \qquad (7.25)$$

and

$$\hat{\sigma}_{[pq]} = (t_{pr}\psi_{rq} - \psi_{pr}t_{rq})/2, \qquad (7.26)$$

respectively. In the special case of isotropic damage ($\psi_{ij} = \psi\delta_{ij}$) we have $\hat{\sigma}_{(pq)} = \psi t_{pq}$ and $\hat{\sigma}_{[pq]} = 0_{pq}$.

An interpretation of the introduced pseudo-net stress tensor (7.8) can be given in the following way. An alternative form of Cauchy's formula (7.1) is

$$dP_i = \sigma_{ji}\,dS_j, \qquad (7.27)$$

where dP_i is the actual force vector (7.3), and according to (6.26b) we can write

$$dP_i = \sigma_{ji}\psi_{jr}^{(-1)}\,d\hat{S}_r, \qquad (7.28a)$$

or inserting (7.15) we find the relation

$$dP_i = \psi_{ip} t_{pr} d\hat{S}_r, \tag{7.28b}$$

which can be multiplied by $\psi_{ki}^{(-1)}$, so that we have

$$\psi_{ki}^{(-1)} dP_i = t_{kr} d\hat{S}_r, \tag{7.29a}$$

or after changing the indices:

$$\psi_{ik}^{(-1)} dP_k \equiv d\tilde{P}_i = t_{ji} d\hat{S}_j. \tag{7.29b}$$

Comparing (7.27) and (7.29b) we see that (7.29b) can be interpreted as Cauchy's formula for the damaged configuration, which is subjected to the pseudo-force $d\tilde{P}_i \equiv \psi_{ik}^{(-1)} dP_k$ instead of to the actual force dP_i.

Because of the nonsymmetric properties of the "net stress tensor" $\hat{\sigma}$ and the operator φ, , i.e.,

$$\hat{\sigma}_{ij} = \tfrac{1}{2}(\hat{\sigma}_{ij} + \hat{\sigma}_{ji}) + \tfrac{1}{2}(\hat{\sigma}_{ij} - \hat{\sigma}_{ji}) \tag{7.30}$$

and

$$\varphi_{ijpq} = \tfrac{1}{2}(\varphi_{ijpq} + \varphi_{ijqp}) + \tfrac{1}{2}(\varphi_{ijpq} - \varphi_{ijqp}), \tag{7.31}$$

respectively, we find, from (7.5a), the decompositions:

$$\sigma_{ij} = \tfrac{1}{4}(\varphi_{ijpq} + \varphi_{ijqp})(\hat{\sigma}_{pq} + \hat{\sigma}_{qp}) + \tfrac{1}{4}(\varphi_{ijpq} - \varphi_{ijqp})(\hat{\sigma}_{pq} - \hat{\sigma}_{qp}), \tag{7.32a}$$

$$\sigma_{ij} = \tfrac{1}{8}(\psi_{ip}\delta_{jq} + \psi_{jp}\delta_{iq} + \psi_{iq}\delta_{jp} + \psi_{jq}\delta_{ip})(\hat{\sigma}_{pq} + \hat{\sigma}_{qp}),$$

$$+ \tfrac{1}{8}(\psi_{ip}\delta_{jq} + \psi_{jp}\delta_{iq} - \psi_{iq}\delta_{jp} - \psi_{jq}\delta_{ip})(\hat{\sigma}_{pq} - \hat{\sigma}_{qp}). \tag{7.32b}$$

Because of (7.6a), the right-hand sides in (7.32a) and (7.32b) are symmetric with respect to the indices i and j. Furthermore, we see the symmetry with respect to the indices p and q. This fact can be seen immediately from (7.5a). In the special case of isotropic damage, i.e., $\psi_{ij} = \psi\delta_{ij}$ or $\hat{\sigma}_{pq} = \hat{\sigma}_{qp}$, the second term of the right-hand side in (7.32b) vanishes. Then, equation (7.32b) is identical to those formulated by Rabotnov (1969).

In a similar way, from (7.5b) we find the decomposition of the net stress tensor $\hat{\sigma}$ into a symmetric and an antisymmetric part:

$$\hat{\sigma}_{ij} = \tfrac{1}{2}(\Phi_{ijpq} + \Phi_{jipq})\sigma_{pq} + \tfrac{1}{2}(\Phi_{ijpq} - \Phi_{jipq})\sigma_{pq}, \tag{7.33a}$$

$$\hat{\sigma}_{ij} = \tfrac{1}{4}(\psi_{ip}^{(-1)}\delta_{jq} + \psi_{iq}^{(-1)}\delta_{jp} + \psi_{jp}^{(-1)}\delta_{iq} + \psi_{jq}^{(-1)}\delta_{ip})\sigma_{pq}$$

$$+ \tfrac{1}{4}(\psi_{ip}^{(-1)}\delta_{jq} + \psi_{iq}^{(-1)}\delta_{jp} - \psi_{jp}^{(-1)}\delta_{iq} - \psi_{jq}^{(-1)}\delta_{ip})\sigma_{pq}. \tag{7.33b}$$

The results given above can be expressed by the damage tensor $\boldsymbol{\omega}$. For instance, from (6.27a) and (6.27b) in connection with (6.19) and because of $\psi_{ij} \equiv \delta_{ij} - \omega_{ij}$ we have

$$\psi_{ijpq} = \delta_{ijpq} - \omega_{kij}\varepsilon_{kpq} \equiv (\varepsilon_{kij} - \omega_{kij})\varepsilon_{kpq}, \tag{7.34a}$$

$$\psi_{ijpq} = (\delta_{ip}\delta_{jq} - \delta_{iq}\delta_{jp})(1 - \omega_{rr}) + (\omega_{ip}\delta_{jq} - \omega_{iq}\delta_{jp}) + (\delta_{ip}\omega_{jq} - \delta_{iq}\omega_{jp}). \tag{7.34b}$$

Furthermore, instead of (7.5a) and (7.32b) we find

$$\sigma_{ij} = \tfrac{1}{2}[\delta_{ip}\delta_{jq} + \delta_{iq}\delta_{jp} - (\omega_{ip}\delta_{jq} + \delta_{iq}\omega_{jp})]\hat{\sigma}_{pq} \tag{7.35a}$$

and

$$\sigma_{ij} = \tfrac{1}{2}(\hat{\sigma}_{ij} + \hat{\sigma}_{ji}) - \tfrac{1}{8}(\omega_{ip}\delta_{jq} + \delta_{iq}\omega_{jp} + \omega_{iq}\delta_{jp} + \delta_{ip}\omega_{jq})(\hat{\sigma}_{pq} + \hat{\sigma}_{qp})$$
$$- \tfrac{1}{8}(\omega_{ip}\delta_{jq} + \delta_{iq}\omega_{jp} - \omega_{iq}\delta_{jp} - \delta_{ip}\omega_{jq})(\hat{\sigma}_{pq} - \hat{\sigma}_{qp}). \tag{7.35b}$$

By using the inverse

$$\psi_{ir}^{(-1)} \equiv \frac{1}{2\det(\psi)}\varepsilon_{rqp}\varepsilon_{ikl}\psi_{pk}\psi_{ql} \tag{7.36}$$

and because of the symmetry $\sigma_{ij} = (\sigma_{ij} + \sigma_{ji})/2$, we find the following relations for the net stress tensor:

$$\hat{\sigma}_{ij} = \frac{1}{2\det(\boldsymbol{\delta} - \boldsymbol{\omega})}[(\delta_{is}\delta_{jt} + \delta_{it}\delta_{js})(1 - \omega_{rr}) + (\omega_{is}\delta_{jt} + \omega_{it}\delta_{js})$$
$$+ \tfrac{1}{2}\varepsilon_{ikl}(\varepsilon_{spq}\delta_{jt} + \varepsilon_{tpq}\delta_{js})\omega_{pk}\omega_{ql}]\sigma_{st}, \tag{7.37a}$$

$$= \frac{1}{\det(\boldsymbol{\delta} - \boldsymbol{\omega})}[(1 - \omega_{rr})\sigma_{ij} + \omega_{ir}\sigma_{rj} + \tfrac{1}{2}\varepsilon_{ikl}\varepsilon_{spq}\omega_{pk}\omega_{ql}\sigma_{sj}], \tag{7.37b}$$

$$= \frac{1}{\det(\boldsymbol{\delta} - \boldsymbol{\omega})}\{[1 - J_1(\boldsymbol{\omega}) - J_2(\boldsymbol{\omega})]\sigma_{ij} + [1 - J_1(\boldsymbol{\omega})]\omega_{ir}\sigma_{rj} + \omega_{ir}^{(2)}\sigma_{rj}\}, \tag{7.37c}$$

where

$$J_1(\boldsymbol{\omega}) \equiv \delta_{ij}\omega_{ji}, \qquad J_2(\boldsymbol{\omega}) \equiv \tfrac{1}{2}(\omega_{ij}\omega_{ji} - \omega_{ii}\omega_{jj}) \tag{7.38a,b}$$

are invariants of the damage tensor $\boldsymbol{\omega}$.

Finally, we consider Cauchy's stress equations of equilibrium,

$$\sigma_{ji,j} = 0_i, \tag{7.39}$$

in the absence of the body forces. Then by using transformation (7.4), we have the equilibrium equations in the net stresses:

$$\hat{\sigma}_{ri}\psi_{jr,j} + \psi_{jr}\hat{\sigma}_{ri,j} = 0_i. \tag{7.40}$$

The symmetry of Cauchy's stress tensor ($\sigma_{ij} = \sigma_{ji}$) resulting from moment equilibrium yields the condition

$$\psi_{ip}\hat{\sigma}_{pj} = \psi_{jq}\hat{\sigma}_{qi} \quad \text{or} \quad \hat{\sigma}_{ij} = \psi_{iq}^{(-1)}\psi_{jp}\hat{\sigma}_{pq}, \tag{7.41a,b}$$

which states, that the net stress tensor is nonsymmetric. From (7.41b) we find the decomposition into a symmetric part and an antisymmetric one:

$$\hat{\sigma}_{ij} = \tfrac{1}{4}(\psi_{iq}^{(-1)}\psi_{jp} + \psi_{ip}^{(-1)}\psi_{jq})(\hat{\sigma}_{pq} + \hat{\sigma}_{qp}) + \tfrac{1}{4}(\psi_{iq}^{(-1)}\psi_{jp} - \psi_{ip}^{(-1)}\psi_{jq})(\hat{\sigma}_{pq} + \hat{\sigma}_{qp}). \tag{7.42}$$

For the isotropic damage case ($\psi_{ij} = \psi\delta_{ij}$), (7.42) is equal to the decomposition

$$\hat{\sigma}_{ij} = (\hat{\sigma}_{ji} + \hat{\sigma}_{ij})/2 + (\hat{\sigma}_{ji} - \hat{\sigma}_{ij})/2,$$

i.e., the net stress tensor is symmetric ($\hat{\sigma}_{ij} = \hat{\sigma}_{ji}$) in this special case only.

VIII. Comparison with Own Experiments

Apart from experimental data taken from literature, the author and his coworkers have also taken results from their own experimental measurements in order to examine the validity of a mathematical model, for instance, experiments by Waniewski (1984, 1985), Betten and Waniewski (1986, 1989, 1990, 1991, 1995), and Betten *et al.* (1990, 1995), to name but a few. In the following some of these experiments are explained in more detail.

To justify the *simplified theory* based on the *mapped stress tensor* (2.25) many tests were performed by Betten and Waniewski (1989) on thin-walled tubes (Figure 8) of austenitic, chromium-nickel steel at 873 K and of pure copper at 573 K.

The tubes are loaded under combined tension, torsion, and internal pressure. The anisotropy of the material is entirely involved in the fourth-

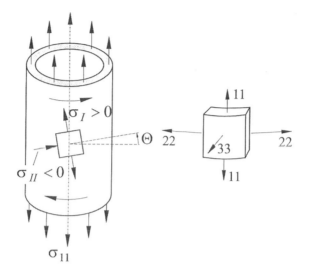

FIG. 8. Thin-walled tube loaded under combined tension, torsion, and internal pressure.

rank tensor $\boldsymbol{\beta}$ in (2.25), the components of which are related to the experimental data. From these experiments we could illustrate that the *simplified theory* is very useful to describe the creep process under non-proportional multiaxial load paths, assuming different inclination θ between load directions and the orthonormal frame of the material (Figure 8). In the following this *simplified theory* will be illustrated in some detail.

In the linear representation

$$\tau_{ij} - \beta_{ijkl}\sigma_{kl},\tag{2.25}$$

the fourth-rank tensor $\boldsymbol{\beta}$ transforms the *anisotropic* stress state $\boldsymbol{\sigma}$ of the *actual material* on the *equivalent isotropic* stress state $\boldsymbol{\tau}$ of the *fictitious material* (Betten, 1981a,b). The irreducible *basic* invariants of the *mapped stress tensor* (2.25) are given, by analogy with (2.15), as

$$S_v(\tau) \equiv \operatorname{tr} \boldsymbol{\tau}^v \equiv \tau_{kk}^{(v)}; \qquad v = 1, 2, 3.\tag{8.1}$$

Alternatively, the set of the three irreducible *principal* invariants

$$J_1(\tau) \equiv \tau_{ii}, \qquad J_2(\tau) \equiv -\tau_{i[i]}\tau_{j[j]}, \qquad J_3(\tau) \equiv \tau_{i[i]}\tau_{j[j]}\tau_{k[k]}\tag{8.2a,b,c}$$

form the *integrity basis* of the tensor $\boldsymbol{\tau}$ in (2.25). The invariants (8.2b,c) can be determined by performing the operation of alternation and are contained in the set (5.42) as special expressions foer $n = 3$. The multiaxial secondary creep behavior of anisotropic materials can be described by using the invariants (8.1) or (8.2a,b,c) in the creep potential

$$F = F[S_\nu(\tau)] \quad \text{or} \quad F = F[J_\nu(\tau)], \qquad \nu = 1, 2, 3, \qquad (8.3a,b)$$

instead of the invariants of the actual stress tensor $\boldsymbol{\sigma}$ as has been employed in constitutive eq. (2.21).

Assuming *incompressibility* (Betten and Waniewski, 1989), the creep potential has the form

$$F = F[J_2(\tau'), J_3(\tau')], \qquad (8.4)$$

where, by analogy with (4.16b,c),

$$J_2(\tau') \equiv \tau'_{ij}\tau'_{ji}/2 \quad \text{and} \quad J_3(\tau') \equiv \tau'_{ij}\tau'_{jk}\tau'_{ki}/3 \qquad (8.5a,b)$$

are the principal invariants of the deviator

$$\tau'_{ij} = \beta'_{\{ij\}pq}\sigma'_{pq} \qquad (8.6)$$

of the mapped stress tensor (2.25). The fourth-rank tensor

$$\beta'_{\{ij\}pq} = \beta_{ijpq} - \beta_{kkpq}\delta_{ij}/3 \qquad (8.7)$$

is deviatoric with respect to the first index pair $\{ij\}$.

Starting from the creep potential (8.4) one can find the constitutive equation by using the *anisotropic flow rule*

$$d_{ij} = \lambda[\partial F(\tau')/\tau_{pq}]J_{pqij} \equiv \dot{\gamma}_{pq}J_{pq\{ij\}} \qquad (8.8)$$

where Jacobi's matrix is defined as

$$J_{pqij} \equiv \partial\tau'_{pq}/\partial\sigma_{ij} = \beta'_{pq\{ij\}}, \qquad (8.9)$$

and $\dot{\gamma}_{pq}$ are the coordinates of the *steady-state creep rate tensor* to be specified in the fictious state.

In a fictitious creep state, defined by (8.6) and by the equality of the equivalent isochoric creep rates in both states,

$$\dot{\gamma} \equiv d, \qquad (8.10)$$

we have, by analogy of Norton's creep law (4.1b), the uniaxial fictious relation

$$\dot{\gamma} = K^*\tau^m \equiv d. \qquad (8.11)$$

The equivalent fictitious isotropic creep stress τ in (8.11) can be determined by the hypothesis of the equivalent dissipation rate $\dot{D}$. Thus, in connection with (8.10), we require

$$\tau\dot{\gamma} = \tau d \overset{!}{=} \sigma'_{ij} d_{ji} \equiv \dot{D}. \tag{8.12}$$

From the flow rule (8.8), combined with the relations (8.11) and (8.12), Betten (1981a,b) finally obtains the tensorial constitutive equation

$$d_{ij} = \Phi\beta'_{pq\{ij\}}\left(\frac{\partial F}{\partial J_2}\tau'_{pq} + \frac{\partial F}{\partial J_3}\tau''_{pq}\right), \tag{8.13}$$

in which the scalar function Φ is defined by

$$\Phi \equiv \frac{1}{2}K^*\left\{3\bigg/\left[\left(\frac{\partial F}{\partial J_2}\right)_V + \frac{1}{3}\left(\frac{\partial F}{\partial J_3}\right)_V\right]\right\}^{(m+1)/2}$$
$$\times\left[\frac{\partial F}{\partial J_2}J_2(\tau') + \frac{3}{2}\frac{\partial F}{\partial J_3}J_3(\tau')\right]^{(m-1)/2}. \tag{8.14}$$

The index V, appended to the round brackets in (8.14), indicates the equivalent fictitious stress state $(\tau_{ij})_V \equiv \text{diag}\{\tau_{11} \equiv \tau, 0, 0\}$. Contrary to (8.7), the tensor

$$\beta'_{pq\{ij\}} = \beta_{pqij} - \beta_{pqkk}\delta_{ij} \tag{8.15}$$

in (8.13) is deviatoric with respect to the second index pair $\{ij\}$. By analogy of (5.33) the tensor

$$\tau''_{pq} \equiv \partial J_3(\tau')/\partial\tau_{pq} = \tau'_{pr}\tau'_{rq} - 2J_2(\tau')\delta_{pq}/3 \tag{8.16}$$

in (8.13) is the deviator of the square of the reduced stress $\tau'_{pq} \equiv \tau_{pq} - \tau_{rr}\delta_{pq}/3$.

Inserting (8.13), together with (8.14), in (8.12) we obtain the rate of dissipation of creep energy:

$$\dot{D} = \left[2\frac{\partial F}{\partial J_2}J_2(\tau') + 3\frac{\partial F}{\partial J_3}J_3(\tau')\right]\Phi. \tag{8.17}$$

In the isotropic special case, characterized by $\beta_{pqij} = (\delta_{pi}\delta_{qj} + \delta_{pj}\delta_{qi})/2$, $K^* \to K$, $m \to n$, $\tau'_{ij} \to \sigma'_{ij}$, and $\tau''_{ij} \to \sigma''_{ij}$, the constitutive equation (8.13), together with (8.14), immediately lead to the corresponding tensorial generalization of Norton's creep law in Section IV.A.

To verify the mapped stress tensor concept for incompressible solids, Betten and Waniewski (1989) assumed the creep potential $F = J_2(\tau') = \tau^2/3$

TABLE 6
EXPERIMENTAL DATA DUE TO BETTEN AND WANIEWSKI (1989)

Material	T	m	K^*	ω_I	ω_{II}	ω_{III}
Austenitic steel	873 K	2.80	1.05e-11	1.11	0.90	1.17
Pure copper	573 K	5.44	1.5e-13	1.09	0.92	1.20

of Mises type and described the orthotropic case by the tensor (8.7) specified as

$$\beta'_{\{ij\}pql} \equiv \tfrac{1}{2}(\omega_{ip}\omega_{jq} + \omega_{iq}\omega_{jp}) - \tfrac{1}{3}\omega_{pq}^{(2)}\delta_{ij}, \tag{8.18}$$

where $\omega_{ij} = \omega_{ji}$ is a second-order tensor of anisotropy with the principal values ω_I, ω_{II}, ω_{III}. These values and the essential creep parameters K^*, m involved in the constitutive equation (8.13) are related to experimental data. Note that the tensor ω_{ij} in (8.18) should not be confused with the damage tensor in (2.34a,b). Some experimental results obtained by Betten and Waniewski (1989) are listed in Table 6.

Plastic prestrain can produce anisotropy in materials that are initially isotropic. This anisotropy strongly influences the creep behavior of such materials. Nonlinear constitutive equations with two argument tensors for the secondary creep behavior have been proposed, based on the representation theory of tensor functions:

$$^{\theta}d_{ij} = f_{ij}(\boldsymbol{\sigma}', \mathbf{A}') = \frac{1}{2}\sum_{\lambda,\mu=0}^{2}\varphi_{[\lambda,\mu]}(M_{ij}'^{[\lambda,\mu]} + M_{ji}'^{[\lambda,\mu]}), \tag{8.19}$$

where $\lambda, \mu = 0, 1, 2$, several deviators

$$M_{ij}'^{[\lambda,\mu]} := \sigma_{ik}'^{(\lambda)}A_{kj}'^{(\mu)} - \tfrac{1}{3}\sigma_{pr}'^{(\lambda)}A_{rp}'^{(\mu)}\delta_{ij} \tag{8.20}$$

have been introduced by Betten (1987c), so that the condition of incompressibility ($^{\theta}d_{kk} \equiv 0$) is *a priori* fulfilled. The two argument tensors in (8.19) are the Cauchy stress deviator $\boldsymbol{\sigma}'$ and an appropriately defined second-rank tensor characterizing the *plastic predeformation* as a function of different plastic prestrain paths θ:

$$^{p}e_{ij} := A_{ij} = \begin{pmatrix} A_{11} & A_{12} & 0 \\ A_{12} & -\tfrac{1}{2}A_{11} & 0 \\ 0 & 0 & -\tfrac{1}{2}A_{11} \end{pmatrix} \equiv A_{ij}'. \tag{8.21}$$

The Mises equivalent plastic prestrain ${}^P e_V = {}^P e_M$ can be expressed by the coordinates A_{11} and A_{12} of the tensor (8.21) as:

$$
{}^P e_M := \sqrt{2 A'_{ij} A'_{ji}/3} \equiv \sqrt{A_{11}^2 + 4 A_{12}^2/3}. \tag{8.22}
$$

The creep strain rates ${}^\theta d_{ij}$ in (8.19) depends on the equivalent plastic prestrain (8.22) and the angle

$$
\theta = \tfrac{1}{2} \arctan(A_{12}/A_{11}), \tag{8.23}
$$

where $\theta = 0$ and $\theta = \pi/4$ characterize the plastic prestrain induced by tension and torsion, respectively.

A few experimental tests have been carried out by Betten and Waniewski (1986, 1995) on round specimens of Inconel 617. First, the strain-induced anisotropy was produced at room temperature by several combinations of tension and torsion. After that, the specimens were loaded uniaxially under creep conditions at 1223 K in helium-gas-atmosphere.

For rolled sheet metal, the internal structure generally presents three orthogonal planes of symmetry; the macroscopic behavior is thus ortho-tropic. The initial orthotropy of the material is modified according to the orientation of the principal directions of the irreversible deformation with respect to the rolling direction; this is the phenomenon of anisotropic hardening. The multiaxial anisotropic creep behavior of rolled sheet metals is analyzed within an invariant formulation of the secondary creep consti-tutive equations by Betten *et al.* (1990), developed in the framework of the theory of tensor function representation, as very briefly illustrated in the following.

We consider an orthotropic sheet steel with privileged directions **u**, **v**, and **w** defining the *transverse*, the *rolling*, and the *normal directions*, respectively, where

$$
|\mathbf{u}| = |\mathbf{v}| = |\mathbf{w}| = 1 \quad \text{and} \quad \mathbf{u} \perp \mathbf{v} \perp \mathbf{w} \perp \mathbf{u} \tag{8.24}
$$

are basic vectors of the symmetry of the material. Since **u**, **v**, and **w** are unit vectors and mutually orthogonal, because of (8.24) we have the relation

$$
u_i u_j + v_i v_j + w_i w_j = \delta_{ij}. \tag{8.25}
$$

Thus, we can discard the dyadic product $w_i w_j$ of the set (8.25) in favor of the unit tensor $\boldsymbol{\delta}$, and take the following basic set of *structural tensors* into account:

$$\mathbf{M} = \mathbf{u} \otimes \mathbf{u} \quad \text{or in index notation:} \quad M_{ij} = u_i u_j, \qquad (8.26a)$$

$$\mathbf{N} = \mathbf{v} \otimes \mathbf{v} \quad \text{or in index notation:} \quad N_{ij} = v_i v_j. \qquad (8.26b)$$

In the frame of material's symmetry $(\mathbf{u}, \mathbf{v}, \mathbf{w})$, the *structural tensors* $\mathbf{M}$ and $\mathbf{N}$ are specified by

$$M_{ij} = \begin{pmatrix} 1 & 0 & 0 \\ 0 & 0 & 0 \\ 0 & 0 & 0 \end{pmatrix} \quad \text{and} \quad N_{ij} = \begin{pmatrix} 0 & 0 & 0 \\ 0 & 1 & 0 \\ 0 & 0 & 0 \end{pmatrix}. \qquad (8.27a,b)$$

We suppose that the sheet metal has been subjected to an irreversible multiaxial creep process. For the secondary creep rate $\mathbf{d}$ and for orthotropic material, the constitutive equation has a form

$$d_{ij} = f_{ij}(\boldsymbol{\sigma}, \mathbf{M}, \mathbf{N}), \qquad (8.28)$$

which can be represented by the minimum polynomial

$$d_{ij} = \frac{1}{2} \sum_{\lambda,\mu,\nu=0}^{2} \varphi_{[\lambda,\mu,\nu]} (G_{ij}^{[\lambda,\mu,\nu]} + G_{ji}^{[\lambda,\mu,\nu]}), \qquad (8.29)$$

where the *tensor generators* are formed by matrix products:

$$G_{ij}^{[\lambda,\mu,\nu]} = M_{ip}^{(\lambda)} \sigma_{pq}^{(\mu)} N_{qj}^{(\nu)}, \qquad \lambda, \mu, \nu = 1, 2, 3. \qquad (8.30)$$

The $\varphi_{[\lambda,\mu,\nu]}$ coefficients in (8.29) are scalar-valued functions of the integrity, the elements of which can be found by forming all the irreducible traces of the matrix products (8.30):

$$G_{rr}^{[\lambda,\mu,\nu]} = M_{ip}^{(\lambda)} \sigma_{pq}^{(\mu)} N_{qi}^{(\nu)}, \quad \text{where} \begin{cases} \lambda, \mu, \nu = 1, 2 \\ \text{and} \\ \lambda = 0, \mu = 0 \Rightarrow \nu = 1, 2, 3 \\ \mu = 0, \nu = 0 \Rightarrow \lambda = 1, 2, 3 \\ \nu = 0, \lambda = 0 \Rightarrow \mu = 1, 2, 3 \end{cases} \qquad (8.31)$$

For orthotropic materials the *integrity basis*, proposed in (8.31), may contain redundant elements because of the orthogonal condition (8.25) as has been discussed in more detail by Betten *et al.* (1990). An application of the tensor function theory (8.28) has been specified to derive the constitutive equation in the case of the plane, multiaxial creep tests, performed at elevated temperature, on cruciform flat specimens (Figure 9) subjected to the

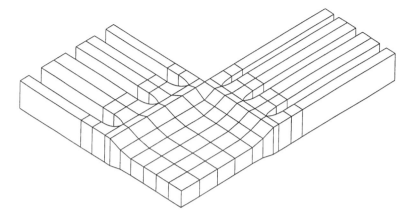

FIG. 9. Biaxial tension cruciform specimen.

two independent mutually perpendicular tension stress states. The biaxial tension method has been improved by Betten *et al.* (1990) in order to determine complex stress state in specimens of orthotropic rolled sheet metal. The shape of the specimen, for instance, the biaxial tension cruciform specimen in Figure 9, has been taken into account with respect to the uniform stress distribution in the central part of the specimen.

In conclusion, the research of Betten *et al.* (1990) has shown that the presented theoretical model for multiaxial anisotropic creep behavior of rolled sheet metal proves that the tensor function representation is a powerful and efficient tool for constructing nonlinear inelastic constitutive equations. Our analysis of off-axes tests on orthotropic solids in the inelastic range has been confirmed by experimental results.

Other experiments have been carried out by Betten *et al.* (1995) in order to predict the influence of creep history, e.g., predamage and preloading, on the further creep behavior after changing the loading direction. From preloading specimens (loading direction x_1) smaller specimens were cut along in several directions ${}^\theta x_1$ and then loaded under creep conditions (Figure 10). The creep tests were performed on flat specimens of austenitic steel X 8 Cr Ni Mo Nb 16 16 at 953 and 973 K and ferritic steel 13 Cr Mo 4 4 at 803 and 823 K.

The experimental results have been utilized in order to determine the material parameters in the tensorial *constitutive equation*,

$$d_{ij} = f_{ij}(\boldsymbol{\sigma}, \mathbf{D}) = \frac{1}{2} \sum_{v,\mu=0}^{2} \psi_{[v,\mu]}(\sigma_{ik}^{(v)} D_{kj}^{(\mu)} + D_{ik}^{(\mu)} \sigma_{kj}^{(v)}), \qquad (4.19)$$

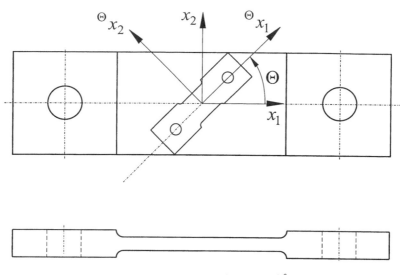

FIG. 10. Loading directions x_1 and $^\theta x_1$.

and *evolutional equation,*

$$\overset{\circ}{D}_{ij} = g_{ij}(\boldsymbol{\sigma}, \mathbf{D}) = \frac{1}{2} \sum_{v,\mu=0}^{2} \eta_{[v,\mu]}(\sigma_{ik}^{(v)} D_{kj}^{(\mu)} + D_{ik}^{(\mu)} \sigma_{kj}^{(v)}), \tag{8.32}$$

where $^\circ$ denotes the *Jaumann derivative.* In contrast with (2.34a,b), the damage state is characterized in (4.19) and (8.32) by the tensor **D** defined in (2.40). In view of polynomial representations of *constitutive* (4.19) and *evolutional* (8.32) *equations* it is more convenient to use the tensor **D** as an argument tensor than the tensorial damage variable **ω** as argued in Section II.

The research of Betten *et al.* (1995) has shown that the *experimental method,* illustrated in Figure 10, is very suitable for finding internal parameters of damaged materials and that the *tensor function theory* has become a powerful tool in order to formulate nonlinear *constitutive* and *evolutional equations* for anisotropic damaged materials in a state of multiaxial stress.

A continuum damage mechanics model for the dislocation creep response associated with the growth of parallel planar mesocracks in initially isotropic materials has been developed by Betten *et al.* (1998). This model describes simultaneously different damage development in tension, compression, torsion, and damage-induced anisotropy, as well as different creep properties in tension, compression, and torsion. The proposed constitutive equation for creep and the damage growth equation contain joint invariants of the stress tensor and the second-order damage tensor constructed by Betten (1982b, 1983a). The material parameters required in the proposed equations have been determined by Betten *et al.* (1998) based on the following experiments:

1. Creep behavior in the primary stage of a tinanium alloy VT 9 at 673 K under uniaxial tension, uniaxial compression or pure torsion;

2. Secondary creep behavior of aluminum alloy AK 4-1 T at temperature of 473 K again under tension, compression, and torsion; and

3. Damage accumulation in two materials in the tertiary creep phase, i.e., again aluminum alloy AK 4-1 T at 473 K and, furthermore, titanium alloy OT 4 at 748 K under proportional and nonproportional loading.

Comparison of the theory with experimental results showed satisfactory agreement.

Extensive experimental research is currently being carried out by Betten *et al.* (1999) concerning the creep behavior of materials with different damage in tension and compression, namely, the incorporation of the microstructural creep characteristics of polycrystalline materials in the damage model and, furthermore, the description of creep behavior of initially anisotropic materials (submitted for publication).

Some tensile tests have been recently carried out by the author and coworkers on aluminum alloy AA 7075 T 7351 at room temperature, as illustrated in Figures 11a and 11b. If $E_n = (1 - \omega)E_0$ is defined as the *effective* elasticity modulus of the damaged material, the values of the damage parameter may be derived from measurements of E_n according to $\omega = 1 - E_n/E$. The curves in Figure 11c can be interpreted as the evolution of damage with permanent strains of loading/unloading tests. The linear decreasing of the elasticity modulus with damage, $E_n = (1 - \omega)E_0$, is based on the assumption of the hypothesis of *strain equivalence*, while the relation $\omega^* = 1 - (E_n/E_n)^{1/2}$ in Figure 11c can be derived, if the hypothesis of *energy equivalence* is assumed (Chow and Lu, 1992; Skrzypek and Ganczarski, 1999).

Note that the symbols $\triangle\triangle\triangle$ and $+++$ in Figure 11c indicate calculated points from the formula $\omega = 1 - E_n/E_0$ and $\omega^* = 1 - (E_n/E_0)^{1/2}$, respectively, where E_i, $i = 0, 1, 2, \ldots, n$, are experimental data taken from Figure 11b. The solid curves in Figure 11c are the best approximations to the calculated points based on *cubic Bézier-splines*.

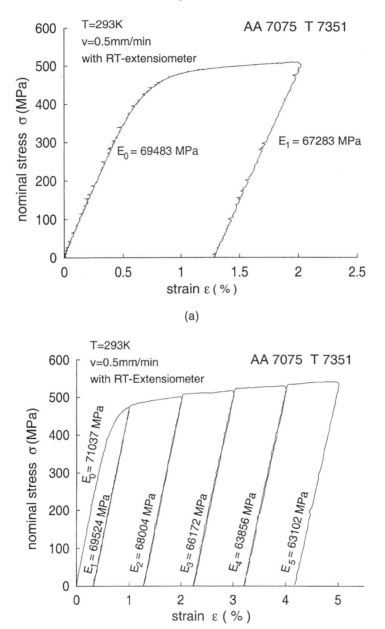

FIG. 11. (a) Tensile test on aluminum alloy at room temperature. (b) Loading/unloading tests on aluminum alloy at room temperature.

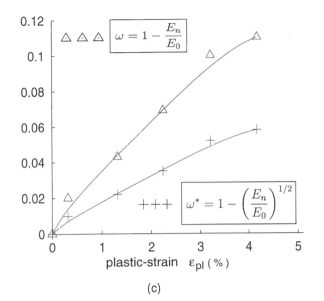

FIG. 11. (c) Evolution of damage with permanent strains of loading/unloading tests in Fig. 11b.

In Figure 12a the Ramberg–Osgood (1943) stress-strain curve (4.45a) and the modified relation (4.45b) are compared with the experimental results from Figure 11a. One can see that the Ramberg–Osgood relation does not agree with the tension test on aluminum alloy. Thus, a modification is necessary (Betten, 1989a).

Finally, the *hardening* of aluminium alloy AA 7075 T 7351 at room temperature can be expressed by the simple formula $\sigma = \sigma_F + k\varepsilon_{pl}^n$ as illustrated in Figure 12b, where the yield stress $\sigma_F = 350\,MPa$, the hardening coefficient $k = 323\,MPa$, and the hardening exponent $n = 0.17$ have been determined based on the experimental results from Figure 11a by using the *nonlinear Maquardt–Levenberg algorithm* with MAPLE. This algorithm has also been applied in order to determine the parameters in Figure 12a.

For engineering applications it is very important to generalize nonlinear material laws found in experiments to multiaxial states of stress (Betten, 1989a). This can be achieved by using interpolation methods for tensor functions (Betten, 1984c), as has been pointed out in detail in Section IV. For instance, the Ramberg–Osgood (1943) stress-strain relation (4.45a) is generalized in Section IV.D to a tensor-valued constitutive equation (4.46) or (4.47). The modified relation (4.45b), which agrees better with the tension

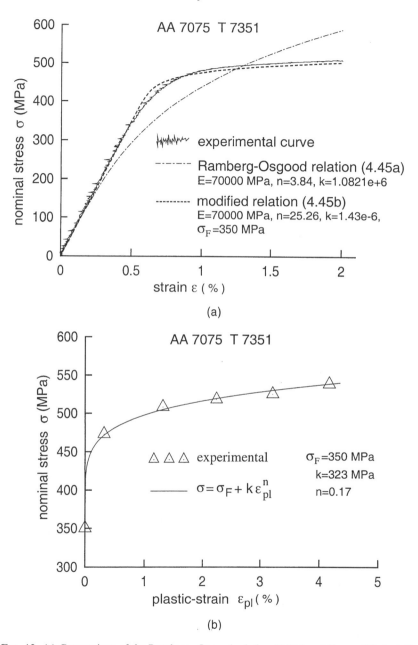

(a)

(b)

FIG. 12. (a) Comparison of the Ramberg–Osgood relation (4.45a) and the modified relation (4.45b) with the experimental curve from Figure 11a. (b) Hardening parameters k and n for aluminum alloy at room temperature.

test on aluminum alloy than the Ramberg–Osgood relation (Figure 12a), can be generalized to multiaxial states of stress in the same way.

References

Allirot, D., and Boehler, J. P. (1982). Yielding and failure of transversely isotropic solids — Part I. Experiments. *Res. Mechanica* **4**, 97–113.

Andrade, E. N. da Costa, (1910). The viscous flow in metals and allied phenomena. *Proc. Roy. Soc. London A,* **84**, 1–12.

Astarita, G. (1979). Why do we search for constitutive equations? Presented at the Golden Jubilee Meeting of the Society of Rheology, Boston, MA, 1979.

Avula, X. J. R. (1987). Mathematical modelling. In *Encyclopedia of Physical Science and Technology* **7**, 719–728.

Betten, J. (1973). Die Traglasttheorie der Statik als mathematisches Modell. *Schweizerische Bauzeitung* **91**, 6–9.

Betten, J. (1975a). Zur Verallgemeinerung der Invariantentheorie in der Kriechmechanik. *Rheol. Acta* **14**, 715–720.

Betten, J. (1975b). Bemerkungen zum Versuch von Hohenemser. *Z. Angew. Math. Mech.* **55**, 149–158.

Betten, J. (1975c). Zum Traglastverfahren bei nichtlinearem Stoffgesetz. *Ing.-Archiv.* **44**, 199–207.

Betten, J. (1976). Plastische Anisotropie und Bauschinger-Effekt–allgemeine Formulierung und Vergleich mit experimentell ermittelten Fließortkurven. *Acta Mechanica* **25**, 79–94.

Betten, J. (1981a). Creep theory of anisotropic solids. *J. Rheology* **25**, 565–581. Presented at the Golden Jubilee Meeting of the Society of Rheology, Boston, MA, 1979.

Betten, J. (1981b). Representation of constitutive equations in creep mechanics of isotropic and anisotropic materials. In *Creep in Structures* (A. R. S. Ponter and D. R. Hayhurst, eds.), pp. 179–201. Springer-Verlag, Berlin.

Betten, J. (1982a). Integrity basis for a second-order and a fourth-order tensor. *Int. J. Math. Math. Sci.* **5**, 87–96.

Betten, J. (1982b). Net-stress analysis in creep mechanics. *Ingenieur Archiv.* **52**, 405–419. Presented at the Second German–Polish Symposium on Inelastic Solids and Structures, Bad Honnef, 1981.

Betten, J. (1982c). Pressure-dependent yield behaviour of isotropic and anisotropic materials. In *Deformation and Failure of Granular Materials* (P. V. Vermeer and H. J. Luger, eds.), pp. 81–89. A. A. Balkema, Rotterdam.

Betten, J. (1983a). Damage tensors in continuum mechanics. *J. Mécanique Théorique Appliquée* **2**, 13–32. Presented at Euromech Colloquium 147 on Damage Mechanics, Paris-VI, Cachan, 1981.

Betten, J. (1983b). Damage tensors and tensors of continuity in stress analysis. *Z. Naturforschung* **38a**, 1383–1390.

Betten, J. (1984a). Constitutive equations of isotropic and anisotropic materials in the secondary and tertiary creep stage. In *Creep and Fracture of Engineering Materials and Structures* (B. Wilshire and D. R. J. Owen, eds.), Part II, pp. 1291–1305. Pineridge Press, Swansea.

Betten, J. (1984b). Interpolation methods for tensor functions. In *Mathematical Modelling in Science and Technology* (X. J. R. Avula *et al.*, eds.), pp. 52–57. New York: Pergamon Press.

Betten, J. (1985a). The classical plastic potential theory in comparison with the tensor function theory. *Engineering Fracture Mechanics* **21**, 641–652. Presented at the Int. Symposium on Plasticity Today, Udine, 1983.

Betten, J. (1985b). *Elastizitäts- und Plastizitätslehre*, 2nd ed. Vieweg-Verlag, Braunschweig, Wiesbaden.

Betten, J. (1986a). Beitrag zur tensoriellen Verallgemeinerung einachsiger Stoffgesetze. *Z. Angew. Math. Mech.* **66**, 577–581.

Betten, J. (1986b). Applications of tensor functions to the formulation of constitutive equations involving damage and initial anisotropy. *Engng. Frac. Mech.* **25**, 573–584. Presented at the IUTAM Symposium on Mechanics of Damage and Fatigue, Haifa and Tel Aviv, Israel, 1985.

Betten, J, (1987a). Irreducible invariants of fourth-order tensors. In *Mathematical Modelling in Science and Technology* (X. J. R. Avula *et al.*, eds.), pp. 29–33. Pergamon Journals Limited, Exeter.

Betten, J. (1987b). Tensor functions involving second-order and fourth-order argument tensors. In *Applications of Tensor Functions in Solid Mechanics* (J. P. Boehler, ed), Springer-Verlag, Berlin.

Betten, J. (1987c). *Tensorrechnung für Ingenieure.* (B. G. TeubnerVerlag, ed.), Stuttgart.

Betten, J. (1988a). Applications of tensor functions to the formulation of yield criteria for anisotropic materials. *Int. J. Plasticity* **4**, 29–46.

Betten, J. (1988b). Mathematical modelling of materials behaviour. In *Mathematical Modelling in Science and Technology* (X. J. R. Avula *et al.*, eds.), pp. 702–708. Pergamon Press, Oxford.

Betten, J. (1989a). Generalization of nonlinear material laws found in experiments to multi-axial states of stress. *Eur. J. Mech. A Solids* **8**, 325–339. Presented at Euromech Colloquium 244 on Experimental Analysis of Nonlinear Problems in Solid Mechanics, Poznan, Poland, 1988.

Betten, J. (1989b). Recent advances in mathematical modelling of materials behaviour. Presented as a general lecture at the Seventh Int. Conf. on Mathematical and Computer Modelling, Chicago, 1989, published in the proceedings (X. J. R. Avula, ed.).

Betten, J. (1991a). Applications of tensor functions in creep mechanics. In *Creep in Structures* (M. Życzkowski, ed.), pp. 3–22. Springer-Verlag, Berlin. Presented as a "general lecture" at the IUTAM Symposium Cracow, Poland, 1988.

Betten, J. (1991b). Recent advances in applications of tensor functions in solid mechanics. *Adv. Mechanics (Uspechi Mechaniki)* **14**, 79–109.

Betten, J. (1992). Applications of tensor functions in continuum damage mechanics. *Int. J. Damage Mechanics* **1**, 47–59.

Betten, J. (1993). *Kontinuumsmechanik.* Springer-Verlag, Berlin.

Betten, J. (1998). Anwendungen von Tensorfunktionen in der Kontinuumsmechanik anisotroper Materialien. *Z. Angew. Math. Mech.* **78**, 507–521, Hauptvortrag (Invited Plenary Lecture) auf der 75, GAMM Tagung in Regensburg, 1997.

Betten, J. (2001). The eigenvalue problem of a fourth-order tensor (in preparation).

Betten, J., and Borrmann, M. (1984). Einfluß der plastischen Kompressibilität und des Strength-Differential-Effektes auf das Fließverhalten von Sinter- und Polymerwerkstoffen. *Rheol. Acta* **23**, 109–116.

Betten, J., and Butters, T. (1990). Rotationssymmetrisches Kriechbeulen dünnwandiger Rreiszylinderschalen im primären Kriechbereich. *Forschung im Ingenieurwesen* **56**, 84–89.

Betten, J., and Helisch, W. (1995a). Integrity bases for a fourth-rank tensor. In *IUTAM Symp. Anisotropy, Inhomogeneity, and Nonlinearity in Solid Mechanics* (D. F. Parker and A. H.

England, eds.), pp. 37–42. Kluwer Academic Publishers, Netherlands.

Betten, J., and Helisch, W. (1995b). Simultaninvarianten bei Systemen zwei- und vierstufiger Tensoren. *Z. Angew. Math. Mech.* **75**, 753–759.

Betten, J., and Helisch, W. (1996). Tensorgeneratoren bei Systemen von Tensoren zweiter und vierter Stufe. *Z. Angew. Math. Mech.* **76**, 87–92.

Betten, J., and Meydanli, S. C. (1990). Materialgleichungen, DFG Report Be 766/12-2, *Z. Angew. Math. Mech.* **75** (1995), 181–182.

Betten, J., and Waniewski, M. (1986). Einfluß der plastischen Anisotropie auf das sekundäre Kriechverhalten inkompressibler Werkstoffe. *Rheol. Acta* **25**, 166–174.

Betten, J., and Waniewski, M. (1989). Multiaxial creep behaviour of anisotropic materials. *Arch. Mech.* **41**, 679–695. Presented at the Fourth Polish–German Symposium on Mechanics of Inelastic Solids and Structures, Mogilany, Poland, 1987.

Betten, J., and Waniewski, M. (1990) Stress-path influence on multiaxial creep behaviour due to multiple load changes. Presented at the 4th Int. Conf. on Creep and Fracture, Swansea, 1990, published in the proceedings (B. Wilshire, ed).

Betten, J., and Waniewski, M. (1991). Biaxial tension creep test of rolled sheet-metals. *Arch. Mech.* **43**. Presented at III Sympozjum nt. Zagadnién Pelzania Materialow, Bialystok, Poland, 1989.

Betten, J., and Waniewski, M. (1995). Tensorielle Stoffgleichungen zur Beschreibung des anisotropen Kriechverhaltens isotroper Stoffe nach plastischer Verformung. *Z. Angew. Math. Mech.* **75**, 831–845.

Betten, J., and Waniewski, M. (1998). The strain path dependence of multiaxial cyclic hardening behaviour. *Forschung im Ingenieurwesen* **64**, 231–244.

Betten, J., Borrmann, M., and Butters, T. (1989). Materialgleichungen zur Beschreibung des primären Kriechverhaltens innendruckbeanspruchter Zylinderschalen aus isotropem Werkstoff. *Ingenieur-Archiv.* **60**, 99–109.

Betten, J., Breitbach, G., and Waniewski, M. (1990). Multiaxial anisotropic creep behaviour of rolled sheet-metals. *Z. Angew. Math. Mech.* **70**, 371–379.

Betten, J., El-Magd, E., Meydanli, S. C., and Palmen, P. (1995). Anisotropic damage growth under multi-axial stress (theory and experiments). *Arch. Appl. Mech.* **65**, 110–120, 121–132.

Betten, J., Sklepus, S., and Zolochevsky, A. (1998). A creep damage model for initially isotropic materials with different properties in tension and compression. *Engng. Fracture Mechanics* **59**, 623–641.

Betten, J., Sklepus, S., and Zolochevsky, A. (1999). A microcrack description of creep damage in crystalline solids with different behaviour in tension and compression. *Int. J. Damage Mechanics* **8**, 197–232.

Bodner, S. R., and Hashin, Z. (eds.) (1986). *Mechanics of Damage and Fatigue*. Pergamon Press, New York.

Boehler, J. P. (ed.). (1987). *Applications of Tensor Functions in Solid Mechanics*. Springer-Verlag, Wien, New York.

Boehler, J. P., and Sawczuk A. (1977). On yielding of oriented solids, *Acta Mechanica* **27**, 185–206.

Brown, S. G. R., Evans, R. W., and Wilshire, B. (1986). Exponential descriptions of normal creep curves. *Scripta Metallurgica* **20**, 855–860.

Chaboche, J. L. (1984). Anisotropic creep damage in the framework of continuum damage mechanics. *Nucl. Eng. Des.* **79**, 304–319.

Chow, C. L., and Lu, T. J. (1992). An analytical and experimental study of mixed-mode ductile fracture under nonproportional loading. *J. Damage Mechanics* **1**, 191–236.

Chrzanowski, M. (1973). *The Description of Metallic Creep in the Light of Damage Hypothesis and Strain Hardening*. Diss. Hab., Politechnika Krakowska, Krakow.

Chrzanowski, M. (1976). Use of the damage concept in describing creep-fatigue interaction under prescribed stress. *Int. J. Mech. Sci.* **18**, 69–73.

Edward, G. H., and Ashby, M. F. (1979). Intergranular fracture during power-law creep. *Acta Metall.* **27**, 1505–1518.

Evans, H. E. (1984). *Mechanisms of Creep Rupture.* Elsevier, London.

Fitzgerald, J. E. (1980). A tensorial Hencky measure of strain and strain rate for finite deformations. *J. Appl. Phys.* **51**, 5111–5115.

Goel, R. P. (1975). On the creep rupture of a tube and a sphere. *J. Appl. Mech. Trans. ASME* **43**, 625–629.

Hayhurst, D. R., Trampczynski, W. A., and Leckie, F. A. (1980). Creep rupture and non-proportional loading. *Acta Metall.* **28**, 1171–1183.

Helisch, W. (1993). *Invariantensysteme und Tensorgeneratoren bei Materialtensoren zweiter und vierter Stufe.* Dr.-Ing. Dissertation, RWTH-Aachen.

Ikegami, K. (1975). An historical perspective of the experimental study of subsequent yield surfaces for metals. *J. Soc. Mat. Sci.* **24**, Part 1: 491–505, Part 2: 709–719.

Ilschner, B. (1973). *Hochtemperatur-Plastizität.* Springer-Verlag, Berlin.

Jakowluk, A. (1993). *Process of Creep and Fatigue in Materials.* Wydawnictwa Naukowo-Techniczne, Warszawa. (in Polish)

Johnson, A. E. (1960). Complex-stress creep of metals. *Metallurg. Rev.* **5**, 447–506.

Kachanov, L. M. (1958). On the time to failure under creep conditions (in Russian). *Izv. Akad. Nauk USSR Otd. Tekh. Nauk* **8**, 26–31.

Kachanov, L. M. (1986). *Introduction to Continuum Damage Mechanics.* Martinus Nijhoff Publishers, Dordrecht.

Krajcinovic, D. (1983). Constitutive equations for damaging materials. *J. Appl. Mech.* **50**, 355–360.

Krajcinovic, D. (1996). *Damage Mechanics.* Elsevier North-Holland, Amsterdam.

Krajcinovic, D., and Lemaitre, J. (eds.). (1987). *Continuum Damage Mechanics.* Springer-Verlag, Wien, New York.

Leckie, F. A., and Hayhurst, D. R. (1977). Constitutive equations for creep rupture. *Acta Metallurg.* **25**, 1059–1070.

Leckie, F. A., and Ponter, A. R. S. (1974). On the state variable description of creeping materials. *Ing.-Archiv.* **43**, 158–167.

Litewka, A., and Hult, J. (1989). One parameter CDM model for creep rupture prediction. *Eur. J. Mech. A Solids* **8**, 185–200.

Litewka, A., and Morzyńska, J. (1985). On yielding and fracture of damaged materials. In *Structural Mechanics in Reactor Technology* (J. Stalpaert, ed.), pp. 281–286. North-Holland, Amsterdam.

Litewka, A., and Morzyńska, J. (1989). Theoretical and experimental study of fracture for a damaged solid. *Res. Mechanica* **27**, 259–272.

Mazilu, P., and Meyers, A. (1985). Yield surfaces description of isotropic materials after cold prestrain. *Ing.-Archiv.* **55**, 213–220.

Meydanli, S. C. (1994). *Tensorielle Beschreibung der Kriechschädigungen in anisotropen Werkstoffen unter besonderer Berücksichtigung des experimentellen Befundes.* Dr.-Ing. Dissertation, RWTH-Aachen.

Monkman, F. C., and Grant, N. J. (1956). An empirical relationship between rupture life and minimum creep rate in creep-rupture tests. *Proc. ASTM* **56**, 593–620.

Murakami, S. (1983). Notion of continuum damage mechanics and its application to aniso-tropic creep damage theory. *J. Engng. Mater. Technol.* **105**, 99–105.

Murakami, S. (1987). Progress of continuum damage mechanics. *JSME Int. J.* **30**, 701–710.

Murakami, S., and Ohno, N. (1981). A continuum theory of creep and creep damage. In *Creep in Structures* (A. R. S. Ponter and D. R. Hayhurst, eds.), pp. 422–444. Springer-Verlag, Berlin.

Murakami, S., and Sawczuk, A. (1981). A unified approach to constitutive equations of inelasticity based on tensor function representations. *Nucl. Eng. Des.* **65**, 33–47.

Murakami, S., Sanomura, Y., and Saitoh, R. (1986). Formulation of Cross-Hardening in Creep and its Effect on the Creep Damage Process of Copper. *J. Engng. Mater. Technol.* **108**, 167–173.

Odquist, F. K. G., and Hult, J. (1962). *Kriechfestigkeit metallischer Werkstoffe*. Springer-Verlag, Berlin.

Onat, E. T. (1986). Representation of mechanical behavior in the presence of internal damage. *Engng. Frac. Mech.* **25**, 605–614.

Phillips, A., and Das, P. K. (1985). Yield surfaces and loading surfaces of aluminium and brass: An experimental investigation at room and elevated temperatures. *Int. J. Plasticity* **1**, 89–109.

Rabotnov, Y. N. (1968). Creep rupture. In *Appl. Mech. Conf.* (M. Hetenyi and H. Vincenti, eds.), pp. 342–349. Stanford University, Palo Alto.

Rabotnov, Yu. N. (1969). *Creep Problems in Structural Members*. North-Holland, Amsterdam.

Ramberg, W., and Osgood, W. R. (1943). Description of stress-strain curves by three parameters. NACA Technical Note No. 902.

Rice, J. R. (1970). On the structure of stress-strain relations for time-dependent plastic deformations in metals. *Trans. ASME J. Appl. Mech.* **37**, 728–737.

Riedel, H. (1987). *Fracture at High Temperatures*. Springer-Verlag, Berlin.

Rivlin, R. S. (1970). An introduction of non-linear continuum mechanics. In *Non-linear Continuum Theories in Mechanics and Physics and Their Applications* (R. S. Rivlin, ed.), pp. 151–309. Edizioni Cremonese, Rome.

Sawczuk, A., and Anisimowicz, M. (1981). Tensor functions approach to creep laws after prestrain. In *Creep in Structures* (A. R. S. Ponter and D. R. Hayhurst, eds.), pp. 220–223. Springer-Verlag, Berlin.

Sawczuk, A., and Trampczýnski, W. A. (1982). A theory of anisotropic creep after plastic pre-straining. *Int. J. Mech. Sci.* **24**, 647–653.

Sedov, L. I. (1966). *Foundations of the Non-Linear Mechanics of Continua*. Pergamon Press, Oxford.

Skrzypek, J., and Ganczarski, A. (1999). *Modeling of Material Damage and Failure of Structures*. Springer-Verlag, Berlin.

Sobotka, Z. (1984). *Tensorial Expansions in Non-Linear Mechanics*. Academia Nakladatelstvi Ceskoslovenské, Akademie VED, Praha.

Spencer, A. J. M. (1971). Theory of invariants. In *Continuum Physics* (A. C. Eringen, ed.), pp. 239–353. Academic Press, New York.

Spencer, A. J. M. (1987). Polynomial invariants and tensor functions. In *Applications of Tensor Functions in Solid Mechanics* (J. P. Boehler, ed.), pp. 141–201. Springer-Verlag, Wien, New York.

Waniewski, M. (1984). The influence of direction and value of plastic prestrain on steady-state creep rate using the combined isotropic-kinematic hardening rule. *Rozpr. Inz.* **23**, 4.

Waniewski, M. (1985). A simple law of steady-state creep for material with anisotropy induced by plastic prestraining. *Ing.-Archiv.* **55**, 368–375.

Zheng, Q.-S. (1994). Theory of representation for tensor functions — a unified invariant approach to constitutive equations. *Appl. Mech. Rev. (AMR)* **47**, 545–587.

Zheng, Q.-S., and Betten, J. (1995a). On the tensor function representations of 2nd-order and 4th-order tensors: Part I. *Z. Angew. Math. Mech.* **75**, 269–281.

Zheng, Q.-S., and Betten, J. (1995b). The formulation of constitutive equations for fibre–reinforced composites in plane problems: Part II. *Arch. Appl. Mech.* **65**, 161–177.

Zheng, Q.-S., and Betten, J. (1997). The formulation of elastic and plastic response for cubic crystals. *Appl. Mech. Eng.* **2**, 171–186.

Zheng, Q.-S., Betten, J., and Spencer, A. J. M. (1992). The formulation of constitutive equations for fibre-reinforced composites in plane problems: Part I. *Arch. Appl. Mech.* **62**, 530–543.

Życzkowski, M. (1981). *Combined Loadings in the Theory of Plasticity*. PWN-Polish Scientific Publishers, Warszawa.

Author Index

Subject Index

ISBN 0-12-002037-8

90065 >

9 780120 020379